Linear Algebra for Physics

Nikolaos A. Papadopoulos · Florian Scheck

Linear Algebra for Physics

 Springer

Nikolaos A. Papadopoulos
Institute of Physics (WA THEP)
Johannes Gutenberg University of Mainz
Mainz, Germany

Florian Scheck (Deceased)
Mainz, Germany

ISBN 978-3-031-64907-3   ISBN 978-3-031-64908-0 (eBook)
https://doi.org/10.1007/978-3-031-64908-0

This Springer imprint is published by the registered company Springer Nature Switzerland AG
The registered company address is: Gewerbestrasse 11, 6330 Cham, Switzerland

If disposing of this product, please recycle the paper.

πάντες ἄνθρωποι τοῦ εἰδέναι
ὀρέγονται φύσει.

All humans by nature desire to know.

—Aristotle
Metaphysics A 1.980a21

καὶ ὅσῳ δὴ ἂν περὶ προτέρων τῷ
λόγῳ καὶ ἁπλουστέρων, τοσούτῳ
μᾶλλον ἔχει τὸ ἀκριβές (τοῦτο δὲ
τὸ ἁπλοῦν ἐστίν)

To the extent that [a science] is concerned
with what is prior in account and simpler, to
that extent the more exactness it has (and this
is what simplicity does).

—Aristotle
Metaphysics M 3.1078a9-11

*To Constantin Carathéodory
and Hermann Weyl*

Preface

What is a matrix? What is linear algebra? Most of those who open this book already have an idea or perhaps a good understanding of both.

When one first comes across linear algebra, one may naively think that it is a standalone area of mathematics, distinct from the rest. On the other hand, matrices seem to simply be rectangular arrays of numbers and the reader may think that they already know everything there is to know about them. However, once the reader has gone through the first chapters of this book, they will experience that matrices are omnipresent throughout the book, and it will gradually become clear that matrices and linear algebra are two sides of the same coin. In fact, matrices are what you can do with them, and linear algebra itself is ultimately the mathematical theory of matrices.

A physicist constantly uses coordinates, which essentially are also matrices. This fact is particularly pleasing for a physicist and could greatly facilitate the access and understanding of linear algebra.

So why is linear algebra important for physicists?

The well-known mathematician Raoul Bott stated that 80% of mathematics is linear algebra. According to our own experience with physics, we would state that almost 90% of mathematics in physics is linear algebra. Furthermore, according to our experience with physics students, the most challenging subject in mathematics for Bachelor students is linear algebra. Students usually have hardly any problem with the rest of mathematics, such as, for example, calculus which is already known from school. The important challenge of linear algebra seems to be that it is underestimated, both from the curriculum point of view and from the students themselves. The reason for this underestimation is probably the widespread idea that, with linear algebra being "linear", it is trivial and plain, easy to learn and use.

Our intention is therefore, among others, to contribute with this book to ameliorate this asymmetric relation between linear algebra and the rest of mathematics.

Finally, we would like to add that with the advent of Gauge theory, structures connected to symmetries became much more important for physics than they ever were at any time before. This means that virtuosity in handling linear algebra is needed since linear structures are one of the chief instruments in symmetries.

Mainz, Germany   Nikolaos A. Papadopoulos
                                                                          Florian Scheck

Acknowledgments

Our first and foremost Acknowledgment goes to Christiane Papadopoulos, who tirelessly and devotedly prepared the LaTeX manuscript with extreme efficiency, and beyond. We also extend our gratitude to our students. One of us, N.P., particularly thanks the numerous physics students who, over the past few years, have contributed significantly to our understanding of the role of Linear Algebra in physics through their interest, questions, and bold responses in several lectures in this area. Our thanks also go to many colleagues and friends. Andrés Reyes contributed significantly to determining the topics to be considered in the early stages of the book. Our physics colleague, Rolf Schilling, provided extensive support and highly constructive criticism during the initial version of the manuscript. Similar appreciation is extended to Christian Schilling for several selected chapters. We vividly remember how much we learned about current mathematics many years ago from Matthias Kreck. The numerous discussions with him about mathematics and physics have left traces in this book. The same applies to Stephan Klaus, Stephan Stolz, and Peter Teichner. One of us, N.P., has benefited greatly from discussions with Margarita Kraus and Hans-Peter Heinz. Equally stimulating have been the discussions and collaborations with mathematician Vassilios Papakonstantinou, which have lasted for decades until today. We would like to express our heartfelt gratitude to mathematician Athanasios Bouganis, whose advice in the final stages of the book was crucial. Last but not least, we would like to thank mathematician Daniel Funck from Durham University, who significantly improved not only the linguistic quality of the book but also its overall content with great care and dedication.

We also thank the staff of Springer Nature and, in particular, Ute Heuser, who strongly supported this endeavor.

About This Book

In this book, we present a full treatment of linear algebra devoted to physics students both undergraduate and graduate since it contains parts which are relevant for both. Although the mathematical level is similar to the level of comparable mathematical textbooks with definitions, propositions, proofs, etc., here, the subject is presented using the vocabulary corresponding to the reader's experience made in his lectures. This is achieved by the special emphasis given to the role of bases in a vector space. As a result, the student will realize that indices, as many as they may be, are not enemies but friends since they give additional information about the mathematical object we are using.

The book begins with an introductory chapter, the second chapter, which provides a general overview of the subject and its relevance for physics. Then, the theory is developed in a structured way, starting with basic structures like vector spaces and their duals, bases, matrix operations, and determinants. After that, we recapitulate the role of indices in linear algebra and give a simple but mathematically accurate introduction to tensor calculus.

The subject material up to Chap. 8 may be considered as the elementary part of linear algebra and tenor calculus. Detailed discussion about eigenvalues and eigenvectors is followed by Chap. 9 on operators on inner product spaces, which includes, among many other things, a full discussion of the spectral theorem. This is followed by a thorough presentation of tensor algebra in Chaps. 3, 8, and 14 which takes full advantage of the material developed in the first chapters, thus making the introduction of the standard formalism of multilinear algebra nothing else but a déjà vu.

Chapter 1 includes material that is usually left for the appendix. However, as we wanted to highlight the usefulness on this chapter, especially for physicists, we placed it as Chap. 1.

All chapters contain worked examples. The exercises and the hints are destined mainly for physics students. Our approach is in many regards quite different from the standard approach in the mathematical literature. We therefore hope that students of both physics and mathematics will benefit a great deal from it.

Where the organization of the book is concerned, the first eight chapters deal with what we would call elementary linear algebra and is therefore perfectly suitable for bachelor students. It covers what is commonly needed in everyday physics. The remaining chapters give a perspective and allow insights into what is interesting and important beyond this. Hence the subjects of Chap. 9 up to the last one can be considered as the advanced linear algebra part of the book.

Everything is written from a physicist's perspective but respecting a stringent mathematical form.

Contents

Chapter 1
The Role of Group Action

In this chapter, we shortly present some prerequisites for the book concerning mathematical structures in calculus and geometry.

We introduce and discuss in detail, especially for the benefit of physicists, the notion of quotient spaces in connection with equivalence relations.

The last two sections deal with group actions, which are the mathematical face of what we meet as symmetries in physics.

This chapter could be considered as appendix, but we set it at the beginning of the book to point out its significance. For physicists, it can be skipped on the first reading.

1.1 Some Prerequisites: Mathematical Structures

In this chapter, we are dealing informally with the various mathematical structures needed in physics. Some of these structures will often be introduced without proof and used intuitively, as in the literature in physics. It is a fact that in physics, we often have to rely on our intuition, sometimes more than we would like to. This may cause difficulties, not only for mathematically oriented readers. We try to avoid this as much as possible in the present book. Therefore, we will rely here on our intuition no more than necessary, and be precise enough to avoid misunderstandings.

We treat set theory as understood; we here discuss only the notion of quotient space. Quotient spaces appear in many areas of physics, but in most cases they are not recognized as such. In this chapter, we will concentrate on general and various aspects of group actions and the revision of some essential definitions. In this context, we also introduce the notion of an equivariant map that respects the group actions of the input and output spaces in a compatible way.

Usually, we call a set with some structure a space. So we can talk about topological spaces, metric spaces, Euclidean and semi-Euclidean spaces, affine spaces, vector

© The Author(s), under exclusive license to Springer Nature Switzerland AG 2024
N. A. Papadopoulos and F. Scheck, *Linear Algebra for Physics*,
https://doi.org/10.1007/978-3-031-64908-0_1

spaces, tensor spaces, dual spaces, etc. Almost every set we meet in physics also has a manifold structure. This is the case in the spaces mentioned above. Some examples of manifolds that the reader already knows are the three-dimensional Euclidean space we live in, any two-dimensional smooth surface, and any smooth curve (one-dimensional manifold) we may think of or see in it. We may also think of manifolds in any dimension $k \leqslant n$ in $\mathbb{R}^n$ which is also the simplest manifold we can have in n-dimensions.

In this book, we will talk freely about manifolds (smooth manifolds) without defining them, and we expect the reader to know at least intuitively what we refer to.

1.2   Quotient Spaces

We start with some general remarks that might be quite useful for many readers. Equivalence relations first appear in real life when we want to talk about objects that are not absolutely the same but show clear similarities. In this way, we can observe rough structures more clearly and more precisely. We get new sets, often with far fewer elements. In mathematics, such a set is called a quotient set and the elements of this quotient set are called equivalence classes. Each such element, called an equivalence class, is itself a special subset of the originally given set.

We could consider as an example from real life the set of the inhabitants of the European Union. A possible equivalence relation exists if we consider the inhabitants of each European state as equivalent. In this case, the quotient set is the set of European states and the elements are the corresponding states Germany, France, Cyprus and so on.

It would be surprising if we could not apply things that are produced in life constantly and often unconsciously, in mathematics, and in particular in linear algebra too. So, in the following, we describe this phenomenon and some consequences of it, here in the framework of mathematical formalism.

In mathematics, it is well-known that we can construct new sets with set operations like, for example, union and intersection. However, building quotient spaces is a much more complex operation than obtaining a new set or a new manifold by union or intersection. This happens when we have to talk not about equal elements but about equivalent elements. Here we use the equivalence relation to construct a new topological space or a new manifold. As this approach to constructions of quotient spaces may seem highly abstract, we have to be much more precise than we usually would be in the standard physics literature (at least at the beginning) and not rely, entirely on our physical or geometric intuition. Interestingly enough, we use many quotient spaces in physics intuitively and often without being aware of the precise mathematical situation. A prominent example in special relativity is when we have to consider our three-dimensional Euclidean space of a given four-dimensional space of events (spacetime points). As is well-known, a point belonging to a Euclidean space within this setup, is not a point but a straight line in the four-dimensional space of events. This is, for example, the set of all the events at all times of a free moving point particle. Mathematically speaking, all the events on this straight line are equivalent elements (events) of the spacetime and form an equivalent class or simply a class or a coset. For example, the space point "London" is the equivalence class of all events along a straight line. The set of all such equivalent classes or cosets (parallel straight lines of free point particles) fills the whole spacetime. This set of equivalent classes is called quotient space. So it is evident that the three-dimensional Euclidean space is described by a quotient space of the four-dimensional spacetime.

Such a construction is a mathematically precise approach to obtaining new spaces out of given ones. Even if it seems quite abstract, for this procedure we need only elementary mathematics from set theory. The usefulness of this formalism in many applications justifies its introduction here. Later, we shall also use this approach to show precisely that the basis dependent (component-wise with indices) tensor formalism is equivalent to the coordinate-free formalism. In this sense, tensor formalism as mostly used in physics, can also be considered as a coordinate-free formulation. Therefore, in this book, we try to use the coordinate-free formalism and the tensor formalism at the same level and take advantage of both.

We first remember the definition of an equivalence relation:

Definition 1.1 Equivalence relation.

On a set X an equivalence relation is a relation, that concerns a subset $R \subset X \times X$, but we usually write instead of $(x, x') \in R : x \sim x'$.

The relation $\sim$ (or $\sim_R$) is an equivalence relation if it is:

(i) reflexive: $x \sim x$;
(ii) symmetric: $x \sim x' \Leftrightarrow x' \sim x$;
(iii) transitive: $x \sim y$ and $y \sim z \Rightarrow x \sim z$.

Definition 1.2 Equivalence class and quotient space.

We call the subset

$$[x] := \{x' : x' \sim x, x \in X\} \subset X$$

the equivalence class or coset of x relative to the equivalence relation $\sim$. The new set of equivalence classes which we call quotient space of X, determined by the equivalence relation $\sim$, is given by:

$$X/\sim := \{[x] : x \in X\}.$$

We have a natural, surjective map,

$$\pi : X \to X/\!\sim$$
$$x \mapsto \pi(x) := [x]$$

which may also be called canonical map, canonical projection or quotient map.

Remark 1.1 Equivalence relation and disjoint decomposition.

It is important to notice that any equivalence relation of X induces a disjoint decomposition of X (called a partition) given by the fibers of $\pi^{-1}([x]) \subset X$. We may represent this symbolically in Fig. 1.1. This will also be demonstrated in Sect. 2.6 in connection with quotient vector spaces.

It is easy to realize intuitively that a given partition of X also introduces an equivalence relation on X.

At this point some explanations are necessary.

All the equivalent points of x in X are also given geometrically by the fibers of π:

$$\pi^{-1}(\pi(x)) = \pi^{-1}([x]) \subset X.$$

So we have again

$$\pi^{-1}([x]) = \{x' : x' \sim x, x \in X\} \subset X.$$

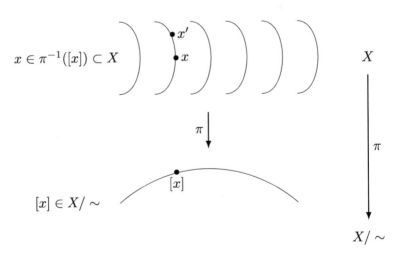

Fig. 1.1 Disjoint decomposition of X

We may call any element $x' \in \pi^{-1}([x]) \in X$ a representative of the class $[x]$. If the different elements x and x' are for example different because they have different properties, we ignore these different features and consider x and x' as essentially identical. So we may identify all the other equivalent elements of x with x. We so obtain a new object, a new element $[x]$ and a new set $X/\!\sim = \{[x], \dots\}$, both with completely different properties as x and $X = \{x, \dots\}$.

It is clear that $[x]$ is not an element of X, so we have $[x] \notin X$. The set of representatives of $[x]$, the fiber $\pi^{-1}([x])$, is a subset of X (i.e., $\pi^{-1}([x]) \subset X$). It is a slight misuse of notation when sometimes we mean $\pi^{-1}([x])$ and write "$[x] \subset X$". Note again that for the element $[x] \in X/\!\sim$ we may use different names: equivalent class, coset, or fiber (i.e., $(\pi^{-1}([x]))$). All this is demonstrated symbolically in Fig. 1.1.

In many cases, the quotient map π is defined in such a way that X and $X/\!\sim$ have the same algebraic (or geometric) structure. In this case, π is a homomorphism relative to the relevant structure. See for example the proposition in Sect. 2.6.

We are now going to give two very simple and essential geometrical examples of quotient spaces. The first example corresponds to the special relativity case mentioned earlier in this section. The second one corresponds to a pure geometric case.

Example 1.1 Rays in $\mathbb{R}^2$ as quotient space.
  We consider the following subset:

$$X = \mathbb{R}^2 - \{\vec{0}\} = \{x = \vec{\xi} : \vec{\xi} \in \mathbb{R}^2 \quad \text{and} \quad \vec{\xi} \neq \vec{0}\}.$$

So, as shown in Fig. 1.2, this is the two-dimensional plane without the zero point. We denote with $\mathbb{R}^+$ the positive numbers $\mathbb{R}^+ := \{\xi \in \mathbb{R} : \xi > 0\}$.
  For each $x \in X$, the x-ray $A(x)$ is given by $A(x) = \mathbb{R}^+ x$.

Fig. 1.2 Rays in $\mathbb{R}^2$

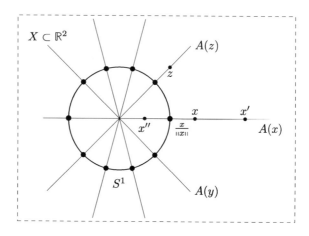

We now define that equivalence relation $\sim$ in X which is given by the x-rays: For $x \in X$, let $A(x)$ be the set of points in X which are in relation with x:

$$x' \sim x \;\Leftrightarrow\; x' \in \mathbb{R}^+ x = A(x). \tag{1.1}$$

We denote the equivalence class or coset of x by $[x]$. This can also be defined formally like this (for $X = \mathbb{R}^2 - \{0\}$):

$$[x] := \{x' \in X : x' \sim x\}. \tag{1.2}$$

In Eq. (1.1), we see that $[x]$ is the set $A(x)$, the ray which contains x.

Note also that by this construction we obtain a disjoint decomposition of X (see Fig. 1.2). Obviously the following holds:

$$\text{From}\quad x'' \sim x \quad\text{follows}\quad A(x'') = A(x).$$

We further see that for two equivalence classes the following is true: either they are equal or they are disjoint. This means that for $A(x)$ and $A(z)$ either $A(x) = A(z)$ or $A(x) \cap A(z) = \emptyset$. The quotient space $X/\sim$ is given by the set of rays:

$$X/\sim = \{[x]\} = \{A(x) : x \in X\}.$$

We can also describe this by a set of suitable representatives. For example, we may choose the circle with radius one $S^1 = \{x \in X : \|x\| = 1\}$ and the bijective map:

$$\Phi : \quad X/\sim \longrightarrow S^1$$
$$A(x) \longmapsto \frac{x}{\|x\|}.$$

So we determined for every equivalence class $[x] \equiv A(x)$ one and only one representative, the point $\frac{x}{\|x\|} \in S^1 \subset \mathbb{R}^2$. That is for each ray a single point on the circle and so we get the bijection:

$$X/\sim \cong S^1.$$

There are of course innumerable such bijections which characterize the quotient space, $X/\sim$, but S^1 seems to be the most pleasant.

1.3   Group Actions

There is hardly an area in mathematics or theoretical physics where groups and, particularly, group actions on manifolds are irrelevant. Symmetries are present in nature, they lie at the heart of the laws of nature. Group actions are the mathematical face of what we call symmetries in physics. The notion of a group arises naturally if we consider the bijective maps from a given set to itself. The composition of two bijective maps is, again, a bijective map. The identity is immediately present, and it is evident that for every bijective map, its inverse exists. Additionally, composition is associative. Throughout this book, we would like to refer to an action of a group element, or a bijective map, as a *transformation*. It is plausible that the set of all transformations (bijective maps) in a given set leads to the notion of a group:

Definition 1.3 Group.
A group is a set G and a binary operation.

$$G \times G \longrightarrow G,$$
$$(a, b) \longmapsto a * b \equiv ab$$

with the following axioms:

(i) associativity: $(a * b) * c = a * (b * c)$;
(ii) existence of a neutral element e with: $e * a = a * e = a$;
(iii) existence of inverse elements: for every $a \in G$ there exists an element $a^{-1} \in G$ such that $a * a^{-1} = a^{-1} * a = a$.
A group is abelian or commutative if $a * b = b * a$.

We believe that groups would be useless in physics if there were no actions or realizations of groups. Indeed, everywhere in physics, we have actions of groups. Therefore, we introduce and apply group actions from the beginning:

Definition 1.4 Left group action.
The group G is acting on the set X (from the left).

$$\phi : G \times X \longrightarrow X$$
$$(g, x) \longmapsto \Phi(g, x) \equiv gx$$

and the following two conditions are valid:

(i) "Compatibility": for $g, h \in G$ $h(gx) = (hg)x$ and
(ii) "Identity": $ex = x$ $(\phi(e, x) = id_X(x))$.

This means that $e \in G$ acts as the identity-map

$$id_X : X \longrightarrow X$$
$$x \longrightarrow id_X(x) = x.$$

Even if it seems quite trivial, it turns out that the identity map, id_X, is a very important map in mathematics and physics.

The map ϕ above gives two families of partial maps which we denote by $(\phi_g)_G$ and $(\phi_x)_X$.

$$\phi_g : X \longrightarrow X$$
$$x \longmapsto \phi_g(x) := \phi(g, x)$$

and

$$\phi_x : G \longrightarrow X$$
$$g \longmapsto \phi_x(g) := \phi(g, x).$$

We may call ϕ_g a g-transformation and ϕ_x an x orbit maker! There are two more corresponding maps $\hat{\phi}$ and $\tilde{\phi}$. Using $Trf(X) \equiv$ bijective (X) for bijective maps in X, we have:

$$\hat{\phi} : G \longrightarrow Trf(X)$$
$$g \longmapsto \hat{\phi}(g) := \phi_g : X \longrightarrow X.$$

The map, $\hat{\phi}$, converts an abstract group element into a transformation ϕ_g in X. If we denote the set of all maps between X and Y by $Map(X, Y) = \{f, \dots\}$,

$$f : X \longrightarrow Y$$
$$x \longrightarrow f(x) = y.$$

If we have $X = Y$, we write $Trf(X) \subset Map(X, X)$. The second map $\tilde{\phi}$ which corresponds to ϕ_x is given by

$$\tilde{\phi} : X \longrightarrow Map(G, X)$$
$$x \longmapsto \tilde{\phi}(x) := \phi_x : G \longrightarrow X.$$

The map, $\tilde{\phi}$, converts the point x into the map ϕ_x, a kind of an "orbit maker"! According to our convention, $Trf(X)$ is the set of bijections on X so that we have $Trf(X) \equiv Bij(X)$. The official name for $Trf(X)$ is the symbol $S(X)$, the group of all permutations in X. The map, $\hat{\phi}$, is a group homomorphism (G-homomorphism):

$$\hat{\phi}(gh) = \hat{\phi}(g)\hat{\phi}(h).$$

This follows immediately from the G-action. For each $x \in X$:

$$\left(\hat{\phi}(h) \circ \hat{\phi}(g)\right)(x) = \phi_h \circ \phi_g(x) = \phi_h\left(\phi_g(x)\right)$$
$$= \phi_h(gx) = hgx = \phi_{hg}(x) = \hat{\phi}(hg)(x).$$

So we have $\hat{\phi}(h) \circ \hat{\phi}(g) = \hat{\phi}(hg)$.

As we see, the abstract group multiplication * corresponds in this consistent way to the composition of transformations on X. The map, $\hat{\phi}$, is a realization of the group G as a particular G-transformation group on X.

There is also an analog action on X from the right hand side (right group action).

$$\psi : X \times G \longrightarrow X$$
$$(x, g) \longmapsto \psi(x, g) = \psi_g(x) = xg$$

with

$$\hat{\psi} : G \longrightarrow Trf(X)$$
$$g \longmapsto \hat{\psi}(g).$$

The map, $\hat{\psi}$, is now an antihomomorphism:

$$\hat{\psi}(hg) = \hat{\psi}(g) \circ \hat{\psi}(h)$$

So we have

$$(\hat{\psi}(g) \circ \hat{\psi}(h))(x) = \psi_g \circ \psi_h(x) = \psi_g(\psi_h(x)) = \psi_g(\psi(xh)) = \psi(xhg) = xhg =$$
$$= \psi_{hg}(x) = \hat{\psi}(hg)(x) \Rightarrow$$
$$\Rightarrow \hat{\psi}(g) \circ \hat{\psi}(h) = \hat{\psi}(hg).$$

If there is no danger of confusion, we may also write anonymously $L_g := \phi_g$ and $R_g := \psi_g$. For this reason, given the map

$$G \times X \to X,$$

we may call the set

$$X : \text{left } G\text{-space}$$

since we consider, as indicated by $G \times X$, a left action. The group G acts from the left. In the case

$$Y \times G \to Y,$$

we may call the set

$$Y : \text{right } G\text{-space}$$

since we consider as indicated by $Y \times G$, a right action. The group G acts from the right.

It is important to realize that left and right actions correspond to two different maps. In particular, as was shown above, a left action leads to a homomorphism $\hat{\phi}(h) \circ \hat{\phi}(g) = \hat{\phi}(hg)$. A right action leads to an antihomomorphism $\hat{\Psi}(g) \circ \hat{\Psi}(h) = \hat{\Psi}(hg)$.

Comment 1.1 *The meaning of the right action.*

We would like to point out again that R_g is an antihomomorphism. We have for example $R_g(ab) = R_g(b)R_g(a)$ and $L_g(ab) = L_g(a)L_g(b)$. Here we meet a tricky technical point in the notation. If the action leads to a homomorphism as above with $\hat{\phi}$, it is commonly referred to as a left action. In the case of an antihomomorphism, as with $\hat{\Psi}$, we are taking about a right action. This is why we write for example:

$$\Psi' : X \times G \longrightarrow X$$
$$(x, g) \longmapsto \Psi'(x, g) := g^{-1}x.$$

Since this leads to an antihomomorphism, Ψ' is a right action even if the group element g^{-1} acts on x, as we see, from the left. We may also indicate this fact by writing $\bar{R} := L_{g^{-1}}$. We will need this fact later.

The left or right action is also relevant for what follows. It is obvious that for every given point $x_0 \in X$, the left group action leads for example to a left orbit which we denote by Gx_0 and which is the subset of X given by:

$$Gx_0 = \{gx_0 : g \in G\}.$$

A subgroup of G, J_{x_0} (that is, $J_{x_0} \leqslant G$) is connected with the orbit Gx_o at the position x_o. This subgroup characterizes entirely the orbit, naturally together with G. This leads to the following definition.

Definition 1.5 Isotropy group.
   The isotropic subgroup of G, with respect to x_0, is given by $J_{x_0} := \{g \in G : gx_0 = x_0\}$.

J_{x_0} is often called isotropy group or stability group or even stabilizer subgroup of G with respect to x_0. Different terms for the same thing sometimes indicate their importance.

We will now give a definition of a few other essential types of action relevant to some aspects of linear algebra and physics in general, for example, gravity and especially cosmology.

Definition 1.6 Transitive action. The group, G, acts transitively on X if X is an orbit of G.

So we have $Gx_0 = X$.
Equivalently, for any x and x' in X there exists a $g \in G$ such that $x' = gx$.

Definition 1.7 Effective action.
   The group, G, acts effectively if the only element of G that fixes every $x \in X$ is e. That is, if $g \in G$ has $gx = x \ \forall x \in X$, then $g = e$.

In other words, only the neutral element e is acting on X as the identity.

Definition 1.8 Free action. The group, G, acts freely on X if for all $x_0 \in X$, the isotropy group J_{x_0} is trivial (i.e., $J_{x_0} = \{e\}$). This means that, in other words, G acts on X without fixed points.

Remark 1.2 On homogeneous spaces.

   Generally, any orbit (Gx_o) is a homogeneous space. It is determined by the group G and a specific subgroup H of G. More precisely, the structure of the orbit Gx_o is given by the quotient space G/H, so that

$$Gx_o \cong G/H.$$

So we may consider the quotient space G/H as a model of the orbit Gx_o in the same sense as $\mathbb{R}^n$ may be considered as the model of a real vector space V with $\dim V = n$. This is the first application of Sect. 1.2. The relevant equivalence relation is given below.
   Note that the orbit Gx_o which corresponds to the action of the group G on X and which in general is not a group, is described by the two groups G and H with $H \leqslant G$. It turns out that the second group H is given, as expected, by the data of the orbit Gx_o and is precisely the stability group J_{x_o}. So we have $H := J_{x_o}$.

Given G and H as above, in order to obtain the quotient space G/H, we have to consider the right action of H on G:

$$G \times H \longrightarrow G,$$
$$(g, h) \longmapsto gh \equiv R_h g.$$

It is clear that H acts freely on G. This follows directly from the group axioms and Definition 1.8 (Free action): If $gh = g$, we also have $g^{-1}gh = g^{-1}g$. Since $g^{-1}g = e$, it follows that $eh = e$ and $h = e$ and so, the isotropy group $J_g = \{e\}$ is trivial. This means that H acts freely on G.

We now define the following equivalence relation.

$$g' \underset{H}{\sim} g :\leftrightarrow g' = gh.$$

So we have

$$[g] := \{g' : g' = gh, \ h \in H\} \equiv gH.$$

This means that g' and g are in the same (right) H orbit gH in G.

This equivalence relation leads to the quotient space, as defined in Sect. 1.2:

$$G/H := \{[g] = gH : g \in G\} \tag{1.3}$$

Here the cosets $[g]$ are H orbits in G. Since the action of H an G is free, every such H orbit in G is bijectively related to the subgroup H. So we get $gH \underset{bij}{\cong} H$ for all $g \in G$ and G/H consists of all such H orbits. For this reason G/H is also called H orbit space in G and we may draw symbolically (Fig. 1.3).

This is one further example of a quotient space. Figure 1.3 and Eq. (1.3) also show explicitly that the quotient space G/H is given by the set of all positions gH, $g \in G$ that the subgroup H takes by the natural G-action. This means nothing else but that G/H by itself is an orbit of the G-action. Therefore G/H is a homogeneous space. At the same time, G/H is generally a model for every G orbit Gx_o in X.

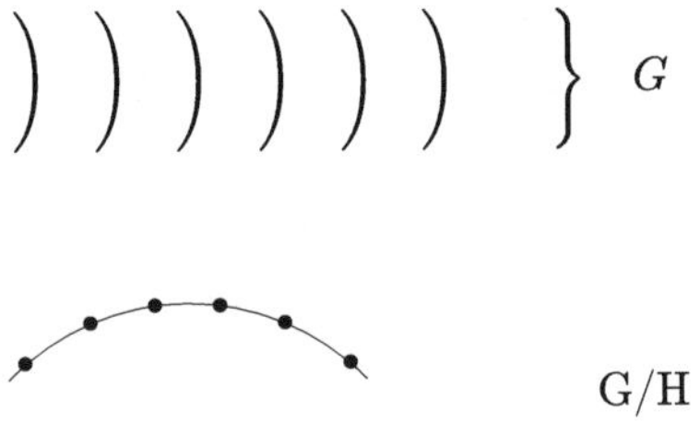

Fig. 1.3 H orbit space in G

G

G/H

Summing up the above discussion, we can state that this type of quotient space leads to a kind of universal relation: Every G orbit in X has the same structure and is called a homogeneous space. So for a given G orbit $Gx_o \subseteq X$ there exists a subgroup H of G (in fact, this subgroup H is the isotropy group J_{x_0}) such that

$$Gx_o \cong G/H.$$

This shows the importance of quotient spaces. Numerous applications in theoretical physics make use of them, for example in cosmology, in various aspects in symmetries, in quantum mechanics, and in numerous cases in linear algebra.

Remark 1.3 Transitive and free G-action.

There is still one further important fact. If we have simultaneously a transitive and a free G-action on X, then X is bijective to the group G (see Definitions 1.6 and 1.8. So we get:

$$X \underset{bij}{\cong} G.$$

By assumption, X is not a group but has the same "number" of elements, that is, the same cardinality, as the group G.

We use this fact in Sect. 3.2. If we work with all (linear) coordinate systems simultaneously, it signifies that we are de facto coordinate-free. This is true for linear algebra as well as for tensor calculus.

This proves that tensor calculus, in the basis dependent component formulation, is completely equivalent to and not less valuable than the corresponding basis free formulation.

1.4 Equivariant Maps

We consider two (left) G-spaces X and Y:

$$\phi : G \times X \to X \text{ and } \psi : G \times Y \to Y$$

and the map

$$F : X \to Y.$$

The interesting case occurs when we demand F to commute with the group actions ϕ and ψ. This leads to the notion of an equivariant map:

Definition 1.9 F is an equivariant map if $F \circ \phi_g = \psi_g \circ F$ holds for all $g \in G$.

Note that the interesting maps between groups are the group homomorphisms in the same sense. Here we have G-spaces, X and Y, and the interesting maps now are the equivariant maps.

This can also be expressed by the following commutative diagram:

$$
\begin{array}{ccc}
X & \xrightarrow{\ F\ } & Y \\
\phi_g \downarrow & & \downarrow \psi_g \\
X & \xrightarrow{\ F\ } & Y
\end{array}
$$

In the case $X = Y$, we have $F \circ \phi_g = \psi_g \circ F$.

Summary

This chapter should normally be in the appendix. However, because we wanted to highlight the usefulness of this chapter, especially for physicists, we placed it as chapter one.

Firstly, we extensively motivated and discussed the significance of the quotient space. Its relationship to the corresponding equivalence relation was elucidated. The connection between the underlying set of the quotient space and the canonical surjection was also clearly highlighted graphically and with a typical example.

Group actions are fundamental in physics, considering the role of symmetries and the fundamental forces of physics. They are also relevant at every step in linear algebra. Therefore, this section became necessary, even though it may appear quite challenging upon first reading. The various aspects and definitions allow for a clear distinction between the different structures that arise in linear algebra.

Finally, the very brief section on equivariant maps was a useful addition, although it will only become relevant in the third chapter.

References

1. W.M. Boothby, *An Introduction to Differentiable Manifolds and Riemannian Geometry* (Academic Press, 1986)
2. T. Bröcker, T. Tom Dieck, *Representations of Compact Lie Groups* (Springer, 2013)
3. M. DeWitt-Morette, C. Dillard-Bleick, Y. Choquet-Bruhat, *Analysis, Manifolds and Physics* (North-Holland, 1978)

4. M. Göckeler, Th. Schücker, *Differential Geometry, Gauge Theories, and Gravity* (Cambridge University Press, 1989)
5. K.-H. Goldhorn, H.-P. Heinz, M. Kraus, *Moderne mathematische Methoden der Physik*. Band 1 (Springer, 2009)
6. K. Jänich, *Topologie* (Springer, 2006)
7. J.M. Lee, *Introduction to Smooth Manifolds. Graduate Texts in Mathematics* (Springer, 2013)
8. S. Roman, *Advanced Linear Algebra* (Springer, 2005)
9. C. Von Westenholz, *Differential Forms in Mathematical Physics* (Elsevier, 2009)

.

Chapter 2
A Fresh Look at Vector Spaces

We start at the level of vector spaces, and we first consider quite generally a vector space as it is given only by its definition, an abstract vector space.

We have to investigate, to compare, and use vector spaces to describe, whenever this is possible, parts of the physical reality. The most appropriate way is to use maps that are in harmony, that is, compatible with a vector space structure. We have to use linear maps, also called vector space homomorphisms.

It turns out that the physical reality demands additional structures which we have to impose on an abstract vector space. The most prominent structure of this kind is a positive definite scalar product (special symmetric bilinear form), the inner product, and in this way we obtain an inner product vector space or a Euclidean vector space which is strongly connected with our well-known (affine) Euclidean space. We have, of course, also semi-Euclidean vector spaces where the scalar product is no more positive definite.

It is interesting that instead of adding, as in Sect. 2.3 with a symmetric bilinear form, we could also "subtract" structures from vector spaces. This means we can consider a vector space a "special" manifold, a linear manifold which is usually called affine space.

2.1 Vector Spaces

The discussion in Sect. 1.3 allows us to define a vector space that emphasizes the point of view of group action.

In preparation of this first approach, we have to define what a field is, for example the real and complex numbers. These are also called scalars and are used essentially to stretch vectors, an operation which is also called scaling. We already know what a group is. In some sense, a group G contains a perfect symmetry structure. It is characterized by one operation, by the existence of a neutral element e with $eg = ge = e$, for all g in G and by the property that for every g in G the inverse g^{-1} exists.

© The Author(s), under exclusive license to Springer Nature Switzerland AG 2024
N. A. Papadopoulos and F. Scheck, *Linear Algebra for Physics*,
https://doi.org/10.1007/978-3-031-64908-0_2

Hence we have $gg^{-1} = g^{-1}g = e$. We often also write $e = 1$. In the case that G is commutative (abelian) and if the operation is additive, we usually write $e = 0$. If the operation is multiplicative, we write $e = 1$.

On the way to determining what a field is, it seems useful to make a stop and first define what a ring is.

Definition 2.1 Ring.

A ring is a set with two operations, $+$ (addition) and $\cdot$ (multiplication), so that the following properties hold:

(i) Let $(R, +)$ be an abelian additive group with neutral element, 0, and the inverse of $\alpha \in R$ given by $-\alpha$. Hence we have $\alpha + (-\alpha) = \alpha - \alpha = 0$;

(ii) $(R, \cdot)$ The multiplication is associative for $\alpha, \beta, \gamma \in R$:

$$\alpha \cdot (\beta \cdot \gamma) = (\alpha \cdot \beta) \cdot \gamma.$$

The neutral element of multiplication is called unit 1 $(\alpha \cdot 1 = 1 \cdot \alpha = \alpha)$;

(iii) The distributive laws hold, for $\alpha, \beta, \gamma \in R$:

$$\alpha \cdot (\beta + \gamma) = \alpha \cdot \beta + \alpha \cdot \gamma;$$
$$(\alpha + \beta) \cdot \gamma = \alpha \cdot \gamma + \beta \cdot \gamma.$$

A ring R is called commutative if $\alpha \cdot \beta = \beta \cdot \alpha$ for all $\alpha, \beta \in R$.

In this case, both operations, addition as well as multiplication, are commutative. In general, a ring needs not be commutative and in fact, it does not have to have inverse elements for each of its elements.

As we shall see, a field is a commutative ring with one more condition, the existence of inverse elements, for multiplication too.

Definition 2.2 Field (1).

A commutative ring $\mathbb{K}$ is called a field if each nonzero element has a mul-tiplicative inverse.

This means that for every $\alpha \in \mathbb{K}$ with $\alpha \neq 0$, there exists an element α^{-1} so that $\alpha \cdot \alpha^{-1} = \alpha^{-1} \cdot \alpha = 1$. It is easy to recognize the following equivalent definition.

Definition 2.3 Field (2).

A field is a set $\mathbb{K}$ with two operations, addition $(+)$ and multiplication $(\cdot)$ so that

(i) $(\mathbb{K}, +)$ is a commutative group with neutral element 0;

(ii) $(\mathbb{K} \setminus 0, \cdot)$ is a commutative group with neutral element 1;

(iii) The distributive laws hold: For all $\alpha, \beta, \gamma \in \mathbb{K}$, we have

$$\alpha \cdot (\beta + \gamma) = \alpha \cdot \beta + \alpha \cdot \gamma;$$
$$(\alpha + \beta) \cdot \gamma = \alpha \cdot \beta + \beta \cdot \gamma.$$

In the case of a vector space, we have a new situation. We have to consider two different sets. The main set is the vector space $V = \{u, v, w, \dots\}$ and the second set is the field $\mathbb{K} = \{\alpha, \beta, \gamma, \dots\}$. $(V, +)$ is an additive group. In addition, we see that the second operation is an external operation between $\mathbb{K}$ and V. To be more precise: we get $(\mathbb{K}, \cdot)$, the multiplicative group of the field $\mathbb{K}$ which acts on V. The notion of group action was discussed in Sect. 1.3 in general terms. This means here that every scalar $\lambda \in \mathbb{K}, \lambda \neq 0$ can expand or shrink every element (vector) of V. All this is summarized in the following definition.

Definition 2.4 Vector space.

A vector space is an abelian group (additive) V on which the field $\mathbb{K}$ acts through its multiplicative group.

More precisely, we denote by $\mathbb{K}$ the field of real or complex numbers ($\mathbb{K} \in \{\mathbb{R}, \mathbb{C}\}$). A vector space over $\mathbb{K}$ is the set V with two operations:
Addition:

$$+ : V \times V \longrightarrow V,$$
$$(v, w) \longmapsto v + w.$$

Scalar multiplication:

$$\mathbb{K} \times V \longrightarrow V,$$
$$(\alpha, x) \longmapsto \alpha v,$$

with the following properties:

(i) V is an abelian group;
(ii) $\mathbb{K}$ acts on V by
      $a(\beta v) = (\alpha\beta)v$ for all $\alpha, \beta \in \mathbb{K}$ and $v \in V$;
      $1v = v$ for all $v \in V$ and
      $\alpha v = v\alpha$;
(iii) The distributive law holds:
      $(\alpha + \beta)v = \alpha v + \beta v$ and $\alpha(v + w) = \alpha v + \alpha w$.

Part (ii) of the definition refers to the group action, as discussed rather generally in Sect. 1.3.

We restrict ourselves from the beginning throughout this book to finite-dimensional vector spaces. Only a few examples are referring to nonfinite-dimensional vector spaces. We use greek letters like $\alpha, \beta, \xi, \lambda$ for the scalars since we want to underline the scalar action on vectors.

In physics, the two fields $\mathbb{R}$ and $\mathbb{C}$ play the leading role; therefore, we restrict $\mathbb{K}$ to these two fields.

Although $\mathbb{R}$ and $\mathbb{C}$ vector spaces correspond to the same linear structure, in some cases they have different properties, for example in the spectral theorems, which are very important in physics. That is why it is necessary to distinguish them.

It is interesting to realize that the scalar action $\mathbb{K}$ on the abelian group V creates the vector space we know, an object which is very "compact" and at the same time very flexible. This is due to the existence of its basis and with it the notion of dimension, especially in the finite-dimensional case we consider here. Thus every vector space V is entirely characterized by its dimension and the scalar action on it.

As we saw, all the above spaces, rings, fields, and vector spaces, start with an abelian group and after that another operation is introduced. In linear algebra, for square matrices and linear operators in a vector space, we can further define another operation. This results in an algebra. In accordance with our procedure in this section, there are two ways to arrive at an algebra. We can either start from a ring and then introduce a scalar multiplication or start with a vector space and then introduce a vector multiplication. In the following definition, we consider both.

Definition 2.5 Algebra.
An algebra over the field $\mathbb{K}$ is a set A, together with three operations: addition, multiplication, and scalar multiplication for which the following holds:

(i)   The set A is a vector space over $\mathbb{K}$ under addition and scalar multiplication.
(ii) The set A is, in addition, a ring under addition and multiplication.
(iii) For a scalar $\lambda \in \mathbb{K}$ and $a, b \in A$,

$$\lambda(ab) = (\lambda a)b = a(\lambda b).$$

holds.

Here we see explicitly that an algebra is a vector space in which we can take the product of vectors. Or, equivalently, we see that an algebra is a ring in which we can multiply each element by a scalar.

2.1.1 Examples of Vector Spaces

There are thousands of examples of vector spaces. We prefer to discuss examples for real vector spaces. These are more commonly used at the beginning of studies in physics than complex vector spaces. Real vector spaces are quite familiar to physicists and this facilitates to a great extend the understanding of the structures we discuss here.

In what follows, we are using an obvious notation, trying to avoid cumbersome definitions and pathological situations. In addition, we introduce the reader to the notation we use as systematically as possible.

Example 2.1
$$V = \mathbb{R}^0 := \{0\}.$$

This is the simplest but trivial vector space we have. Here, and in most examples below, the verification of the vector space-axioms are very straightforward.

Example 2.2
$$V = \mathbb{R} = \mathbb{R}^1 = \{v \in \mathbb{R}\}.$$

This is the simplest (nontrivial) vector space we can have. Both, scalars α and vectors v, are real numbers ($\alpha, v \in \mathbb{R}$).

Example 2.3
$$V = \mathbb{R}^2 = \mathbb{R} \times \mathbb{R} := \{x = \begin{bmatrix} \xi^1 \\ \xi^2 \end{bmatrix} : \xi^1, \xi^2 \in \mathbb{R}\}. \tag{2.1}$$

We may also write $x = e_1\xi^1 + e_2\xi^2$ with $e_1 = \begin{bmatrix} 1 \\ 0 \end{bmatrix}$ and $e_2 = \begin{bmatrix} 0 \\ 1 \end{bmatrix}$. We would like to clarify that multiplication is commutative, that is, we have (Definition 2.4) for example $e_1\xi^1 = \xi^1 e_1$.

This is the simplest typical example of a vector space. The scalars ξ^1, ξ^2 are the components, coefficients, coordinates of the vector x. $\mathbb{R}^2$ may also be seen as the coordinate plane. It is clear that the data of the vector $x \in \mathbb{R}^2$ is the list of length two (ξ^1, ξ^2) $\xi^1, \xi^2 \in \mathbb{R}$, but we choose, as is common practice for good reasons, to present this as a column which is a 2×1 matrix, as indicated in Eq. (2.1) with square brackets. We assume, as usual in physics, that the reader learnt very early to use matrices. In physics, we also like arrows! Here too, and we shall use them, especially when we want to emphasize that these vectors belong to the standard vector space $\mathbb{R}^n$ (here $n = 2$) written as columns. For this reason we freely use both notations, $x \in \mathbb{R}^2$ and $x \equiv \vec{\xi} = \left[\begin{smallmatrix}\xi^1 \\ \xi^2\end{smallmatrix}\right]$, $\xi^1, \xi^2 \in \mathbb{R}$. Here, the symbol "$\equiv$" indicates that we use a different notation for the same thing.

Comment 2.1 Lists and Matrices.

We write as usual a 2-tuple or a list of length 2, horizontally, with round brackets, symbolically like $(*, *)$, with a comma. We write a row in $\mathbb{R}^2$ which is a 1×2-matrix also horizontally but without comma (because this is the common way to write matrices), and with square brackets symbolically like $[**]$ in analogy to a column $\left[\begin{smallmatrix}*\\ *\end{smallmatrix}\right]$ in $\mathbb{R}^2$ as in Eq. (2.1).

On the other hand, if we want to write a vertical list of length 2 simply as data, and want to distinguish this list from the corresponding 2×1-matrix, we write it with round brackets but necessarily without comma, symbolically like $\left(\begin{smallmatrix}*\\ *\end{smallmatrix}\right)$. We proceed similarly in the case of $\mathbb{R}^n$.

This difference between a list and a matrix seems at first to be quite pedantic. But in linear algebra, whenever we consider linear combinations of vectors in $\mathbb{R}^n$ or vectors in V, it is better to talk of a linear combination with respect to a list of length n rather than of a linear combination of the entries of an $1 \times n$-matrix. It is clear that in most cases we identify lists with matrices. For example, we identify the standard basis in $\mathbb{R}^n$ which is the list$(e_1, \cdots, e_n)$ with the matrix $[e_1 \cdots e_n] = \mathbb{1}_n$:

$$(e_1, \cdots, e_n) = [e_1 \cdots e_n],$$

and similarly we may do the identification (symbolically)

$$\begin{pmatrix}*\\ \vdots \\ *\end{pmatrix} = \begin{bmatrix}*\\ \vdots \\ *\end{bmatrix}.$$

Example 2.4

$$V = \mathbb{R}^n = \left\{ x = \begin{bmatrix} \xi^1 \\ \vdots \\ \xi^i \\ \vdots \\ \xi^n \end{bmatrix} : \xi^1, \dots, \xi^n \in \mathbb{R} \right\}. \tag{2.2}$$

This is an extension of the Example 2.3 for all $n \in \mathbb{N} := \{1, 2, 3, \dots\}$. n is, as we know, the dimension of the vector space $\mathbb{R}^n$. The precise definition of the dimension is given in Sect. 3.1, after knowing precisely what a bases in a vector space is.
If we define

$$i \in I(n) := \{1, 2, \dots, n\},$$

we may write, as usually in various situations,

$$x = \vec{\xi} \equiv (\xi^i)_n \equiv (\xi^i).$$

The symbol "$\equiv$" here indicates only a different notation of the same object. We write the list of numbers $\xi^1, \xi^2, \dots, \xi^n$ as a column of length or size n, a $n \times 1$ matrix, written as a vertical list (without commas of course). It is clear that in general we cannot distinguish a $n \times 1$ matrix (column) from a vertical list of length n.

In physics, the vector space $\mathbb{R}^n$ is extremely relevant and well-known, especially for $n = 3$. The vector space $\mathbb{R}^3$ is the model for our Euclidean space which we may denote by E^3 in order to distinguish it from $\mathbb{R}^3$. It should be clear that the Euclidean space E^3 with its elements, the points $p \in E^3$ (we denote elements of E^3 with "p", "q", etc.), is not a vector space, and is therefore different to $\mathbb{R}^3$ whose elements we denote with "$\vec{x}$" and "$\vec{\xi}$", and so on. As we know, E^3 is a homogeneous space whereas $\mathbb{R}^3$ is not. The presence of the element $\vec{0} = \begin{bmatrix} 0 \\ 0 \\ 0 \end{bmatrix} \in \mathbb{R}^3$, the neutral element ($\vec{0} + \vec{\xi} = \vec{\xi}$), makes $\mathbb{R}^3$ nonhomogeneous. It is clear that the points of E^3 are not numbers and that we cannot add points. Nevertheless, we consider $\mathbb{R}^3$, not only in physics, as a perfect model for E^3. This enables us, among other things, to do calculations choosing a bijective correspondence ($p \leftrightarrow \vec{\xi}$) between the points p in E^3 and the three numbers ξ^1, ξ^2, ξ^3. The three numbers ξ^1, ξ^2, ξ^3 describe the position of the point p. This allows for example to formulate Newton's axioms in coordinates and to perform all the calculations we need in Newtonian mechanics. Similarly, $\mathbb{R}^n$ is the model of E^n for $n \in \mathbb{N}$.

Comment 2.2 Identification of vectors with points.

Very often in physics and mathematics, we identify $\vec{\xi}$ with p and we also call $\vec{\xi}$ (the three numbers) a point. We furthermore use $\vec{\xi}$ to denote a point p, a position which is not a vector. Here, we have to distinguish the following cases: A translation of a vector leads to the same vector. If we consider a translation of a point, we may think that we obtain another point. In other words, this identification means that we may consider a vector space as a manifold (a linear manifold).

Example 2.5

$$V = \mathbb{R}^{2\times 2} := \{A = \begin{bmatrix} \alpha & \beta \\ \gamma & \delta \end{bmatrix} : \alpha, \beta, \gamma, \delta \in \mathbb{R}\}.$$

This is the vector space of 2×2-matrices. $0 \equiv [0] \equiv \begin{bmatrix} 0 & 0 \\ 0 & 0 \end{bmatrix}$ is the zero. If we write

$$A = \begin{bmatrix} \alpha_1^1 & \alpha_2^1 \\ \alpha_1^2 & \alpha_2^2 \end{bmatrix} = (\alpha_s^i) \text{ with } \alpha_s^i \in \mathbb{R}, \quad i, s \in I(2)$$

and

$$B = (\beta_s^i) \quad C = (\gamma_s^i),$$

the addition is given component-wise, $\alpha_s^i + \beta_s^i = \gamma_s^i$ which is $A + B = C$. The scalar multiplication is given by

$$\mathbb{R} \times \mathbb{R}^{2\times 2} \longrightarrow \mathbb{R}^{2\times 2},$$

$$(\alpha, B) \longmapsto \alpha B := \begin{bmatrix} \alpha\beta_1^1 & \alpha\beta_2^1 \\ \alpha\beta_1^2 & \alpha\beta_2^2 \end{bmatrix}.$$

Example 2.6

$$V = \mathbb{R}^{2\times 3} := \left\{ A = \begin{bmatrix} \alpha_1^1 & \alpha_2^1 & \alpha_3^1 \\ \alpha_1^2 & \alpha_2^2 & \alpha_3^2 \end{bmatrix} = (\alpha_s^i) \right\}$$

with $\alpha_s^i \in \mathbb{R}$}, and with $i \in I(2)$, and $s \in I(3)$.

We thereby have correspondingly to the previous example the vector space of 2×3-matrices.

Example 2.7

$$V = \mathbb{R}^{\mathbb{N}} := \{x := (\xi_n)_n \equiv (\xi_1, \xi_2, \xi_3, \ldots), \ \xi_1, \xi_2, \xi_3, \ldots \in \mathbb{R}\}.$$

We here use the notation $Y^X \equiv \text{Map}(X, Y)$ for the maps $\text{Map}(X, Y) = \{f : X \to Y\}$ between the sets X and Y. V is the vector space of sequences and we can interpret the sequence x as the map

$$x : \mathbb{N} \longrightarrow \mathbb{R}$$
$$n \longmapsto x(n) := \xi_n \in \mathbb{R}.$$

Zero is $0 = (0, 0, \ldots)$. Addition and scalar multiplication are given componentwise. For $z = x + y$ and $\alpha \cdot x$ ($y = (\eta_n)_n$, $z = (\zeta_n)_n$), we write:

$$z(n) := (x + y)(n) := x(n) + y(n)$$

and

$$(\alpha \cdot x)(n) := \alpha x(n).$$

This means that $\zeta_n := (\xi + \eta)_n := \xi_n + \eta_n$ and $(\alpha \cdot x)_n := \alpha \xi_n$. Finally, we write again $z = x + y$ and αx. Note that for $\mathbb{R}^n$, we can write in the above notation $\mathbb{R}^n \equiv \mathbb{R}^{I(n)}$.

Example 2.8

$$V = \mathbb{R}^{(\mathbb{N}_0)} := \{\alpha := (\alpha_0, \alpha_1, \ldots, \alpha_m) : \alpha_0, \alpha_1, \ldots \alpha_m \in \mathbb{R} \text{ and}$$
$$m \in \mathbb{N}_0 = \{0, 1, 2, \ldots\}\}.$$

V is the vector space of the interrupted sequences. Zero, addition, and scalar multiplication are given as in the previous example. It turns out that this vector space is equivalent to the space of polynomials denoted by

$$\mathbb{R}[x] = \{\alpha(x) := \sum_{k=0}^{m} \alpha_k x^k, m \in \mathbb{N}_0\}.$$

Then we get the vector isomorphism (without proof)

$$\mathbb{R}^{(\mathbb{N}_0)} \cong \mathbb{R}[x].$$

Example 2.9

$$V = \mathbb{R}^X \equiv \text{Map}(X, \mathbb{R}) \equiv \mathcal{F}(X) := \{f, g, \dots\},$$

with f given by

$$f : X \longrightarrow \mathbb{R}$$
$$x \longmapsto f(x).$$

Zero, addition and scalar multiplication are in analogy to the Example 2.7. We therefore have for zero the constant map $\hat{0}$.

$$\hat{0} : X \longrightarrow \mathbb{R},$$
$$x \longmapsto \hat{0}(x) := 0 \in \mathbb{R}$$

for all $x \in X$,

$$(f + g)(x) := f(x) + g(x) \quad \text{and}$$
$$(\alpha f)(x) := \alpha f(x).$$

Example 2.10

$$V = C^0(X).$$

V is the set of all continuous functions. Zero, addition, and scalar multiplication are defined as in the previous Example 2.9. From analysis, we know that $C^0(X)$ is a vector space: For example, the addition of two continuous functions is a continuous function too.

Example 2.11
$$V = C^1(X).$$

The set of differentiable functions. Analogously to the previous Examples 2.9 and 2.10, we know that $C^1(X)$ has a vector space structure too.

Example 2.12
$$V = \mathcal{L}(X).$$

The set of integrable functions in X. We may take as an example the interval $X = [-1, 1] \in \mathbb{R}$. $\mathcal{L}(X)$ is also a vector space since addition and scalar multiplication of integrable functions produce integrable functions.

Example 2.13
$$V = Sol(A, 0).$$

The set of solutions (Sol) of a homogeneous linear equation given by a matrix A.
Suppose $A \in \mathbb{K}^{1 \times n}$, is the row

$$A = [\alpha_1 \alpha_2 \cdots \alpha_n] \equiv (\alpha_s)_n \qquad \alpha_s \in \mathbb{R}, \quad s \in I(n)$$

and $Sol(A, 0)$ is the solution to the equation

$$\alpha_1 \xi^1 + \alpha_2 \xi^2 + \cdots + \alpha_n \xi^n = 0. \tag{2.3}$$

If we use the Einstein convention for the summation,

$$\sum_{s=1}^{n} \alpha_s \xi^s = \alpha_s \xi^s. \tag{2.4}$$

Equation 2.3 takes the form
$$\alpha_s \xi^s = 0 \tag{2.5}$$

or, with $x = \vec{\xi}$ as matrix equation, the form

$$Ax = 0. \tag{2.6}$$

The set of all solutions $Sol(A, 0)$ of Eq. 2.3, is a vector space: If x and y are solutions of Eq. 2.3, the sum and the scalar product are also solutions of Eq. 2.3. If $x, y \in Sol(A, 0) : Ax = 0, Ay = 0$, then it follows directly that

$$A(x + y) = Ax + Ay = 0 + 0 = 0,$$
$$A(\lambda x) = \lambda Ax = \lambda 0 = 0 (\lambda \in \mathbb{R})$$

are also valid.

Remark 2.1 $\mathbb{R}$- and $\mathbb{C}$-vector spaces.

If we replace $\mathbb{R}$ by $\mathbb{C}$ in all the above examples, nothing changes formally in the existing structures. Therefore, the discussion applies equally well to real and complex vector spaces.

Comment 2.3 Linear combinations.

All the above equations, Eqs. (2.3) to (2.6), contain linear combinations. As we shall see in Sect. 3.1 and Remark 3.1, it might not be exaggerated to claim that linear combinations are the most important operation in linear algebra. The precise definition of a linear combination is given in Sect. 3.1, Definition 3.3.

The next example can have a very sobering effect on physicists!

Example 2.14 What is a vector?.

We consider the circle with radius r in $\mathbb{R}^2$. See Fig. 2.1.

$$S^1 = \{x = \begin{bmatrix} \xi \\ \eta \end{bmatrix} \in \mathbb{R}^2 : \sqrt{\xi^2 + \eta^2} = r > 0\}.$$

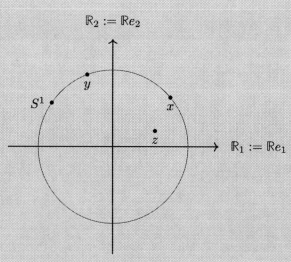

Fig. 2.1 Circle S^1 in $\mathbb{R}^2$

We choose for example the elements $x, y \in S^1 \subset \mathbb{R}^2$ and $z \in \mathbb{R}^2$ but $z \notin S^1$. We may ask: Are x and y vectors? The answer is not clear because the question is not clear. The elements of S^1, x and y, are obviously not vectors since S^1 is not a vector space. The elements x, y, z of $\mathbb{R}^2$ are obviously vectors since $\mathbb{R}^2$ is a vector space. To decide whether an element is a vector or not, we need to know where it belongs to. The elements of a vector space are, of course, vectors. The elements of a circle are not vectors, we call them points. This clarifies the paradox that x and y are not vectors and at the same time x and y are vectors.

Comment 2.4 What is a point?

All this reminds us of Euclid's approach to geometry. Euclid does never tell us what a point is, he only says how the points behave towards each other. If we compare this with Comment 2.2, we see the difference. We considerate elements on the circle S^1, x and y not as vectors but, since S^1 is not a vector space, as points in S^1.

At this stage, it is natural to address subsets of V which have the same structure as the vector space V. This means that a subset U should be by itself also a vector space. In this sense, we may say that for the subset U of V, we would like to stay in the vector space category, and we write $U \leqslant V$. This leads to the following definition.

Definition 2.6 Subspace.
A subset U of V is called subspace of V if U is also a vector space (with the same structure).

Remark 2.2 Criterion for a subspace.

A necessary and sufficient condition (a criterion) for a subset $U \subseteq V$ to be a subspace of V ($U \leqslant V$), is given if for $\lambda \in \mathbb{K}, u, v \in U$; $0, u + v, \lambda u \in U$ holds. This means in detail:

(i)   The neutral element 0 belongs to U;
(ii) U is closed under addition;
(iii) U is closed under scalar multiplication.

2.1.2   Examples of Subspaces

Example 2.15 $\{0\}$ and V are clearly subspaces of V.

Example 2.16

Given $V = \mathbb{R}^2 = \mathbb{R} \times \mathbb{R}$.

(i)   The x-axis $U_1 := \mathbb{R} \times \{0\} = \begin{bmatrix} 1 \\ 0 \end{bmatrix} \mathbb{R} = \{ \begin{bmatrix} x \\ 0 \end{bmatrix} : x \in \mathbb{R} \}$ and
      the y-axis $U_2 := \{0\} \times \mathbb{R} = \begin{bmatrix} 0 \\ 1 \end{bmatrix} \mathbb{R} = \{ \begin{bmatrix} 0 \\ x \end{bmatrix} : x \in \mathbb{R} \}$,
      are subspaces of V and we write $U_1, U_2 \leqslant V$;
(ii) For every v different from zero ($v \neq 0$), the set $U_v := \mathbb{R}v = \{ x \in \mathbb{R}^2 :$
      $x = \lambda v, \lambda \in \mathbb{R} \}$ is a subspace of V;
(iii) The set of solutions of the equation $\alpha_1 \xi^1 + \alpha_2 \xi^2 = 0$ with $A = [\alpha_1, \alpha_2] \in$
      $\mathbb{R}^{1 \times 2}$ is a subspace of V. Assuming that $A \neq \underset{\sim}{0} \equiv [00]$, we may write for

      the above set of solutions $U = Sol(A, \vec{0})$

$$U = \mathbb{R}v_0 \quad \text{with} \quad v_0 = \begin{bmatrix} \alpha_2 \\ -\alpha_1 \end{bmatrix}.$$

Example 2.17 Standard subspaces in $\mathbb{R}^3$.

Given that $V = \mathbb{R}^3 = \mathbb{R}_1 \times \mathbb{R}_2 \times \mathbb{R}_3$, $U_{1,2}$, $U_{1,3}$, $U_{2,3}$ and U_0 are subspaces of V.

(i) $U_{1,2} = \mathbb{R}^2 \times \{0\} = \left\{ \begin{bmatrix} x \\ y \\ 0 \end{bmatrix} : x, y, \in \mathbb{R} \right\} \leqslant V$;

(ii) $U_{2,3} = \left\{ \begin{bmatrix} 0 \\ y \\ z \end{bmatrix} : y, z \in \mathbb{R} \right\} \leqslant V$;

(iii) $U_{1,3} = \left\{ \begin{bmatrix} x \\ 0 \\ z \end{bmatrix} : x, z \in \mathbb{R} \right\} \leqslant V$;

(iv) $U_0 = Sol(A, \vec{0})$, the solution of the equation $A\vec{x} = \vec{0}$ with $A = [\alpha_1 \alpha_2 \alpha_3] \in \mathbb{R}^{1 \times 3}$ is also subspace of V.

Example 2.18 Given that $V = \text{Map}(\mathbb{R}, \mathbb{R})$, $C^1(\mathbb{R})$ and $C^0(\mathbb{R})$ are subspaces of V, we may write

$$C^1(\mathbb{R}) \leqslant C^0(\mathbb{R}) \leqslant V.$$

Comment 2.5 Union of two subspaces.

Similarly, it is natural to require for the union $U_1 \cup U_2$ of the two subspaces U_1 and U_2 the same structure as for V. But we may immediately realize that this is, in general impossible. (We may choose $u_1, u_2 \in U_1 \cup U_2$ so that $u_1 + u_2 \notin U_1 \cup U_2$.) The notion of the sum $U_1 + U_2$ and $u_1 + u_2 \in U_1 + U_2$ is defined in Sect. 2.7, Definition 2.19. Here, there exists uniquely, as we shall see in Sect. 2.7, the smallest subspace of V, which contains the set $U_1 \cup U_2$.

Example: In $V = \mathbb{R}^2$ we take $U_1 := \mathbb{R}_1 = \{(x, 0) : x \neq 0, x \in \mathbb{R}\}$ and $U_2 = \mathbb{R}_2 := \{(0, y), y \neq 0, y \in \mathbb{R}\}$. In this case it is clear that for the example $u_1 = (1, 0)$, and $u_2 = (0, 2)$, we get $u_1, u_2 \in U_1 \cup U_2$ but $u_1 + u_2 = (1, 2) \notin U_1 \cup U_2$. It is evident that $U_1 \cup U_2$ is not a vector space nor even a subspace of $V = \mathbb{R}^2$. Geometrically speaking, this is perfectly clear. The vector $(1, 2)$, for example, belongs neither to the x-axis nor to the y-axis, and consequently not to their union. To simplify our notation, we write here $u_1 = (1, 0)$ instead of $u_1 = [\begin{smallmatrix} 1 \\ 0 \end{smallmatrix}]$.

Remark 2.3 Intersection of two subspaces.

It is interesting to notice what is easy to see here, that the intersection $U_1 \cap U_2$ (opposite to $U_1 \cup U_2$) is a subspace of V. This is also plausible because the intersection is a much stronger condition than the union.

2.2 Linear Maps and Dual Spaces

It is essential and beneficial to use the maps that conserve the mathematical structure for every mathematical structure. Here, these so-called homomorphisms are the linear maps.

Definition 2.7 Linear maps and $\mathrm{Hom}(V, V')$.
A map f from V to V' $f : V \to V'$ is linear (or homomorph) if

(i) $f(u + v) = f(u) + f(v)$ $\longrightarrow$ additivity ;
(ii) $f(\alpha v) = \alpha f(v)$ $\longrightarrow$ homogeneity
or, equivalently, if $f(\alpha v + \beta v) = \alpha f(v) + \beta f(v)$ holds.

$\mathrm{Hom}(V, V')$ is the set of all linear maps from V to V'.

A few further comments about linear maps: We would like to remind the reader that we do not assume that they come upon this definition for the first time and the same holds for most definitions and many propositions and theorems in this book. But we are convinced that essential and fundamental facts have to be repeated and thus this is by no means a loss of time. It is further an excellent opportunity to fix our notation. In this sense, we remember that a subspace U of V (denoted by $U \leqslant V$ to distinguish it from the symbol $\subseteq$ (subset)) is a vector space in its own right by the restriction of addition and scalar multiplication to U.

2.2.1 Examples of Linear Maps

Example 2.19 The zero map $f_0 \in \mathrm{Hom}(V, V')$.

The value of f_0 is zero at each $v \in V$:

$$f_0 : V \longrightarrow V',$$
$$v \longmapsto f_0(v) := 0' \in V'.$$

f_0 is linear since $f_0(\lambda v) = f_0(u + v) = 0'$. We have

$$f_0(\alpha v_1 + \beta v_2) = 0' = 0' + 0' = f_0(\alpha v_1) + f_0(\beta x_2).$$

Example 2.20 The identity map $f = id$.

This is given by

$$f : V \longrightarrow V,$$
$$v \longmapsto f(v) := v.$$

f is linear since $f(\alpha v_1 + \beta v_2) = \alpha v_1 + \beta v_2 = \alpha f(v_1) + \beta f(v_2)$.

Example 2.21 The map $f_\lambda = \lambda id, \lambda \in \mathbb{R}$ is linear.

It is linear since

$$f_\lambda(\alpha v_1 + \beta v_2) = \lambda(\alpha v_1 + \beta v_2)$$
$$= \alpha \lambda v_1 + \beta \lambda v_2$$
$$= \alpha f_\lambda(v_1) + \beta f_\lambda(v_2).$$

Maps similar to f_λ in the form $f_\lambda : U \to U$ for $U \leqslant V$ are widely used in quantum mechanics and in physics wherever symmetries are relevant.

Example 2.22 A parallel projection map $f = P_U$.

For $V = \mathbb{R}^3$, we define

$$U := \left\{ \begin{bmatrix} \xi^1 \\ \xi^2 \\ 0 \end{bmatrix} : \xi^1, \xi^2 \in \mathbb{R} \right\} \quad \text{and}$$

$$W := \left\{ \begin{bmatrix} 0 \\ 0 \\ \xi \end{bmatrix} : \xi \in \mathbb{R} \right\}.$$

It is apparent that we have $V = U + W$ as we shall see later in Definition 2.19 and Sect. 2.8. For each $v \in V$, $v = u + w$, $u \in U$ and $w \in W$.

$$f \;:\; V \longrightarrow V$$
$$v = u + w \longmapsto f(v) := u.$$

The linearity is clear, and thus the image of f is $imf = f(V) = U$. Note that for this projection, we have to use both the subspace U, and the subspace W. A parallel projection corresponds to the usual parallelogram rule when adding two vectors. Here, we did not use the dot product in $\mathbb{R}^3$ which means that we do not have to use orthogonality.

In physics, we use for the most part orthogonal projections. These are, as we will show later, connected with the inner product: here the dot product (see Sect. 10.4). In this case, we only need to know the space $U = imf$. The corresponding complement (here W) is given by the orthogonality.

Example 2.23 $f = f_A$, a matrix induced map.

For $V = \mathbb{R}^n$ with $n = 2$, we define $v = \vec{\xi} = \begin{bmatrix} \xi^1 \\ \xi^2 \end{bmatrix}$ and $A = \begin{bmatrix} \alpha_1^1 & \alpha_2^1 \\ \alpha_1^2 & \alpha_2^2 \end{bmatrix} \in \mathbb{R}^{2\times 2}$. The matrix A induces the map:

$$f : V \longrightarrow V$$
$$\begin{bmatrix} \xi^1 \\ \xi^2 \end{bmatrix} \longmapsto \begin{bmatrix} \alpha_1^1\xi^1 + \alpha_2^1\xi^2 \\ \alpha_1^2\xi^1 + \alpha_2^2\xi^2 \end{bmatrix} = \begin{bmatrix} \alpha_s^1\xi^s \\ \alpha_s^2\xi^s \end{bmatrix}.$$

Using the Einstein convention for the summation, we obtain componentwise, with $s \in I(2) = \{1, 2\}$:

$$f\left(\vec{\xi}\right)^1 = \alpha_s^1\xi^s \text{ and } f\left(\vec{\xi}\right)^2 = \alpha_s^2\xi^s.$$

With the matrix form, the two above equations lead to a single one:

$$f(\vec{\xi}) = A\vec{\xi}.$$

A quite straightforward approach allows to see the linearity of $f \equiv f_A$. With $\lambda, \mu \in \mathbb{R}$, we obtain:

$$f(\lambda\vec{\xi} + \mu\vec{\eta}) = A(\lambda\vec{\xi}) + A(\mu\vec{\eta}) = \alpha_s^i(\lambda\xi^s) + \alpha_s^i(\mu\eta^s) =$$
$$= \lambda\alpha_s^i\xi^s + \mu\alpha_s^i\eta^s = \lambda f(\vec{\xi}) + \mu f(\vec{\eta}).$$

Example 2.24 Differentiation $D := \frac{d}{dx}$.

For $V = C^1$, whose elements we denote by f, we define the map

$$D : C^1(\mathbb{R}) \longrightarrow C^0(\mathbb{R})$$

$$f \longmapsto Df := \frac{df}{dx}.$$

As is commonly known from analysis, D is a linear map ($\alpha, \beta \in \mathbb{R}$):

$$D(\alpha f_1 + \beta f_2) = \frac{d}{dx}(\alpha f_1 + \beta f_2) = \alpha \frac{df_1}{dx} + \beta \frac{df_2}{dx} = \alpha Df_1 + \beta Df_2.$$

Example 2.25 Integration.

For $V = \mathcal{L}^1(X) = \{f\}$, the space of integrable functions in the interval $X = [0, 1] \subset \mathbb{R}$. We consider the integral

$$J : \mathcal{L}^1(x) \longrightarrow \mathbb{R},$$

$$f \longmapsto J(f) := \int_X f(x)dx.$$

Once again we know from analysis that J is a linear map:

$$J(\alpha f_1 + \beta f_2) = \int_X (\alpha f_1 + \beta f_2)dx =$$

$$= \int_X \alpha f_1 dx + \int_X \beta f_2 dx =$$

$$= \alpha \int_X f_1 dx + \beta \int_X f_2 dx =$$

$$= \alpha J(f_1) + \beta J(f_2).$$

From all these examples of vector spaces and linear maps, we may learn that if we have thousands of vector spaces, we expect millions of linear maps. In addition, taking into account the above examples, we may observe that the image set of vector spaces by linear maps is again a vector space.

Remark 2.4 Some useful well-known facts about linear maps.

A homomorphism $f \in \text{Hom}(V, V')$ is called an isomorphism if f is bijective: $f \in Iso(V, V')$. A homomorphism is called an endomorphism or a linear operator if $V = V' : f \in \text{End}(V) := \text{Hom}(V, V)$, or is called an automorphism if $V = V'$ and f is bijective: $f \in Aut(V)$. In consequence, we have the obvious hierarchy of spaces.

$$Iso(V, V') \subset \text{Hom}(V, V') \leqslant \text{Map}(V, V').$$

The above notation with the symbol "$\leqslant$", indicates that the set Hom (V, V') and Map(V, V') are vector spaces. The set $Iso(V, V')$ is not a vector space since, for example, the sum of two such bijective maps is not necessarily bijective.

In the case $V = V'$ if we take the composition as multiplication,

$$gf := g \circ f \text{ for } f, g \in \text{End}(V),$$

it is not difficult to see that $\text{End}(V)$ is a ring. Hence with the above composition, we have an additional multiplicative operation in $\text{End}(V)$. It is isomorphic to a matrix ring, but we have not yet seen this (see Sect. 3.3, Corollary 3.8 and Comment 3.3). Similar connections of linear maps to matrices are dominant throughout linear algebra. As we shall see, with the help of bases we can entirely describe, that is, represent linear maps and their properties by matrices. In this sense, if we ask what linear algebra is, we may simply state that linear algebra is the theory of matrices.

Definition 2.8 Kernel and image.
The kernel and image of a linear map f are given by:

$$\ker f := \{v \in V : f(v) = 0_{V'}\} \text{ and } \text{im } f := f(V).$$

For $v' \in V'$, the preimage (or fiber) of $v' \in V'$ is given by

$$f^{-1}(v') := \{v \in V : f(v) = v'\}.$$

The kernel is often called null space and the image is called range.
We note some direct conclusion from the definition:

Remark 2.5 Some important properties of linear maps.

Let $f : V \to V'$. Then:

- $f(0_V) = 0_{V'}$.
- If U and W are subspaces of V and V' respectively, (i.e., $U \leq V$ and $W \leq V'$), then $f(U) \leq V'$, $f^{-1}(W) \leq V$ and $\ker f$, $\operatorname{im} f$ are subspaces of V and V' as well.
- If f is an isomorphism, then $f^{-1} : V' \to V$ is an isomorphism, that is, f^{-1} is also linear, as f.
- f injective $\Leftrightarrow \ker f = 0_V$ and f surjective $\Leftrightarrow \operatorname{im} f = V'$.
- If $w \in \operatorname{im} f$, then for every $v_0 \in f^{-1}(w)$, $f^{-1}(w) = v_0 + \ker f := \{v_0 + u : u \in \ker f\}$.

Furthermore, if we know what the dimension of a vector space is (see Definition 3.7), we can also give a number for f which we call rank: $\operatorname{rank}(f) := \dim(\operatorname{im} f)$.

Linear maps themselves build vector spaces. The most prominent example is the dual space of V denoted by V^*. This contains all linear functions from V to $\mathbb{K} : V^* := \operatorname{Hom}(V, \mathbb{K})$.

Definition 2.9 The dual space V^*.
For a given vector space V over $\mathbb{K}$, the dual space of V is given by $V^* = \operatorname{Hom}(V, \mathbb{K})$.

This contains all homomorphisms, here all linear functions from V to $\mathbb{K}$. We denote the elements of V^* preferentially by greek letters, for example $\alpha, \beta, \ldots, \xi, \eta, \theta, \ldots$ and we thus have $V^* = \{\alpha, \beta, \xi, \eta, \theta, \ldots\}$. $\xi \in V^*$ is a linear function:

$$\xi : V \longrightarrow \mathbb{K},$$

$$v \longmapsto \xi(v).$$

with $\xi(\lambda_1, v_1 + \lambda_2 v_2) = \lambda \xi(v_1) + \lambda_2 \xi(v_2)$ with $\lambda_1, \lambda_2 \in \mathbb{K}$. ξ is also called linear form or linear functional. V^* is of course a vector space.

Comment 2.6 On the connection between V and V^*.

Remarkably enough, V^* has the same dimension as V. We consider only finite-dimensional vector spaces throughout this book. It turns out that V and V^* are isomorphic ($V \cong V^*$). But for a given abstract vector space V, no natural isomorphism exists between V and V^*. An isomorphism $\Phi : V \to V^*$ depends on the chosen basis, and there exist many of them. Although V and V^* are isomorphic, these two vector spaces are imperceptibly different. This causes a lot of difficulties, especially in connection with the use of tensors in physics. This is why we are going to deal with it later.

We could cure this problem by introducing a scalar product in the vector space V. A Euclidean or a semi-Euclidean vector space with more structure than an abstract vector space allows to identify the two spaces V and V^*. But we can only fully understand the above mentioned subtlety if we deal explicitly with the dual space V^*, in both cases we take V without and with the inner product. We must admit that we cannot understand tensors if we do not understand broadly the role of the dual space V^*. Therefore we do not avoid, as is usually done in introductions to elementary theoretical physics, but on the contrary, we underline the role of V^* in the present book. As already mentioned, the role of V^* is important to understand tensors.

2.3   Vector Spaces with Additional Structures

We now introduce the most prominent example of an additional structure in an abstract vector space.

Definition 2.10 Inner products.
An inner product on V is given by the function

$$\langle | \rangle : \quad V \times V \longrightarrow \mathbb{K},$$
$$(w, v) \longmapsto \langle w \mid v \rangle.$$

It has the following properties:

(i) linearity in the second slot (argument):
$\langle w \mid u + v \rangle = \langle w \mid u \rangle + \langle w \mid v \rangle$ and $\langle w \mid \lambda v \rangle = \lambda \langle w \mid v \rangle$;
(ii) conjugate symmetry:
$\langle v \mid w \rangle = \overline{\langle w \mid v \rangle}$;
(iii) positive definiteness :
$\langle v \mid v \rangle > 0$ if $v \neq 0$.

In physics, we usually define linearity in the second slot, as above in (i), not in the first one.

For $\mathbb{K} = \mathbb{R}$, we call $(V, \langle | \rangle)$ Euclidean vector space or real inner product space.

For $\mathbb{K} = \mathbb{C}$, we call $(V, \langle | \rangle)$ unitary vector space or complex inner product space or Hilbert space (for finite dimensions). It corresponds usually to a finite-dimensional subspace of the Hilbert space in quantum mechanics.

It follows from this definition by direct calculation,

(a) additivity in the first slot $\langle w + u \mid v \rangle = \langle w \mid v \rangle + \langle u \mid v \rangle$;

(b) and $\langle \lambda w \mid v \rangle = \bar{\lambda} \langle w \mid v \rangle$.

For (a) we have from (i) and (ii)

$$\langle w + u \mid v \rangle = \overline{\langle v \mid w + u \rangle} = \overline{\langle v \mid w \rangle + \langle v \mid u \rangle} = \overline{\langle v \mid w \rangle} + \overline{\langle v \mid u \rangle} = \langle w \mid v \rangle + \langle u \mid v \rangle.$$

For (b) we have by analogy

$$\langle \lambda w \mid v \rangle = \overline{\langle v \mid \lambda w \rangle} = \overline{\langle v \mid w \rangle \lambda} = \overline{\langle v \mid w \rangle} \bar{\lambda} = \langle w \mid v \rangle \bar{\lambda}.$$

The inner product $\langle | \rangle$ is called symmetric in the case of a $\mathbb{R}$ vector space and Hermitian in the case of a $\mathbb{C}$ vector space. As a result, we see altogether that the real inner product is a positive definite symmetric bilinear form. The complex inner product is analogously what is called a Hermitian sesquilinear form positive definite. Note that $\langle v \mid v \rangle$ is real and nonnegative, for a complex vector space too.

Bearing in mind the very important applications in physics, it is instructive to discuss again and separately the situation for a real vector space. This leads to some more definitions for the special case of $\mathbb{K} = \mathbb{R}$ where the scalars are real numbers.

Definition 2.11 The inner product on a $\mathbb{R}$-vector space V.

$\langle | \rangle$ is now a bilinear form, symmetric and positive definite. Bilinear means that if we write $\sigma(x, y) \equiv \langle x \mid y \rangle$ for $x, y \in V$.

(i) σ is a bilinear form:

$$\sigma(x_1 + x_2, y) = \sigma(x_1, y) + \sigma(x_2, y)$$
$$\sigma(x, y_1 + y_2) = \sigma(x, y_1) + \sigma(x, y_2).$$

Therefore σ is linear in both slots;

(ii) σ is symmetric:

$$\sigma(x, y) = \sigma(y, x);$$

(iii) σ is positive definite. Hence we have:

   (a) $0 \leq \sigma(x, x)$ and
   (b) $\sigma(x, x) = 0 \Leftrightarrow x = 0$.

This means:

(a) $\sigma(x, x)$ is never negative,
(b) $\sigma(x, x)$ is never zero if $x \neq 0$.

   This leads to further definitions.

Definition 2.12 Non-degenerate bilinear form.

   σ is called nondegenerate in the second variable whenever $\sigma(x, y) = 0$ for all $x \in V$, then this implies that $y = 0$ too.
   σ is nondegenerate if it is nondegenerate in both variables.

   From (b) it results that σ is nondegenerate. Note that for nondegenerate bilinear forms, $\sigma(x, x)$ can be negative.

Definition 2.13 Semi-Euclidean vector space.
   (V, σ) with a nondegenerate, that is not necessarily positive definite σ, is called a semi-Euclidean vector space.

   The use of inner products leads to different structures and properties for the vectors in V. The most important ones are the norm, the orthogonality, the Pythagorean theorem, the orthogonal decomposition, the Cauchy-Schwarz inequality, the triangular inequalities, and the parallelogram equality.

Definition 2.14 Norm or length, $\|v\|$.

   For $v \in V$, the norm of v denoted by $\|v\|$ is given by $\|v\| := \sqrt{\langle v \mid v \rangle}$.

   This leads to a positive definiteness, $\|v\| = 0 \Leftrightarrow v = 0$ and to a positive homogeneity $\|\lambda v\| = |\lambda| \, \|v\|$. Both are easily recognizable since if $v \in V$, we have $\|v\|^2 = \langle v \mid v \rangle = 0$ if and only if $v = 0$. Similarly, we have for $\lambda \in \mathbb{C}$,

$$\|\lambda v\|^2 = \langle \lambda v \mid \lambda v \rangle = \bar{\lambda} \langle v \mid \lambda v \rangle = \bar{\lambda} \langle v \mid v \rangle \lambda = \bar{\lambda} \lambda \langle v \mid v \rangle = |\lambda|^2 \|v\|^2.$$

Definition 2.15 Orthogonality.
The vectors $u, v \in V$ are orthogonal if $\langle u \mid v \rangle = 0$.
It follows that $0 \in V$ is orthogonal to every vector $\langle 0 \mid v \rangle = 0$.

Theorem 2.1 *The Pythagorean theorem.*
If u and v are orthogonal in V, that is, $\langle u \mid v \rangle = 0$), then we have:

$$\|u + v\|^2 = \|u\|^2 + \|v\|^2.$$

Note that for a real vector space, if we have $\|u + v\|^2 = \|u\|^2 + \|v\|^2$, it follows that u and v are orthogonal:

$$0 = \|u + v\|^2 - \|u\|^2 - \|v\|^2 = \langle u \mid v \rangle + \langle v \mid u \rangle = 2\langle u \mid v \rangle \Rightarrow \langle u \mid v \rangle = 0.$$

This means that $\|u + v\|^2 = \|u\|^2 + \|v\|^2 \Leftrightarrow \langle u \mid v \rangle = 0$.
For a complex vector space, we obtain analogously

$$\|u + v\|^2 = \|u\|^2 + \|v\|^2 \Leftrightarrow \langle u \mid v \rangle + \langle v \mid u \rangle = 0 \Leftrightarrow Re\langle u \mid v \rangle = 0.$$

Definition 2.16 Orthogonal projection and decomposition.
We consider $v, b \in V$ with $b \neq 0$. The orthogonal projection of v onto b is given by

$$P_b v = b \frac{\langle b \mid v \rangle}{\langle b \mid b \rangle}.$$

We define $v_b := b \frac{\langle b \mid v \rangle}{\langle b \mid b \rangle}$, so $v = v_b + v_c$ and $v_c = v - v_b$. A simple calculation shows that $\langle b \mid v_c \rangle = 0$ so v_c is orthogonal to b (Fig. 2.2).
This leads to the orthogonal projection P_b:

$$P_b : \quad V \longrightarrow V,$$

$$v \longmapsto P_b(v) := b \frac{\langle b \mid v \rangle}{\langle b \mid b \rangle} \in \mathbb{K}b.$$

A direct inspection shows that P_b is a linear map and projects the vectors of V orthogonally to the one-dimensional subspace $\mathbb{K}b$. P_b, as any projection operator, is idempotent:

$$P_b^2 = P_b.$$

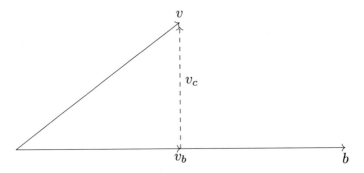

Fig. 2.2 The orthogonal projection

This follows directly from the following: for all $v \in V$,

$$P_b\, P_b v = P_b\, v_b = b\frac{\langle b \mid v_b \rangle}{\langle b \mid b \rangle}$$

$$= \frac{b}{\langle b \mid b \rangle}\langle b \mid v_b \rangle$$

$$= b\frac{1}{\langle b \mid b \rangle}\langle b \mid b \rangle\frac{\langle b \mid v \rangle}{\langle b \mid b \rangle}$$

$$= b\frac{\langle b \mid v \rangle}{\langle b \mid b \rangle}$$

$$= P_b v.$$

This shows that $P_b P_b = P_b$.

At this point, it is interesting to observe that for the projection we may also write $P_b v =\mid b)\frac{\langle b \mid v\rangle}{\langle b \mid b\rangle} = \frac{\mid b\rangle\langle b\mid}{\langle b\mid b\rangle} \mid v)$. This refers to the Dirac notation we use in quantum mechanics (see Sect. 6.4).

If we start with a vector e equal to 1, that is, $\|e\| = 1$, we simply get

$$P_e v = e\langle e \mid v \rangle.$$

If we write $e_b := \frac{b}{\|b\|}$, we obviously get $P_b = P_{e_b}$ since

$$P_b\, v = e_b\langle e_b \mid v \rangle.$$

Note that up to now we used the orthogonal decomposition for the vector v, relatively to the vector $b \neq 0$! We may write

$$id_V\, v = v = v_b + v_c = P_b v + v_c = P_b v + (id_V - P_b)v$$

for all $v \in V$. This leads to the very useful decomposition of the identity id_V into the two projection operations:

$$id_V = P_b + (id_V - P_b).$$

The existence of orthogonal projections is one of the most practical applications of the inner product. See also Sect. 10.4.

As an example, we show this by the proof of the following proposition.

Proposition 2.1 *The Cauchy-Schwarz inequality:* $\forall u, v \in V$

$$| \langle u \mid v \rangle | \leq \|u\| \|v\|.$$

The equality holds if and only if u and v are colinear.

Proof For the inequality, we assume that $u \neq 0$ and we define the projection as $P_u(v) = u \frac{\langle u|v \rangle}{\langle u|u \rangle} =: v_u$ and $v_c := v - v_u$. Hence we have $v = v_u + v_c$. Since

$$\|v\|^2 = \|v_u + v_c\|^2 = \|v_u\|^2 + \|v_c\|^2,$$

it follows that $\|v\| \geq \|v_u\|$ such that

$$\|v\| \geq \|u\| \left| \frac{\langle u \mid v \rangle}{\langle u \mid u \rangle} \right| = \|u\| \frac{| \langle u \mid v \rangle |}{\|u\|^2} = \frac{| \langle u \mid v \rangle |}{\|u\|}.$$

$\square$

That is why the Cauchy-Schwarz inequality indicates purely that an orthogonal projection is less than the original. Further, we have if $\|v_c\| = 0 (v_u = v)$, then v and u are colinear.

Corollary 2.1 *Triangular inequality*
$$\|u + v\| \leq \|u\| + \|v\|.$$

Proof This follows essentially from the Cauchy-Schwarz inequality.

$$\|u + v\|^2 = \langle u + v \mid u + v \rangle = \langle u \mid u \rangle + \langle u \mid v \rangle + \langle v \mid u \rangle + \langle v \mid v \rangle =$$
$$= \|u\|^2 + 2Re\langle u \mid v \rangle + \|v\|^2,$$
$$\leq \|u\|^2 + 2 \mid \langle u \mid v \rangle \mid + \|v\|^2.$$

The Cauchy-Schwarz inequality for $|\langle u \mid v \rangle|$ leads furthermore to the inequality

$$\|u + v\|^2 \leq \|u\|^2 + 2\|u\|\|v\| + \|v\|^2 = (\|u\| + \|v\|)^2.$$

Thus we have $\|u + v\| \leq \|u\| + \|v\|$. $\qquad\qquad\qquad\qquad\qquad\qquad\qquad\square$

2.3.1 Examples of Vector Spaces with a Scalar Product

Example 2.26 $(V, \langle | \rangle) = \mathbb{R}$.

Here V is the most simple nontrivial Euclidean vector space we can have. The canonical or standard scalar product is given by $\langle \alpha \mid \beta \rangle := \alpha\beta$ and the norm by $\|\alpha\| := |\alpha| \geq 0$. Orthogonality leads to the following results:

$$\langle \alpha \mid \beta \rangle = 0 \Leftrightarrow \alpha = 0 \text{ or } \beta = 0;$$
$$\langle \alpha \mid \alpha \rangle = 0 \Leftrightarrow \alpha = 0.$$

The Cauchy-Schwarz and triangular inequalities take the same form, as in elementary analysis:

$$|\langle \alpha \mid \beta \rangle| \leq |\alpha| |\beta| \quad \text{and}$$
$$|\alpha + \beta| \leq |\alpha| + |\beta|.$$

Example 2.27 $(V, \langle | \rangle_\sigma) = (\mathbb{R}^2, \langle | \rangle_\sigma)$.

With $\langle | \rangle_\sigma$ we denote a family of scalar products, like symmetric nondegenerate bilinear forms that make V a Euclidean, in (i) and (ii), or semi-Euclidean vector space, in (iii).

(i) For $\mathbb{R}^2$, with the canonical or standard inner product $\langle | \rangle$, that is, a positive definite scalar product, we have:

If we set $x = \vec{\xi} = \left[\begin{smallmatrix} \xi^1 \\ \xi^2 \end{smallmatrix}\right] = (\xi^s)$, $y = \vec{\eta} = \left[\begin{smallmatrix} \eta^1 \\ \eta^2 \end{smallmatrix}\right] = (\eta^r)$ with $s, r \in I(2) = \{1, 2\}$ and $\xi_s = \xi^s$, $\eta_r = \eta^r \in \mathbb{R}$, the transpose T is given by:

$$V \overset{T}{\longleftrightarrow} V^*,$$

$$x = \vec{\xi} = \begin{bmatrix} \xi^1 \\ \xi^2 \end{bmatrix} \longmapsto x^{\mathsf{T}} = \underset{\sim}{\xi} := [\xi^1 \; \xi^2] = [\xi_1 \; \xi_2],$$

$$\xi^{\mathsf{T}} = \begin{bmatrix} \xi^1 \\ \xi^2 \end{bmatrix} \longleftarrow \underset{\sim}{\xi} = [\xi_1 \; \xi_2].$$

At this point, we introduced a new symbol for good reasons, $\underset{\sim}{\xi} \in \mathbb{R}^{1 \times n}$.

We have to make the difference in our notation between the elements of $\mathbb{R}^2$ and the elements of $(\mathbb{R}^2)^*$. We therefore write

$$x = \vec{\xi} = \begin{bmatrix} \xi^1 \\ \xi^2 \end{bmatrix} \in \mathbb{R}^2$$

and

$$x^{\mathsf{T}} = \underset{\sim}{\xi} = [\xi_1 \; \xi_2] \in (\mathbb{R}^2)^*.$$

Furthermore, in order to use the Einstein summation convention, we denote the coefficients of the linear forms (covectors) by writing the index downstairs: We write for $\xi \in (\mathbb{R}^2)^*$, $\xi = \underset{\sim}{\xi} = [\xi_1, \xi_2]$. Consequently $\underset{\sim}{\xi}$ always corresponds to a row and $\vec{\xi}$ to a column vector in $\mathbb{R}^2$, and similarly for $(\mathbb{R}^n)^*$ and $\mathbb{R}^n$.

It is obvious that T transforms columns to rows and rows to columns. This demonstrates also the vector space isomorphism between $\mathbb{R}^n$ and $(\mathbb{R}^n)^*$. Thus we get :

$$x \longmapsto x^{\mathsf{T}} \longmapsto (x^{\mathsf{T}})^{\mathsf{T}} = x \text{ and}$$

$$\underset{\sim}{\xi} \longmapsto \underset{\sim}{\xi}^{\mathsf{T}} \longmapsto (\underset{\sim}{\xi}^{\mathsf{T}})^{\mathsf{T}} = \underset{\sim}{\xi}.$$

This shows that T is an involution: $T^2 = id$.

With the above preparation, we can express the canonic inner product for $n = 2$ in various forms, as we saw, for example with $x = \vec{\xi}$ and $y = \vec{\eta}$, where we have:

$$\underset{\sim}{\xi} y = x^T y = \langle x \mid y \rangle = \langle \vec{\xi} \mid \vec{\eta} \rangle = \sum_{s=1}^{2} \xi^s \eta^s = \xi^s \delta_{sr} \eta^r = \xi_s \eta^s.$$

The symbol δ_{sr}, the Kronecker symbol, is given by

$$\delta_s^r \equiv \delta_{sr} := \begin{cases} 1 & \text{if } s = r \\ 0 & \text{if } s \neq r. \end{cases}$$

Using the transpose T and the explicit symbols for columns and rows, $x = \vec{\xi}$, $x^{\mathsf{T}} = \underset{\sim}{\xi}$ and $\underset{\sim}{\xi}^{\mathsf{T}} = x$, we may again write for $\langle \,|\, \rangle$:

$$\langle x \mid y \rangle = x^{\mathsf{T}} \mathbb{1} y = x^{\mathsf{T}} y = \underset{\sim}{\xi}\vec{\eta} \text{ with } \mathbb{1} = (\delta_{sr}) = \begin{bmatrix} 1 & 0 \\ 0 & 1 \end{bmatrix}.$$

Here, the $V = \mathbb{R}^2$ orthogonality is no longer trivial, as with $\mathbb{R}$. For a given subspace $U \leq V$, the orthogonal space (perpendicular) $U^\perp$ relative to U, is given by

$$U^\perp = \{v \in V : \langle v \mid u \rangle = 0 \text{ for all } u \in U\}.$$

If we take for example $v_0 \in V$, $v_0 \neq 0$, we can define the subspace

$$U_0 = v_0 \mathbb{R} := \{u : u := \lambda v_0, \lambda \in \mathbb{R}\} \leq V.$$

The orthogonal space $U_0^\perp$, as shown in Fig. 2.3, is given in this case by $w \in V$, $w \neq 0$ with $\langle w \mid v_0 \rangle = 0$. So the subspace $U_0^\perp$ is equal $w\mathbb{R}$ and we may write

$$V = U_0 + U_0^\perp.$$

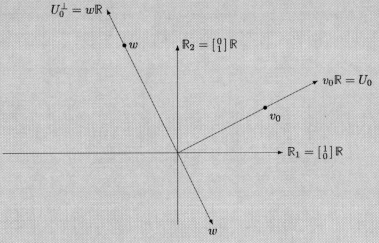

Fig. 2.3 Orthogonality in $\mathbb{R}^2$

(ii) $\mathbb{R}^2$ with an inner product given by the matrix $S = (\sigma_{sr}) = \begin{bmatrix} \sigma_1 & 0 \\ 0 & \sigma_2 \end{bmatrix}$ with σ_1 and σ_2 positive:

$$\langle x \mid y \rangle_\sigma := x^{\mathsf{T}} S y = \xi^s \sigma_{sr} \eta^r.$$

$(V, \langle | \rangle_\sigma)$ is again a Euclidean vector space. All aspects discussed in example (i), apply one to one also here.

(iii) $\mathbb{R}^2$, with a symmetric nondegenerate bilinear form, is given, for example, by the matrix $S = \begin{bmatrix} \sigma_1 & 0 \\ 0 & \sigma_2 \end{bmatrix}$, with σ_1 = positive and σ_2 negative. We can assume that $\sigma_1 = 1$ and $\sigma_2 = -1$. We get $S = \begin{bmatrix} 1 & 0 \\ 0 & -1 \end{bmatrix}$.

$(V, \langle | \rangle_\sigma)$ is a semi-Euclidean vector space. This is our first model for the two-dimensional spacetime of special relativity: In other words, it is the vector space that corresponds to the two-dimensional Minkowski spacetime.

Remark 2.6 The angle.

In this example, we find a very important application of the Cauchy-Schwarz inequality which allows the definition of an angle. If we take $u, v \in \mathbb{R}^2$, u and v not zero, the Cauchy-Schwarz inequality $\langle u \mid v \rangle \leq \|u\| \|v\|$ leads to

$$0 \leq \frac{|\langle u \mid v \rangle|}{\|u\| \|v\|} \leq 1.$$

As a result, we obtain

$$-1 \leq \frac{\langle u \mid v \rangle}{\|u\| \|v\|} \leq 1.$$

This allows the unique determination of a real number $\varphi \in [0, \pi]$, the angle between the two vectors $u, v \in \mathbb{R}^2 - \{0\}$:

$$\cos \varphi := \frac{\langle u \mid v \rangle}{\|u\| \|v\|}.$$

It is clear that this definition applies to any Euclidean vector space V, for any two vectors $u, v \in V - \{0\}$.

Example 2.28 $(V, \langle | \rangle) = (\mathbb{R}^{2 \times 2}, \langle | \rangle_M)$.

For the two matrices

$$A = \begin{bmatrix} \alpha_1^1 & \alpha_2^1 \\ \alpha_1^2 & \alpha_2^2 \end{bmatrix} = (\alpha_s^i) \text{ and } B = (\beta_s^i) \quad i, s \in I(2) = \{1, 2\},$$

an inner product is given by

$$\langle A \mid B \rangle_M := \alpha_1^1 \beta_1^1 + \alpha_2^1 \beta_2^1 + \alpha_1^2 \beta_1^2 + \alpha_2^2 \beta_2^2.$$

If we define the trace of A by

$$tr A := \alpha_1^1 + \alpha_2^2 = \sum_{i=1} \alpha_i^i \equiv \alpha_i^i \in \mathbb{R}$$

and write for the transpose of $A : A^\mathsf{T} = (\alpha(T)_i^s)$ with $\alpha(T)_i^s = \alpha_s^i$, the inner product takes the form:

$$\langle A \mid B \rangle_M := \sum_{i,s} \alpha_s^i \beta_s^i = \alpha(T)_i^s \beta_s^i = tr(A^\mathsf{T} B).$$

The corresponding norm is given by

$$\|A\|_M := (tr A^\mathsf{T} A)^{1/2} = \left(\sum_{i,s} (\alpha_s^i)^2 \right)^{1/2}.$$

It is easy to recognize the isomorphism

$$(\mathbb{R}^{2 \times 2}), \langle | \rangle)_M \cong (\mathbb{R}^4, \langle | \rangle)$$

and to see that $\langle | \rangle_M$ is indeed an inner product in $V = \mathbb{R}^{2 \times 2}$. It is also obvious that all the above relations may be extended to every $n \in \mathbb{N}$. The space $(\mathbb{R}^{n \times n}, \langle | \rangle_M)$, of $n \times n$ matrices, is also a Euclidean vector space.

Example 2.29 $L^2(\mathbb{R})$.

It is interesting to consider the next example, taken from analysis, even if it corresponds to an infinite dimension vector space that we do not discuss in this book. We denote by $L^2(\mathbb{R})$ the vector space of the class of square integrable functions in $\mathbb{R}$, as used in quantum mechanics. Without going into detail, we can give

$$\langle f \mid g \rangle := \int_{\mathbb{R}} f(x)g(x)dx \quad f, g \in L^2(\mathbb{R})$$

and

$$\|f\|_2 := \sqrt{\int_{\mathbb{R}} f(x)^2 dx}.$$

Comment 2.7   Bilinear forms in $V \times W$ and $V^* \times V$.

The definition of a bilinear form φ applies also when we consider two different vector spaces, V and W:

$$\varphi : V \times W \longrightarrow \mathbb{R},$$
$$(v, w) \longmapsto \varphi(v, w) \in \mathbb{R}.$$

The following naturally given bilinear form is a very instructive and remarkable example. We take V and its dual V^* and denote the bilinear form by the symbol $\langle , \rangle$:

$$\langle , \rangle \quad V^* \times V \longrightarrow \mathbb{R},$$
$$(\xi, v) \longmapsto \langle \xi, v \rangle := \xi(v) \in \mathbb{R}.$$

Note that in this case, we write $\langle , \rangle$ and not $\langle | \rangle$ as for the inner product.

We complete this section by presenting a few characteristic examples which now refer to $\mathbb{C}$ vector spaces.

Example 2.30 $(V, \langle | \rangle) = \mathbb{C}$.

For the canonical inner product $\langle | \rangle$, we use the complex conjugation, as usual. If $z, w \in \mathbb{C}, z = x + iy \longmapsto \bar{z} = x - iy, x, y \in \mathbb{R}$, we have

$$\langle z \mid w \rangle := \bar{z}w.$$

The norm takes the form $\|z\| := |z| = (\bar{z}z)^{1/2}$ and the Cauchy-Schwarz inequality is given by $|\bar{z} w| \leqslant |z| |w|$.

Example 2.31 $(V, \langle | \rangle) = (\mathbb{C}^2, \langle | \rangle)$.

For the canonical inner product $\langle | \rangle$, we have to use the Hermitian conjugate of $v \in \mathbb{C}^2$ given by $v^\dagger := \bar{v}^\mathsf{T}$. More explicitly examined, this means that $v = \begin{bmatrix} v^1 \\ v^2 \end{bmatrix}$ with $v^1, v^2 \in \mathbb{C}$, is by definition a column, and $v^\dagger = [\bar{v}_1, \bar{v}_2]$ a row (with $v^1 = \bar{v}_1, v^2 = \bar{v}_2$). The norm is given by

$$\|v\| = (v^\dagger v)^{1/2} = (\bar{v}_1 v^1 + \bar{v}_2 v^2)^{1/2} = (\bar{v}_s v^s)^{1/2} \quad s \in I(2).$$

Example 2.32 $(V, \langle | \rangle) = (\mathbb{C}^{2 \times 2}), \langle | \rangle_M)$.

For $A, B \in \mathbb{C}^{2 \times 2}$, the standard inner product $\langle | \rangle$ is given by

$$\langle A \mid B \rangle_M := tr(A^\dagger B) \text{ and the norm by}$$
$$\|A\|_M := (tr(A^\dagger A))^{1/2}.$$

Remark 2.7 Analogy to vector spaces over $\mathbb{C}$.

All the above relations can undoubtedly be extended from every $\mathbb{R}^n$ to $\mathbb{C}^n$. Using the simplest possible models allowed us to clarify the notation we intend to use systematically throughout the book.

2.4   The Standard Vector Space and Its Dual

If we consider an abstract n-dimensional vector space, it has only the linear structure. We can of course introduce other structures, such as a volume form (see Sect. 7.5) or an inner product, or even a particular basis. The standard vector space $\mathbb{K}^n$ has all of these structures. Therefore, we can say that the vector space $\mathbb{K}^n$ has the maximum structure an n-dimensional vector space can have! The reader should already be familiar with $\mathbb{R}^n$ from a first course in analysis. In Sect. 2.1, Example 2.3, we introduced some elements of this section for the case $n = 2$, so our discussion here can be considered as a review thereof or as an extension to general $n \in N$.

As discussed in Example 2.4, $\mathbb{K}^n$ is the set of all finite sequences of numbers with length n which we may also call n-tuple or list of length or size n. We regard this n-tuple as a column. Any $\mathbb{K}^n$ is given by

$$\mathbb{K}^n = \left\{ x = \begin{bmatrix} \xi^1 \\ \vdots \\ \xi^j \\ \vdots \\ \xi^n \end{bmatrix} : \xi^j \in \mathbb{K}, \, j \in I(n) := \{1, 2, \ldots n\} \right\}.$$

We also use the notation $x = (\xi^j)_n \equiv (\xi^j) \equiv \vec{\xi}$ and we omit the index n when its value is clear. We may say that ξ^j is the jth coordinate or the jth coefficient or the jth component of $x \in \mathbb{K}^n$.

If we apply the standard addition and scalar multiplication by adding and multiplying the corresponding entries, the set $\mathbb{K}^n$ is indeed a vector space since it fulfills all the axioms of the definition of a vector space. $\mathbb{K}^n$ is the standard vector space; it is the model of a vector space with dimension n, and, as is well-known from analysis, it is locally also the model of a manifold!

It is evident that $\mathbb{K}^n$ can be identified with the $n \times 1$-matrices with entries in $\mathbb{K}$. The greatest advantage of $\mathbb{K}^n$ is its canonical basis:

$$E := (e_1, \ldots, e_n) \quad \text{with } e_1 = \begin{bmatrix} 1 \\ 0 \\ \vdots \\ 0 \end{bmatrix}, \ldots e_n = \begin{bmatrix} 0 \\ \vdots \\ 0 \\ 1 \end{bmatrix}.$$

Matrices are closely connected with linear maps. It is well-known that a matrix F with elements in $\mathbb{K}$, may, if we wish, describe a linear map f as a matrix multiplication. This was also demonstrated in Example 2.23. Here we use the letter F to demonstrate the close connection between a matrix and a map, and in our mind we may always identify f with F:

$$f : \mathbb{K}^n \longrightarrow \mathbb{K}^m,$$
$$x \longmapsto f(x) := Fx.$$

This means that we have $F \in \mathbb{K}^{m \times n}$ and $f \in \text{Hom}(\mathbb{K}^n, \mathbb{K}^m)$. Usually, the $m \times n$-matrix F is denoted by $F = (\varphi_{ij})$ with $\varphi_{ij} \in \mathbb{K}$, $j \in I(n) := \{1, \ldots, n\}$, $i \in I(m) := \{1, \ldots, m\}$ and we have $z := f(x)$ and $z^i = \sum_{j=1}^n \varphi_{ij} x^j$, $z^i, x^j \in \mathbb{K}$, as usual in literature. In special relativity, as in tensor analysis, we actually wish for the type of relevant transformation we may perform to be indicated by the position of the indices (up or down). We therefore put initially $\varphi^i_j := \varphi_{ij}$. In addition, it is a great advantage to use the Einstein convention for the summation which is almost implied by this notation. Therefore we first write $F = (\varphi^i_j)$ and we have

$$z^i = \varphi^i_j x^j. \tag{2.7}$$

However, this is not enough for us. Since we always use greek letters for scalars, we set $x = (\xi^j)_n \equiv \vec{\xi}$, $j \in I(n)$, $z = (\zeta^i)_n \equiv \vec{\zeta}$, $i \in I(m)$. In addition, we change the index j into the indices $s, r \in I(n)$ and we write $x = (\xi^s)_n$. In this book, we systematically use this kind of indices and notation, and we may call this "Smart Indices Notation". Hopefully, the reader will soon realize the usefulness of this notation. We therefore write for the Eq. 2.7

$$\zeta^i = \varphi^i_s \xi^s. \tag{2.8}$$

We can verify that f is indeed a linear map in both notations. The matrix $F = (\varphi^i_s)$ contains all the information about the map f.

Before proceeding further, we would like to remark that we restrict ourselves to $\mathbb{R}^n$ for simplicity reasons and because there are direct, relevant connections to the Euclidean geometry, especially in $n = 3$, and to special relativity ($n = 4$). We must mention again that we consider the elements of $\mathbb{R}^n$ as columns. But nevertheless, we need the rows too! If we take the dot product and the matrix multiplication into account, and also take $\xi^s = \xi_s$ and $\eta^s = \eta_s$, we have:

$$\langle | \rangle : \mathbb{R}^n \times \mathbb{R}^n \longrightarrow \mathbb{R},$$

$$(x, y) \longmapsto \langle x \mid y \rangle := \sum_{i=1}^{n} \xi^s \eta^s = \xi_s \eta^s,$$

we may interpret rows ($1 \times n$-matrices) with entries scalars as elements of the dual space of $\mathbb{R}^n$ and write:

$$(\mathbb{R}^n)^* := \left\{ \xi^* \equiv \xi = [\xi_1 \ldots \xi_n] : \xi_i \in \mathbb{R}, i \in \{1, 2, \ldots, n\} \right\}.$$

We used the matrix notation $\xi = [\xi_1 \xi_2 \ldots \xi_n] \in \mathbb{R}^{1 \times n}$ and not the notation for the corresponding (horizontal) list $\widetilde{(\xi_1, \xi_2, \ldots, \xi_n)}$ (see Comment 2.1). In order to use the Einstein convention, we have to write the indices of the coefficients of ξ downstairs and we get, with $y = (\eta^s) = \vec{\eta}$,

$$\xi : \mathbb{R}^n \longrightarrow \mathbb{R},$$

$$y \longmapsto \xi(y) = \xi_s \eta^s = [\xi_1 \cdots \xi_n] \begin{bmatrix} \eta^1 \\ \vdots \\ \eta^n \end{bmatrix}.$$

The Einstein convention is the simplest and nicest way to express matrix multiplications. The same holds for linear combinations. After all, both operations are essentially the same thing. Therefore we can write symbolically in an obvious notation:

$$[* * * \cdots * *] \begin{bmatrix} \vdots \\ \cdots \end{bmatrix} = [* \cdot + * \cdot + * \cdot + \cdots + * \cdot + * \cdot] \tag{2.9}$$

which is

$$[1 \times n][n \times 1] = [1 \times 1].$$

Remark 2.8 A generalization of matrix multiplication.

The above Eq. 2.9 is nothing else but the usual row by column rule for the multiplication of matrices. We assumed tacitly that between the entries "$*$" and "$.$", the multiplication and correspondingly the addition are already defined. At the same time, with this symbolic notation, we also generalized the matrix multiplication, even in the case where the entries are more general objects than scalars. This means for example that if the entries "$*$" and "$.$" are themselves matrices, we obtain also the matrix multiplication of block matrices.

We use the row matrix $[\xi] \in \mathbb{R}^{1 \times n}$ to express the linear map

$$\xi \in \mathrm{Hom}(\mathbb{R}^n, \mathbb{R}) = (\mathbb{R}^n)^*.$$

This leads to the natural identification of $\mathbb{R}^{1 \times n}$ with $(\mathbb{R}^n)^*$. Thus we can call $\xi \in (\mathbb{R}^n)^*$ a linear function, a linear form, a linear functional or a covector, and we may also write:

$$\xi \equiv \underset{\sim}{\xi} \equiv [\xi] \equiv \langle x \mid : \mathbb{R}^n \longrightarrow \mathbb{R},$$

$$y \longmapsto \xi(y) = [\xi]y = \langle x \mid y \rangle.$$

$\langle x \mid \in (\mathbb{R}^n)^*$ corresponds to the Dirac notation and is widely used in quantum mechanics (see also Sect. 6.4).

Furthermore, if we use the transpose, we set $\xi_i = \xi^i$ and we have

$$T : \mathbb{R}^n \longrightarrow (\mathbb{R}^n)^*,$$

$$x \longmapsto x^{\mathsf{T}} = [\xi_1 \ldots \xi_n].$$

Comment 2.8 The use of the transpose $\top$.

We can connect $\mathbb{R}^n$ with $(\mathbb{R}^n)^*$ and in addition we can, if we want, identify $(\mathbb{R}^n)^*$ with $\mathbb{R}^n$. This is an identification we are often doing in physics from the beginning without even mentioning the duality behind it.

However, a note of caution is in order here. In the case of an abstract vector space V, we cannot ignore V^* because there is no basis-independent connection between V and V^*. The situation changes drastically if we introduce an inner product in V. We see this immediately in the case of $\mathbb{R}^n$ since the canonical inner product (dot product as above) already exists here. We have altogether $\underset{\sim}{y}(x) = y^\top x = \langle y \mid x \rangle$. It is clear that here $\xi^i = \xi_i \in \mathbb{R}$ and we write

$$\underset{\sim}{y} \equiv \mid y \rangle \overset{T}{\longmapsto} \langle y \mid \equiv \underset{\sim}{y} \equiv y^\top \in (\mathbb{R}^n)^*.$$

See also the Dirac notation in Sect. 6.4.

2.5 Affine Spaces

In the previous sections, we started with an abstract vector space V given only by its definition, and we introduced the inner product as an additional structure on it. Besides its geometric signification, this other structure facilitates the formalism considerably within linear algebra, which is very pleasant for applications in physics too.

On the other hand, removing some of its structure is necessary when starting with V. It is essential to consider the vector space V as a manifold. This special manifold is also called linear manifold or affine space. The well-known Euclidean space is also an affine space, but its associated vector space is an inner product vector space, a Euclidean vector space. An affine space has two remarkable properties: On the one hand, it is homogeneous and on the other hand, it contains just enough structure for straight lines to exist inside it, just as every every $\mathbb{R}^n$ does. Similarly, any vector space has the same capacity to contain straight lines. But at the same time, a vector space is not homogeneous since its most important element, the zero, is the obstacle to homogeneity. To obtain the associated affine space starting from a vector space, we have to ignore the unique role of its zero, and to ignore that we can add and multiply vectors by scalars. We obtain an affine space with the same elements as the vector space we started with and which we now may call points. The action of a vector space V gives the precise procedure for this construction on a set X, precisely in the sense of group actions in Sect. 1.3. The group which acts on X is the abelian group of V:

> **Definition 2.17** Affine space.
> An affine space X associated to the vector space V is a triple (X, V, τ). We may call the elements of X points and the elements of V vectors or translations (using also the notation $V = T(X)$) with
>
> $$\tau : V \times X \longrightarrow X,$$
> $$(v, x) \longmapsto \tau(v, x) = \tau_v(x) = x + v.$$
>
> The action τ of V on X is free and transitive.

From this follows, according to Comment 2.2 in Sect. 2.1, that V and X have the same cardinality or, in simple words, have the same "number" of elements: For $x_0 \in X$, we have $V \times \{x_0\} \underset{\cong}{\to} X$, and the map

$$V \longrightarrow X,$$
$$v \longmapsto x_0 + v,$$

is bijective.

We can give another equivalent and perhaps more geometric description of an affine space, with essentially the following property:

X a set, $T(X)$ the associated vector space and Δ the map:

$$\Delta : \ X \times X \longrightarrow T(X),$$
$$(x_0, x_1) \longmapsto \overrightarrow{x_0 x_1},$$

such that for $x_0, x_1, x_2 \in X$, the triangular equation $\overrightarrow{x_0 x_2} = \overrightarrow{x_0 x_1} + \overrightarrow{x_1 x_2}$ holds. This means that two points x_0 and x_1 in X determine an arrow $\overrightarrow{x_0 x_1} \in T(X)$ which we may also interpret as a translation. We may consider the straight line through the points x_0 and x_1 as an affine subspace of X. Here, we see explicitly that an affine space can contain straight lines. It is interesting to notice that for every $x_0 \in X$, we have the set

$$T_{x_0} X = \{x_0 + \overrightarrow{x_0 x}, x \in X\}$$

which is the tangent space of X at the point x_0. In this sense, we may regard $T(X)$ as the universal tangent space of X and the letter T could mean not only "translation" but also "tangent" space.

The above construction is equally valid if we start with a vector space $V (X = V)$. The affine space now is the triple $V := (V, T(V), \tau)$. In this case, the arrow $\overrightarrow{v_0 v_1}$ is given by the difference: $\overrightarrow{v_0 v_1} = v_1 - v_0$ and the vector space V is now considered as an affine space! See also Comment 2.2 in Sect. 2.1.

Starting with a vector space V allows to give another, more direct definition of an affine space in V, considered as a subset, not a subspace, of V. This construction is more relevant to linear algebra. It also leads to a new kind of very useful vector spaces, the quotient vector spaces which are discussed in Sect. 2.6.

Definition 2.18 Affine space in V.

U is a vector subspace of V. For a given $v_0 \in V$, an affine space $A = A(v_0) \subset V$, associated to U, is given by the set

$$A = v_0 + U := \{v_0 + u : u \subset U\}.$$

As we saw in Sect. 1.3, we may also write $A(v_0) = U v_0$, that means that the affine space $A(v_0)$ is exactly the orbit of the action U on the vector v_0. Here, the action of U on v_0 is the additive action of the commutative subgroup U (of the commutative group V). This justifies the notation $A(v_0) = U v_0 = \{u + v_0 = v_0 + u : u \in U\} = U + v_0$.

Example 2.33 Affine spaces as solutions of linear equations.

Given a linear map

$$f : V \longrightarrow V',$$

affine spaces appear very naturally as solutions of a linear equation $f(x) = w_0$ or equivalently as a preimage (or fiber) of w_0. As discussed in Remarks 2.4 and 2.5 in Sect. 2.3, ker f and im f are subspaces of V and V' respectively. The preimage of w_0 is given by $f^{-1}(w_0)$.

For $f(v_0) = w_0$ with $w_0 \neq 0$, we may write, as above, taking $U := \ker f$

$$A(v_0) = f^{-1}(w_0) = v_0 + U.$$

In Fig. 2.4, we see the corresponding affine spaces associated to the linear map f. They are all parallel to each other and have the same dimension.

2.6 Quotient Vector Spaces

We know from Sect. 1.2, in quite general terms, what a quotient space is. We now have the opportunity to discuss a very important application which is probably the most important example of a quotient space in linear algebra. In what follows, we

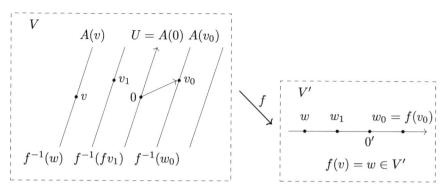

Fig. 2.4 Affine subspaces in V relative to the subspace $U = \ker f < V$ with $w = f(v)$, $w_1 = f(v_1)$, $w_0 = f(v_0)$

would like to stay within the vector space V, we use the notation of Sect. 2.5. We consider the set of all such affine subspaces associated with a given subspace U of V for all points $v \in V$:

$$A(U) := \{A(v) : v \in V\}.$$

As shown in Sect. 2.5, all these affine subspaces are parallel and have the same dimension. Here, it is also intuitively clear, as mentioned in Remark 1.1 in Sect. 1.2, that this gives us a disjoint decomposition of the vector space V.

It turns out that we can introduce a vector space structure on the set $A(U)$ and that we may thus obtain a new vector space out of V and the given subspace U. Therefore we can talk about a vector space $A(U)$ with elements (again vectors of course) which are the affine spaces in V associated with U. The elements of $A(U)$ are by construction equivalent classes or cosets. Every such class is in this case also a vector since, as we will see below, we can introduce naturally a linear structure on $A(U)$. This makes $A(U)$ a vector space and we denote this vector space by V/U. As sets, $A(U)$ and V/U are bijectively connected. $A(U)$ is simply a set as was defined above, and V/U is this set with the vector space structure. Clearly, this new vector space itself cannot be a subspace of V, but we stay, in a more general mathematical language, in the vector space category.

Another way to obtain similar new vector spaces, is to use the notion of equivalence classes as discussed in Sect. 1.2. We notice that the subspace U induces an equivalence class on V:

$$v' \sim v :\Leftrightarrow v' - v \in U.$$

The coset $[v]$ of this equivalence class of v is exactly the additively written orbit of the U action on v and we have for this orbit

$$[v] = Uv = A(v) = v + U.$$

The set of cosets of U in V is given by

$$V/U := \{v + U : v \in V\} \tag{2.10}$$

which we may call the quotient space of V modulo U. It is evident that a bijection of sets $A(U) \cong_{bij} V/U$ holds.

Furthermore, we are now in the position to use the formalism of Sect. 1.2 to introduce a vector space structure on the set $V/U = \{[v]\}$. Consequently, in the end, we will be in the position to add and scalar multiply the affine subspaces, the classes or cosets, associated with U. This means for example that for a given subspace U with dim $= 1$, all straight lines parallel to U behave exactly as vectors of a vector space. This also means that, in general, at the end of our construction, we may expect to have

$$V/U \cong \mathbb{K}^m,$$

and as we will see, with $m = \dim V - \dim U$.

In the two following equivalent figures, we represent the various affine subspaces of $\mathbb{R}^2$ relative to a given subspace $U(U \leq \mathbb{R}^2)$ and the quotient space V/U. Our figures should be self-explanatory with $A(v) = v + U = [v]$.
The x- and y-axes are denoted by $\mathbb{R}_1 = \begin{bmatrix} 1 \\ 0 \end{bmatrix}$ and $\mathbb{R}_2 = \begin{bmatrix} 0 \\ 1 \end{bmatrix}$. $A(v)$, $A(v')$, $A(u_0)$ are elements of V/U. $U = \mathbb{R}\lambda u_0 = A(u_0) = [\vec{0}]$.

Keeping it general, we may also write symbolically for Fig. 2.5 (see Fig. 2.6):
Note that

$$A(v_1) = \pi^{-1}(\pi(v_1)) = \pi^{-1}([v_1]),$$
$$A(v_0) = \pi^{-1}([\vec{0}]),$$
$$A(v) = \pi^{-1}(\pi(v)) = \pi^{-1}([v]).$$

Since all these spaces are parallel and have the same dimension, and are elements of the set $A(U) = V/U$, it seems quite natural to look at the concrete space V/U and try to define the addition and, similarly, the scalar multiplication in the following way:

$$[v] \stackrel{+}{:} [w] := (v + U) + (w + U) = v + w + U,$$
$$\lambda \cdot [v] := \lambda(v + U) := \lambda v + U.$$

We have only to make sure that these operations are well-defined. That is for $[v'] = [v]$ and $[w'] = [w]$, we have $v' + w' + U = v + w + U$, and $\lambda v + U = \lambda v' + U$. Taking into account the definition of the equivalence class (see Definition 1.2), we see immediately that this is indeed the case and that V/U is a concrete vector space. Since we may write the equivalence relation

$$v \sim v' \Leftrightarrow v' - v \in U \Leftrightarrow v' = v + u, \; u \in U \Leftrightarrow [v'] = v + U,$$

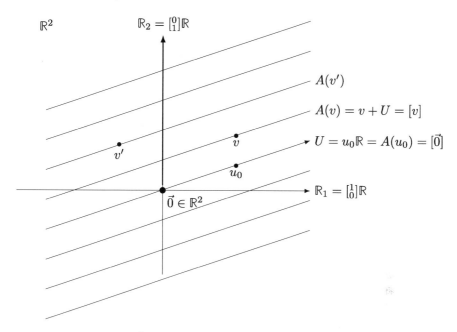

Fig. 2.5 The vector space $\mathbb{R}^2$ with the x- and y-axes and the affine spaces parallel to $U = u\mathbb{R}$

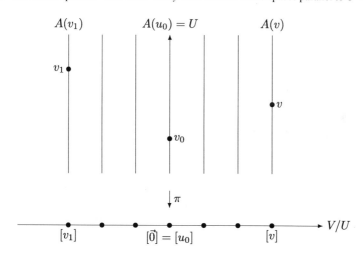

Fig. 2.6 Figure 2.5 symbolically

and equally with w and w', we get

$$v' + w' + U = v + U + w + U + U = v + w + U.$$

This shows that

$$(v' + w') \sim (v + w) \Leftrightarrow [v' + w'] = [v + w] \Leftrightarrow [v]\overset{+}{.}[w] = [v']\overset{+}{.}[w'].$$

Similarly,

$$\lambda v' + U = \lambda(v + U) = \lambda v + \lambda U + U = \lambda vU.$$

This shows that

$$\lambda v' \sim \lambda v \Leftrightarrow [\lambda v'] = [\lambda v] \Leftrightarrow \lambda \cdot [v'] = \lambda \cdot [v].$$

It is interesting to notice that the natural map (canonical map) π,

$$\pi : V \longrightarrow V/U, \quad v \longmapsto \pi(v) := [v],$$

is a linear vector space homomorphism. It is evident since we may write

$$\pi(v + w) = [v + w] = [v]\overset{+}{.}[w] = \pi(v)\overset{+}{.}\pi(w)$$
$$\text{and}$$
$$\pi(\lambda v) = \lambda[v] = \lambda \cdot \pi(v).$$

Hence, we may revert to use "+" instead of "$\overset{+}{.}$":

$$[v] + [w] := [v + w] \text{ and } \lambda[v] := [\lambda v].$$

We herewith established the following very interesting proposition:

Proposition 2.2 *Quotient vector space V/U.*
Let V be a vector space and $U \leqslant V$, then the quotient space $V/U = \{a + U : a \in V\}$ is a vector space with addition and scalar multiplication as follows:

$$V/U \times V/U \longrightarrow V/U,$$
$$(a + U, b + U) \longmapsto (a + b) + U \quad and$$
$$\mathbb{K} \times V/U \longrightarrow V/U,$$
$$(\lambda, a + U) \longmapsto \lambda a + U.$$

The canonical map or quotient map π is given by:

$$\pi : \quad V \longrightarrow V/U,$$
$$v \longmapsto [v] \equiv v + U.$$

Thus, the zero of V/U is $\ker \pi = U \equiv [\vec{0}]$.

The map π is linear and surjective (epimorphism). The dimension formula, a.k.a rank-nullity (see Corollary 3.2 in Sect. 3.3), takes here the form

$$\dim V = \dim U + \dim V/U.$$

Comment 2.9 What is the essence of a vector space?

We now understand even better what a vector space is. The elements of a vector space can be anything. The only one thing we can demand from these elements, is that they behave well; that is, that we can add and multiply by scalars. In consequence, we understand that it is the behavior that matters. Here, we come across the same situation as we met with the Euclidean axioms. We do not know what the essence of a vector is. Nevertheless, it does not matter since we have the definition of a vector space.

Now we can ask ourselves: what are the other benefits of the notion of quotient space for linear algebra? We present two examples that might also be a relevant in physics.

As will be discussed in Sect. 3.5 in Proposition 3.11, for every given subspace U and V there exist a lot of complementary subspaces W in V, satisfying $U \oplus W \cong V$ (see Definition 2.19). In this situation, the following holds: the quotient vector space V/U is isomorphic to W, that is, $W \cong V/U$ and consequently to any complementary subspace of U in V. We can therefore say that V/U represents all these complimentary vector spaces of U.

The second example is the following isomorphism theorem which we give without the proof.

Theorem 2.2 *The first isomorphism theorem:*
Let $f : V \rightarrow V'$ be a linear map. The f induces a canonical isomorphism $\bar{f}$:

$$V/\ker f \underset{\bar{f}}{\cong} \operatorname{im} f.$$

If f is a surjective map, this canonical isomorphism is given by

$$V/\ker f \cong V'.$$

This means that we can describe the vector space V' solely with data of the vector space V.

It is further worthwhile noting that quotient vector spaces play an important role not only in finding and describing new spaces but also in formulating theorems or various proofs in linear algebra. As already mentioned, in physics we also often use quotient spaces intuitively even if we do not apply explicitly the above formalism.

2.7 Sums and Direct Sums of Vector Spaces

We are now going to answer the question that arose in Comment 2.5 of Sect. 2.1. How can we obtain uniquely a subspace of V from a union of two or more subspaces? There is a natural construction that leads to the desired result.

Definition 2.19 Sum of subspaces.
For the given subspaces $U_1, \ldots, U_m$ of V, the sum of $U_1, \ldots, U_m$ is given by the following expression:

$$U_1 + \ldots + U_m := \{u_1 + u_2 + \ldots + u_m : u_i \in U_i, i \in I(m)\}.$$

We may also write

$$U = \sum_{(i=1)}^{m} U_i = U_1 + \cdots + U_m.$$

We can see immediately that U is a subspace of V, that is, $U \leqslant V$. In addition, U is the smallest subspace of V containing the subspaces $U_1, \ldots, U_m$. This is easy to perceive since every other subspace W that contains also $U_1, \ldots, U_m$ must contain something more than U. The following construction seems even more perfect:

Definition 2.20 Direct sum of subspaces.
The sum of the subspaces $U_1, \ldots, U_m$ is a direct sum denoted by

$$U_1 \oplus \cdots \oplus U_m = \bigoplus_{i=1}^{m} U_i$$

if every element of the sum $u \in U_1 + \cdots U_m$ has a unique decomposition as a sum

$$u = u_1 + \cdots u_m \text{ with } u_i \in U_i \quad i \in I(m).$$

Remark 2.9 Direct sum and zero.

Definition 2.20 is equivalent to the following statement. The zero element has a unique decomposition in $U := \bigoplus_{i=1}^{m} U_i$: That is, if $u_1 + \cdots + u_m = 0$, then $u_1 = 0, \ldots, u_m = 0$.

Proof If the zero has a unique decomposition, we may check that if $u = u_1 + \cdots + u_m$, and $u' = w_1 + \cdots + w_m$ are given, we get for the difference $0 = u - u' = (u_1 - w_1) + \cdots + (u_m - w_m)$ and then $u_1 = w_1, \ldots, u_m = w_m$. $\square$

Remark 2.10 Direct sum and linear independence.

In the case of a direct sum, we may say that the list of vector spaces $(U_1, \ldots, U_m)$ is (block) linearly independent too (for the definition, see Sect. 3.1). As a consequence, a direct sum is a form of linear independence.

Comment 2.10 Direct sums and linear maps.

The direct sum decomposition of V,

$$V = U_1 \oplus \cdots \oplus U_m,$$

is very interesting, especially if this decomposition is induced by an operator (endomorphism) $f \in \mathrm{Hom}(V, V) \equiv \mathrm{End}(V)$ since such decompositions characterize the geometric properties of f. As we shall see later, this also leads to a decomposition of every $f \in \mathrm{End}(V)$. In the more general case of a linear map $f \in \mathrm{Hom}(V, V')$, if we choose a basis in V and V', we always obtain the decomposition of the form

$$V = U_1 \oplus U_2 \text{ and } V' = U_2' \oplus U_1'$$

with $U_1 = \ker f$ and $U_2' = \mathrm{im} \, f$. So f is given by

$$U_1 \oplus U_2 \xrightarrow{f} U_2' \oplus U_1'$$

which is a form of the fundamental theorem of linear maps (see Theorem 5.2 in Sect. 5.3).

2.7.1   Examples of Direct Sums

Example 2.34 $V = \mathbb{R}^2$.

As in the examples in 2.16 in Sect. 2.1.2, U_1 and U_2 are the x-axis and y-axis. We have of course $U_1 \cap U_2 = \{0\}$ and we have the direct decomposition of $\mathbb{R}^2$:

$$\mathbb{R}^2 = U_1 \oplus U_2 \equiv (x - axis) \oplus (y - axis).$$

Example 2.35 $V = Pol(5)$.

All real polynomials with degree $\leqslant 5$. If we define the odd polynomials in $Pol(5)$ by U_1, and the even polynomials in $Pol(5)$ by U_2, we get $U_1 \cap U_2 = \{0\}$ and the direct sum

$$Pol(5) = U_1 \oplus U_2.$$

At this point, the question arises concerning a criterion for a sum to be a direct sum. A simple answer is given in the following proposition for $m = 2$.

Example 2.36 Direct sums in $\mathbb{R}^3$.

We consider U_1 and U_2, two arbitrarily chosen one-dimensional subspaces in $\mathbb{R}^3$ with $U_1 \cap U_2 = \{0\}$, and we obtain a two-dimensional subspace V, as shown in Fig. 2.7a.
Similarly, we chose U_1 and U_2, two arbitrarily chosen subspaces in $\mathbb{R}^3$ with $\dim U_1 = 1$, $\dim U_2 = 2$ with $U_1 \cap U_2 = \{0\}$, and we obtain $V = U_1 \oplus U_2 = \mathbb{R}^3$ (Fig. 2.7b).

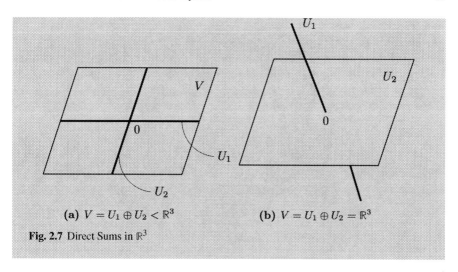

(a) $V = U_1 \oplus U_2 < \mathbb{R}^3$ **(b)** $V = U_1 \oplus U_2 = \mathbb{R}^3$

Fig. 2.7 Direct Sums in $\mathbb{R}^3$

Proposition 2.3 *Direct sum of two subspaces.*

$U + Y$ *is a direct sum if and only if*

$$U \cap Y = \{0\}.$$

Proof If $U + Y$ is a direct sum, we have to show that $U \cap Y = \{0\}$. Suppose $z \in U \cap Y$, with $z \in U$ and $z \in Y$, and $(-z) \in Y$. We get $z + (-z) = 0$ and also $0 + 0 = 0$. The unique representation of 0 leads to $z = 0$ and $(-z) = 0$. It follows that we get $z = 0$ and $U \cap Y = \{0\}$.

If $U \cap Y = \{0\}$, we have to show that $U + Y$ is direct: We have only to show that the decomposition of 0 is unique. If $0 = u + y$ with $u \in U$ and $y \in Y$, we have to show that $u = 0$ and $y = 0$:

$$0 = u + y \Leftrightarrow u = -y,$$
$$\Rightarrow u \in Y,$$
$$\Rightarrow u \in U \text{ and } u \in Y \Rightarrow u \in U \cap Y,$$
$$\Rightarrow u = 0 \text{ and also } y = 0.$$

Thus, the proposition holds. $\qquad\qquad\qquad\qquad\qquad\qquad\qquad\qquad\qquad\qquad\qquad\square$

Remark 2.11 Direct sums for more than two subspaces.

It is of interest to notice, here without proof, that for more than two subspaces, we have the following result: $U_1 + \cdots + U_m$ is a direct sum if and only if

$$U_j \cap \sum_{\substack{i \neq j}}^{m} U_i = \{0\} \text{ for all } j \in I(m) := \{1, 2, \ldots, m\}.$$

This is equivalent to the uniqueness or, generalizing, to the linear independence relation: If $\sum_{i=j}^{m} u_i = 0$ with $u_i \in U_i, i \in I(m)$, then $u_i = 0$ for all $i \in I(m)$.

2.8 Parallel Projections

In an affine space and in every abstract vector space, the notion of parallelism is part of the structure. Consequently we may say that the vectors u and v in V are parallel whenever $u = \lambda v$ for some scalar $\lambda \in \mathbb{K}$. See also Proposition 3.11 in Sect. 3.4 which leads immediately to the definition of a (parallel) projection.

Definition 2.21 (Parallel) Projection.
Given a direct sum $V = U \oplus W$, the linear operator

$$P : V \longrightarrow W \leqslant V,$$
$$v \longmapsto P(v) = w,$$

is given by the relation $v = u + w, u \in U$ and $w \in W$. P is called (parallel) projection along U (see Fig. 2.8).

As we see, parallel projection is essentially the well-known parallelogram rule known in physics. It is clear that this projection depends on both, U and $W : P = P(U, W)$. Furthermore, if we set $U = \ker P$ and $W = \operatorname{im} P$, we have:

$$V = \ker P \oplus \operatorname{im} P$$

and of course $\ker P \cap \operatorname{im} P = \{0\}$.

We see directly that $P|W = id_W$ and $P|U = \hat{0}_U$, where $\hat{0}_U$ is the null operator, holds. Further on, the algebraic characterization of a projection is very interesting:

$$P \text{ is idempotent } P^2 = P.$$

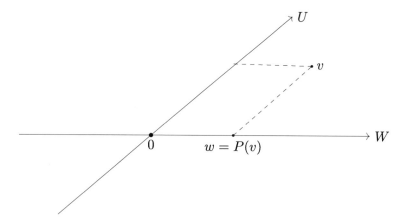

Fig. 2.8 Parallel projection in $\mathbb{R}^2$

Comment 2.11 The importance of projection operators.

The importance of projection operators stems also from the fact that every operator, especially in a complex vector space, always contains essentially projection operators as components. In particular, this is present in connection with the spectral decomposition of operators. In quantum mechanics and symmetries in physics, idempotent operators are particularly important.

2.9 Family of Vector Spaces in Newtonian Mechanics

We now come to a completely different situation in which linear algebra is pointing to analysis, differential geometry, and of course physics. We have to consider not only one vector space, as is usual in linear algebra, but an infinite number of them and they all are identical to each other. They are parametrized by the points of the space we are interested in and which for convenience we take here to be $\mathbb{R}^2$.

2.9.1 Tangent and Cotangent Spaces of $\mathbb{R}^2$

In the following construction, the elements of $\mathbb{R}^2 = \{p, q, \dots\}$ are used as indices. They parametrize the various vector spaces. These are canonically isomorphic to $\mathbb{R}^2$ at the points $p, q \dots \in \mathbb{R}^2$. For this construction, we consider for example the elements $p, q \in \mathbb{R}^2$ as points and the elements $u, v \in \mathbb{R}^2$ as free vectors.

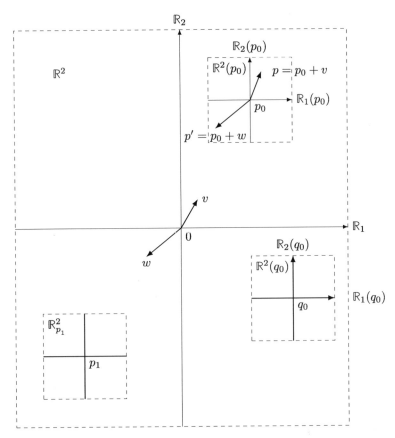

Fig. 2.9 The vector space $\mathbb{R}^2$ with its standard subspaces (x- and y-axes). $\mathbb{R}_1 \equiv \mathbb{R}_1(0) = \mathbb{R}e_1$ and $\mathbb{R}_2 \equiv \mathbb{R}_2(0) = \mathbb{R}e_2$, $T_{p_0}\mathbb{R}^2 \equiv \mathbb{R}^2(p_0) = p_0 + \mathbb{R}^2$: the tangent space at the points p_0, q_0, p_1 and their standard subspaces $\mathbb{R}_1(p_0)$ and $\mathbb{R}_2(p_0)$

As a next step, we fix a point $p_0 \in \mathbb{R}^2$ and we add the vector v so that we have $p_0 + v$, a new point which we consider as the point $p = p_0 + v \in \mathbb{R}^2$. It is usually symbolized by an arrow from p_0 to p. This is demonstrated in Fig. 2.9.

We can do likewise with another vector $w \in \mathbb{R}^2$, and we then obtain another end point $p' = p_0 + w$.

We may do the same with all $v, w, \ldots \in \mathbb{R}^2$. We may think that we thus obtained a new vector space which we denote by

$$\mathbb{R}^2_{p_0} = p_0 + \mathbb{R}^2.$$

We denote $\mathbb{R}^2_{q_0}$, $\mathbb{R}^2_{p_1}$ the tangent spaces at the points q_0 and p_1. Note that in the literature both notations, $T_{p_0}\mathbb{R}^2$ and $\mathbb{R}^n_{p_0}$ are used.

All of the above is illustrated in Fig. 2.9.

It leads to the following definitions.

Definition 2.22 Tangent vector v_{p_0}.

A tangent vector v_{p_0} in $\mathbb{R}^2$ consists of two parts, the point of application p_0 and the vector part v. We may write: $v_{p_0} \equiv p_0 + v = (p_0, v)$ and $w_{p_0} = p_0 + w \equiv (p_0, w)$.

It is clear that we represent p_0 by two numbers (its coordinates) and the vector part v also by two numbers since we are in $\mathbb{R}^2$. Similarly, we may take another point of application q_0 such that:

$$v_{q_0} = q_0 + v \quad \text{or} \quad w_{q_0} = q_0 + w.$$

Without doubt, the tangent vector $v_{q_0} = q_0 + v$ is different from the tangent vector $v_{p_0} = p_0 + v$. The point q_0 is different from p_0, but v_{q_0} and v_{p_0} are parallel to each other.

The tangent vectors v_{p_0} and v_{q_0} are equal if and only if $q_0 = p_0$ and $u = v$.

Remark 2.12 Tangent vectors in physics.

You may ask, where there is a connection with physics. In physics, we met such objects very early! The instantaneous velocity of a point particle moving in $\mathbb{R}^2$ is exactly what we call here a tangent vector. Instantaneous velocity is, as we know, a vector, but it is never alone. The point of its application is the momentary position of the moving particle in $\mathbb{R}^2$.

If we consider a curve α in $\mathbb{R}^2$,

$$\alpha : \mathbb{R} \longrightarrow \mathbb{R}^2,$$
$$t \longmapsto \alpha(t),$$

we get $\dot{\alpha}(0) := (\alpha(0), \frac{d\alpha}{dt}(0)) = (p_0, v)$. Furthermore, we know that the force which may act on this point particle, is a vector which has the same point of application. All this is usually understood in physics first on an intuitive level.

Coming back to tangent vectors in mathematics, we may consider all possible tangent vectors for every fixed p. This means that we fix p and vary the vector $v \in \mathbb{R}^2$. This leads to the following definition:

Definition 2.23 Tangent space at p.
Given a point p of $\mathbb{R}^2$, the set

$$T_p \mathbb{R}^2 \equiv \mathbb{R}_p^2 := \{v_p := (p, v) : v \in \mathbb{R}^2\}$$

consisting of all tangent vectors with p the point of application, is called the tangent space of $\mathbb{R}^2$ at the point p.

Now, we may ask whether $T_p \mathbb{R}^2 \equiv \mathbb{R}_p^2$ is really a vector space. Yes, it is:

$$\text{Addition :} \quad u_p + v_p \quad := p + u + v = p + (u + v) = (p, u + v).$$
$$\text{Scalar multiplication :} \quad \lambda v_p \quad := p + \lambda v = (p, \lambda v).$$

Then $T_p \mathbb{R}^2$ is a vector space and it is clear that $T_p \mathbb{R}^2$ is isomorphic to $\mathbb{R}^2$: ($T_p \mathbb{R}^2 \cong \mathbb{R}^2$):

$$\mathbb{R}^2 \longrightarrow T_p \mathbb{R}^2,$$
$$v \longmapsto (p, v).$$

We further observe that we have such a vector space for every $p \in \mathbb{R}^2$.

$$p \longrightarrow T_p \mathbb{R}^2.$$

We consequently obtain a family of vector spaces canonically isomorphic to $\mathbb{R}^2$ which is parametrized by $p \in \mathbb{R}^2$. If we consider the disjoint union of all these vector spaces, using the symbol $\amalg$ for a disjoint union, we may write:

$$T \mathbb{R}^2 := \coprod_{p \in \mathbb{R}^2} T_p \mathbb{R}^2.$$

This is called the tangent space of $\mathbb{R}^2$ or equivalently, also the tangent bundle of $\mathbb{R}^2$.

Comment 2.12 A tangent bundle is a special vector bundle.

A tangent bundle is a special case of a vector bundle. Vector bundles play an important role in gauge theories. If we take into account the coordinates of $\mathbb{R}^2$ and $T\mathbb{R}^2$, we expect to have:

$$T\mathbb{R}^2 = \mathbb{R}^2 \times \mathbb{R}^2 = \{(p, v) : p \in \mathbb{R}^2, v \in \mathbb{R}^2\}.$$

It is, in this special case, in bijection with the vector space $\mathbb{R}^4$: $T\mathbb{R}^2 \underset{bij}{\cong} \mathbb{R}^4$.

Remark 2.13 Vector space and its dual, tangent space and its dual.

At this point, we have also to remember that an abstract vector space V, as all vector spaces, is never alone. Its dual vector space V^* is always present. We remember for example that the dual of

$$\mathbb{R}^2 = \{\text{columns of length 2}\} = \left\{ \begin{bmatrix} \alpha^1 \\ \alpha^2 \end{bmatrix} : \alpha^1, \alpha^2 \in \mathbb{R}^2 \right\}$$

is

$$(\mathbb{R}^2)^* = \{\text{rows of length 2}\} = \{[\varphi_1, \varphi_2] : \varphi_1, \varphi_2 \in \mathbb{R}^2\}.$$

Note that we write a coefficient (component or coordinate) of a vector (column) in $\mathbb{R}^2$ with the index upstairs but a coefficient of a covector or linear form (row) with the index downstairs. This leads, as expected, for every tangent vector space at p, $T_p\mathbb{R}^2$ to its dual $T_p^*\mathbb{R}^2 = (T_p\mathbb{R}^2)^*$. Similarly, the dual of the tangent bundle $T\mathbb{R}^2$ is denoted by $T^*\mathbb{R}^2$ and called cotangent space or cotangent bundle. In this case too, it is easy to recognize the bijection:

$$T^*\mathbb{R}^2 \underset{bij}{\cong} \mathbb{R}^2 \times (\mathbb{R}^2)^*.$$

2.9.2   Canonical Basis and Co-Basis Fields of $\mathbb{R}^2$

In mathematics and in physics, some aspects of analysis occur when we go from a vector space to a vector bundle (tangent bundle). We have to deal with vector fields and covector fields. We would like to choose, as above, the space $\mathbb{R}^2$ as an example because this facilitates both our notation, and our explanations. Going from vectors to vector fields, we may describe this by the map F which we call a vector field. We use the usual simplified notation:

$$F : \mathbb{R}^2 \longrightarrow \mathbb{R}^2 \times \mathbb{R}^2 (= T\mathbb{R}^2),$$
$$p \longmapsto (p, F(p)).$$

Similarly, we may write for a covector field the map Θ:

$$\Theta : \mathbb{R}^2 \longrightarrow \mathbb{R}^2 \times (\mathbb{R}^2)^* (= T^*\mathbb{R}^2),$$
$$p \longmapsto (p, \Theta(p)).$$

As we see, with F we denote a family of vectors, parametrized by $p \in \mathbb{R}^2$, and with Θ we denote a family of covectors, parametrized by $p \in \mathbb{R}^2$.

We follow the same procedure with the basis and the cobasis of a vector space. We then obtain a basis field and a cobasis field. This means that we get a family of canonical basis vectors, related to $\mathbb{R}^2$.

$$E : p \longmapsto (p; e_1(p), e_2(p)) := (p; (e_1, e_2)).$$

We represent this similarly as in Fig. 2.9. We get similarly a family of canonical cobasis vectors:

$$\mathcal{E} : p \longmapsto (p; \varepsilon^1(p), \varepsilon^2(p)) = (p; (\varepsilon^1, \varepsilon^2)).$$

The covectors $\varepsilon^1(p)$ and $\varepsilon^2(p)$ are the linear forms (covectors, linear functionals, linear functions)

$$\varepsilon^1(p), \varepsilon^2(p) : T_p\mathbb{R}^2 \longrightarrow \mathbb{R}$$

with the duality relation

$$\varepsilon^i(p)(e_j(p)) = \delta^i_j, \quad i, j, \in \{1, 2\}.$$

$(\varepsilon^1(p), \varepsilon^2(p))$ is the dual basis to $(e_1(p), e_2(p))$. Note that both, $[e_1 e_2]$ and $\begin{bmatrix} \varepsilon^1 \\ \varepsilon^2 \end{bmatrix}$ are special 2×2-matrices. Here, we even have $[e_1 e_2] = \mathbb{1}_2$ and $\begin{bmatrix} \varepsilon^1 \\ \varepsilon^2 \end{bmatrix} = \mathbb{1}_2$. The basis vector field E is represented symbolically in Fig. 2.10.

Remark 2.14 Calculus notation.

Traditionally, in calculus, we use another notation, for good reasons, as explained there.

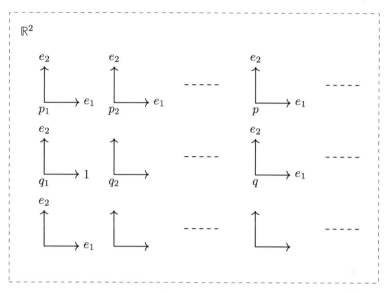

Fig. 2.10 The standard basis field in $\mathbb{R}^2$

For the canonical basis field which corresponds to the Cartesian coordinates $p = (x^1, x^2)$, we write:

$$\frac{\partial}{\partial x^i} \equiv e_i \text{ and } dx^i \equiv \varepsilon^i.$$

The duality relation is given by

$$dx^i \left(\frac{\partial}{\partial x^j} \right) = \frac{\partial x^i}{\partial x^j} = \delta^i_j.$$

This may partially explain the use of the $\frac{\partial}{\partial x^i}$, dx^j notation. Here we used the notation $e_i(p)$, $\varepsilon^i(p)$ because we wanted to emphasize the linear algebra background.

Summary

Beginning with the elementary part of linear algebra, we introduced and discussed the first steps of all necessary mathematical concepts which the physics student should already know and in fact has to use from day one in any theoretical physics lecture. The reader is probably already familiar with most of these concepts, at least in special cases. But we now offered careful definitions and a catalogue of basic properties which will better equip readers to follow the pure physics content of any introductory lecture in theoretical physics. Moreover, this introduction provided the reader a solid foundation to explore linear algebra in the subsequent chapters.

Right from the outset, we introduced the concept of the dual vector space. This aspect is often overlooked in physics lectures, leading to difficulties later on, especially in comprehending tensors.

In physics, abstract vector spaces are rarely encountered. For instance, in Newtonian mechanics, we begin with an affine Euclidean space and immediately utilize its model, the corresponding Euclidean vector space. What we accomplished in this chapter, mathematically speaking, was jumping from an abstract vector space to an affine Euclidean space and then back to an inner product space with significantly more structure than the initial abstract linear structure. The reader learned to manipulate structures within mathematical objects, adding and subtracting structures, a skill utilized throughout the book.

We also introduced the concept of a quotient space. While not typically utilized in physics until the end of master's studies, standard quotient spaces in differential topology and geometry are increasingly relevant in modern fields of physics, such as gravitation and cosmology.

Finally, we address a topic never found in linear algebra books but crucial for readers and one of the most important applications of linear algebra to physics: Newtonian mechanics begins with the concept of velocity, particularly the velocity at a fixed point. This velocity is essentially a vector in a tangential space where motion occurs. Hence, in physics, we often encounter families of isomorphic vector spaces, as discussed at the end of this chapter.

Exercises with Hints

Exercise 2.1 *The zero, the neutral element* 0 *of a vector space* V, *is uniquely defined.* Prove that 0 is the only zero in V.

Exercise 2.2 *The inverse of any vector of a vector space* V *is uniquely defined.* Prove that for any $v \in V$ there is only one additive inverse.

Exercise 2.3 *A scalar times the vector* 0 *gives once more* 0. Prove that $\lambda 0 = 0$ for all $\lambda \in \mathbb{K}$.

Exercise 2.4 *The number* $0 \in \mathbb{K}$ *times a vector gives the zero vector.* Show that for all $v \in V$, $0v = 0_V$.

Exercise 2.5 *The number* -1 *times a vector gives the inverse vector.* Show that for all $v \in V$, $(-\lambda)v = \lambda v$ holds.

Exercise 2.6 *If a scalar times a vector gives zero, then at least one of the two is zero.*
Prove that if $\lambda \in \mathbb{K}\, v \in V$, and $\lambda v = 0$, then either $\lambda = 0$ or $v = 0$.

Exercise 2.7 *A matrix face of the field of complex numbers.*
Consider the set of real matrices given by

$$\mathcal{C} = \{A = \left[\begin{smallmatrix} \alpha_0 & -\alpha_1 \\ \alpha_1 & \alpha_0 \end{smallmatrix}\right] : \alpha_0, \alpha_1 \in \mathbb{R}\}.$$

Show that

(i) if $A, B \in \mathcal{C}$, then $AB \in \mathcal{C}$;
(ii) $AB = BA$;
(iii) if $J = \left[\begin{smallmatrix} 0 & -1 \\ 1 & 0 \end{smallmatrix}\right]$ then $JJ = -\mathbb{1}_2$.
(iv) Check that, as a field, $\mathcal{C}$ is isomorphic to complex numbers

$$\mathcal{C} \cong \mathbb{C}.$$

In the following four exercises, it is instructive to check that all the vector space axioms hold on the given vector space. First, to check the axioms of the commutative group V and then the axioms connected to the $\mathbb{K}$ scalar action on the commutative group V.

Exercise 2.8 Show that $V = \mathbb{K}^n$ is a vector space.

Exercise 2.9 Show that $V = \mathbb{K}^{m \times n}$ is a vector space.

Exercise 2.10 Show that the set of symmetric matrices

$$V = \{S \in \mathbb{R}^{n \times n} : S^{\mathsf{T}} = S\}$$

is a vector space over $\mathbb{R}$, but the set of Hermitian matrices

$$W = \{H \in \mathbb{C}^{n \times n} : H^{\dagger} = H\}$$

is not over $\mathbb{C}$.

Exercise 2.11 Show that the space of functions on the set X,

$$V = \mathbb{R}^X = \mathrm{Map}(X, \mathbb{R}),$$

is a vector space.

Exercise 2.12 *Square matrices have more structure than a vector space. They form an algebra (see Definition 2.3).*
Show that $\mathbb{K}^{n \times n}$ is an algebra.

The next four exercises are dealing with some aspects of subspaces.

Exercise 2.13 *The union of two subspaces of a vector space V is in general not a subspace (see Comment 2.5). However, this exercise shows an exception.*
Prove that the union of two subspaces U and W of V is a subspace of V if and only if one of them is contained in the other.

$$U \leq W \leq V \quad \text{or} \quad W \leq U \leq V.$$

Exercise 2.14 *On the other hand, the intersection of subspaces of V is always a subspace of V (see Remark 2.3).*
Prove that if $U, W \leq V$, then $U \cap W \leq V$.

Exercise 2.15 *A special addition of two subspaces of V.*
Show that when $U \leq V$, then $U + V = V$.

Exercise 2.16 *The vector space complement of a given subspace U of V is not uniquely defined.*
Show that if $U \oplus W_1 = V$ and $U \oplus W_2 = V$, we get in general $W_1 \neq W_2$.

Exercise 2.17 *Linear maps take 0 to $0'$.*
Show that for a linear map $f \in \text{Hom}(V, V')$, then $f(0) = 0'$.

Exercise 2.18 *The null space is a subspace.*
Show that for a map $f \in \text{Hom}(V, V')$, ker f is a subspace of V'.

Exercise 2.19 *The range is a subspace.*
Show that for a map $f \in \text{Hom}(V, V')$, im f is a subspace of V'.

Exercise 2.20 *Criterion for injectivity.*
Show that a linear map $f \in \text{Hom}(V, V')$ is injective if and only if ker $f = \{0\}$.

In the following two examples, we consider very special linear maps: Linear functions, also called linear functionals or linear forms.

Exercise 2.21

$$(V^*)^* = \mathrm{Hom}(V^*, \mathbb{R}).$$

Consider a vector $v \in V$ as a linear form denoted by $v^{\#}$ on V^*, with $v^{\#} \in \mathrm{Hom}(V^*, \mathbb{R})$ is given by

$$v^{\#}: \quad V^* \longrightarrow \mathbb{R}$$
$$\xi \longmapsto v^{\#}(\xi) := \xi(v).$$

Show that $v^{\#}$ is indeed a linear function.

Exercise 2.22 *We consider the vector space of square matrices* $V = \mathbb{R}^{n \times n}$, *and its dual* $V^* = (\mathbb{R}^{n \times n})^*$.
Show that the trace $\mathrm{tr} \in (\mathbb{R}^{n \times n})^*$, given by

$$\mathrm{tr}: \quad (\mathbb{R}^{n \times n}) \longrightarrow \mathbb{R}$$
$$A \longmapsto \mathrm{tr}(A) := \sum_{i=1}^{n} \alpha_i^i$$

is indeed a linear function.

Exercise 2.23 *Properties of the standard inner product on* $\mathbb{C}$:

$$\langle v | u \rangle = \bar{v}^{\mathsf{T}} u = v^{\dagger} u = \sum_{i=1}^{n} \bar{v}^i u^i \quad \text{where} \quad v^i, u^i \in \mathbb{C}.$$

Let $u, v \in \mathbb{C}^n$ and $\lambda \in \mathbb{C}$. Show that

(i) $\overline{\langle v | u \rangle} = \langle u | v \rangle$;
(ii) $\langle \lambda v | u \rangle = \bar{\lambda} \langle v | u \rangle$;
(iii) $\langle v | \lambda u \rangle = \lambda \langle v | u \rangle$.

Exercise 2.24 *Equality in the Cauchy-Schwarz inequality.*
Show that $\langle u | v \rangle = \| v \| \| u \|$ if and only if $u = 0$ or $v = \lambda u$ for some $\lambda \in \mathbb{R}$.

Exercise 2.25 *Equality in the triangular inequality.*
Show that $\| u + v \| = \| u \| + \| v \|$ if and only if either $u = 0$ or if there is a scalar λ with $0 \leq \lambda$ such that $v = \lambda u$.

Exercise 2.26 *The parallelogram law.*
Show that for elements u and v in an inner product vector space,

$$\| u + v \|^2 + \| u - v \|^2 = 2 \| u \|^2 + 2 \| v \|^2.$$

Exercise 2.27 *The polarization identity for a real inner product space V.*
Show that for $u, v \in V$,

$$4\langle u|v\rangle = \|u + v\|^2 - \|u - v\|^2.$$

Exercise 2.28 *Extended polarization identity for a symmetric operator f, $\langle u|fv\rangle = \langle fu|v\rangle$, in a real inner product vector space V.*
Show that

$$4\langle u|fv\rangle = \langle u + v|f(u + v)\rangle - \langle u - v|f(u - v)\rangle.$$

Exercise 2.29 *Expectation values of a zero operator in a complex inner product space V.*
Show that $f \in \text{Hom}(V, V)$ is the operator $\hat{0}$ if and only if $\langle v|fv\rangle = 0$ for all $v \in V$.

Exercise 2.30 *Polarization identity for a complex inner product space V.*
Show that for all $v, w \in V$,

$$4\langle v|w\rangle = \|v + w\|^2 - \|v - w\|^2 - i(\|v + iw\|^2 - \|v - iw\|^2).$$

Exercise 2.31 *Extended polarization identity for an operator f in a complex inner product space V.*
Show that for all $v, w \in V$,

$$4\langle v|fw\rangle = \langle v + w|f(v + w)\rangle - \langle v - w|f(v - w)\rangle -$$
$$-i(\langle v + iw|f(v + iw)\rangle - \langle v - iw|f(v - iw)\rangle).$$

Exercise 2.32 *Affine subset of a subspace in a real vector space V.*
Let $A = a + U$ be an affine subset with U a subspace in V and $\in V$. Show that U is uniquely defined by A.

Exercise 2.33 *Translations in a real vector space V. Definition: Any vector $v \in V$ leads to a translation:*

$$\begin{aligned} T_a : \quad V &\longrightarrow V \\ v &\longmapsto T_a(v) = a + v. \end{aligned}$$

Check the following properties:

(i) $T_0 = id_V$;
(ii) $T_b \circ T_a = T_{b+a}$;
(iii) $T_a^{-1} = T_{-a}$ and hence T_a is bijective;
(iv) When $a \neq 0$, T_a is not a linear map.

Exercise 2.34 *Affine subsets in a real vector space V and translations.*
Let A be a subset of V with $a \in A$. If $T_{-a}(A)$ is a subspace of V, show that A is an affine subset of V and that for any $a' \in A$, $T_{-a'}(A) = T_{-a}(A)$.

Exercise 2.35 *An affine subspace A in a real vector space V contains all its straight lines.*
Show that a subset A in V is an affine subset of V if and only if for all $u, v \in A$ and $\lambda \in \mathbb{R}$, $\lambda v + (1 - \lambda)u \in A$.

Exercise 2.36 *Equivalence proposition for an affine space.*
Let V be a real vector space and $A \subset V$ a subset. Then show that the following assertions are equivalent:

(i) A is an affine subset of V. (There exists a subspace U of V and $a \in V$ such that $A = a + U$);

(ii) There is a vector space V, a linear map $f : V \to V'$, and a vector $a' \in V'$ such that
$$A = f^{-1}(a');$$

(iii) There is a list of vectors $(v_0, \ldots, v_k)$ in A such that

$$A = \{\, v_i \alpha^i \,|\, \alpha^i \in \mathbb{K}\, i \in \{0, \ldots, k\} \text{ and } \sum_{i=0}^{k} \alpha^i = 1\}.$$

Exercise 2.37 *Affine subset in a real vector space V from a list of vectors.*
Show that for a list of vectors $(v_0, \ldots, v_k)$ in V, there exists an affine subset A of V given by:
$$A = \{v_0 + (v_i - v_0)\alpha^i : \alpha^1, \ldots, \alpha^k \in \mathbb{R} \text{ and } i \in I(k)\}.$$

Exercise 2.38 *Quotient space and its basis.*
Let U be a subspace of a vector space V, $(u_1, \ldots, u_r)$ a basis of U, and $(v_1 + U, \ldots, v_l + U)$ a basis of V/U. Show that $(u_1, \ldots, u_k, v_1, \ldots, v_l)$ is a basis of V.

Exercise 2.39 *Quotient space and vector space complement.*
Let U be a subspace of the vector space V. Show that any complement W of U such that $V = U \oplus W$, has the following properties::

$$\dim W = \dim V/U.$$

The natural surjection $V \to V/U$ gives an isomorphic

$$v \to v + U,$$

$$W \cong V/U.$$

Exercise 2.40 *Isomorphism theorem for linear maps.*
Let f be a linear map $f \in \mathrm{Hom}(V, V')$ and V, V' vector spaces. Show that f induces an isomorphism:
$$\bar{f} : V/\ker f \xrightarrow{\cong} \operatorname{im} f$$

and when f is surjective:

$$\bar{f} : V / \ker f \xrightarrow{\cong} V'.$$

References and Further Reading

1. S. Axler, *Linear Algebra Done Right* (Springer Nature, 2024)
2. W.M. Boothby, *An Introduction to Differentiable Manifolds and Riemannian Geometry* (Academic Press, 1986)
3. S. Bosch, *Lineare Algebra* (Springer, 2008)
4. G. Fischer, B. Springborn, *Lineare Algebra. Eine Einführung für Studienanfänger*. Grundkurs Mathematik (Springer, 2020)
5. S.H. Friedberg, A.J. Insel, L.E. Spence, *Linear Algebra*. (Pearson, 2013)
6. K.-H. Goldhorn, H.-P. Heinz, M. Kraus, *Moderne mathematische Methoden der Physik. Band 1* (Springer, 2009)
7. S. Hassani, *Mathematical Physics: A Modern Introduction to its Foundations* (Springer, 2013)
8. K. Jänich, *Mathematik 1. Geschrieben für Physiker* (Springer, 2006)
9. N. Jeevanjee, *An Introduction to Tensors and Group Theory for Physicists* (Springer, 2011)
10. N. Johnston, *Introduction to Linear and Matrix Algebra* (Springer, 2021)
11. M. Koecher, *Lineare Algebra und analytische Geometrie* (Springer, 2013)
12. G. Landi, A. Zampini, *Linear Algebra and Analytic Geometry for Physical Sciences* (Springer, 2018)
13. J.M. Lee, *Introduction to Smooth Manifolds*. Graduate Texts in Mathematics (Springer, 2013)
14. J. Liesen, V. Mehrmann, *Linear Algebra* (Springer, 2015)
15. P. Petersen, *Linear Algebra* (Springer, 2012)
16. S. Roman, *Advanced Linear Algebra* (Springer, 2005)
17. B. Said-Houari, *Linear Algebra* (Birkhäuser, 2017)
18. F. Scheck, *Mechanics. From Newton's Laws to Deterministic Chaos* (Springer, 2010)
19. A.J. Schramm, *Mathematical Methods and Physical Insights: An Integrated Approach* (Cambridge University Press, 2022)
20. B.F. Schutz, *Geometrical Methods of Mathematical Physics* (Cambridge University Press, 1980)
21. G. Strang, *Introduction to Linear Algebra* (SIAM, 2022)
22. R.J. Valenza, *Linear Algebra, An Introduction to Abstract Mathematics* (Springer, 2012)

Chapter 3
The Role of Bases

In this chapter, we discuss in detail the basics of linear algebra. The first important concepts in a vector space are linear combinations of vectors and related notions like generating systems, linear independent and linear dependent systems. This leads directly to the central notions of bases of vector spaces and their dimension.

Bases allow us to perform concrete calculations for vectors needed in physics, such as assigning a list of numbers, the coordinates. This enormous advantage has its price. The representation of an abstract vector by a list of numbers depends on the chosen basis. Any theoretical calculation we do should obviously not depend on the choice of a basis. We discuss in detail the satisfactory but demanding solution to this problem in Sect. 3.2 which can be skipped on a first reading.

We then demonstrate a suitable choice of basis for the representation of linear maps, and we discuss the origin of tensors in an elementary manner. Finally, we provide an important application for physics, and show that the transition from Newtonian mechanics to Lagrangian mechanics is nothing but the transition from a linear dependent to a linear independent system.

3.1 On the Way to a Basis in a Vector Space

The multiplicative action of $\mathbb{K}$ on an abelian group V, as given in Definition 2.4, is what makes the abelian group the vector space V we know. We see immediately that for every fixed vector $v \neq 0$, we can obtain, by a scalar multiplication, all vectors of the one-dimensional subspace $v\mathbb{K} = \mathbb{K}v$ in the direction v. Analogously, for every direction (or equivalently, every one-dimensional subspace), we need to fix only one vector to describe the corresponding subspace. It may further seem a surprise that within linear algebra, we often only need a finite number of such vectors to describe all elements of V which contain all these one-dimensional subspaces. This leads to the existence of a basis in V.

© The Author(s), under exclusive license to Springer Nature Switzerland AG 2024
N. A. Papadopoulos and F. Scheck, *Linear Algebra for Physics*,
https://doi.org/10.1007/978-3-031-64908-0_3

As mentioned above, one main reason for the existence of a basis stems from the scalar action of $\mathbb{K}$ which gives the possibility of scaling the elements of a basis and thus to obtain, after adding, all the elements of the given vector space V.

In addition, we need an abstract and fundamental property that characterizes a basis: the notion of linear independence or linear dependence (see Definition 3.4). Before we can understand this, however, one must first understand the notions of linear combinations and span.

Definition 3.1 Linear combination and span.
The following expression is a linear combination of vectors in V.

$$\alpha^1 v_1 + \alpha^2 v_2 + \cdots + \alpha^k v_k$$

with the coefficients or scalars $\alpha^i \in \mathbb{K}$ and the vectors $v_i \in V$, $i \in I(k)$ and $k \in \mathbb{N}$. $\mathrm{span}(v_1, \ldots, v_k)$ denotes the set of all such linear combinations:

$$\mathrm{span}(v_1, \ldots, v_k) := \left\{ \sum_{i=1}^{n} \alpha^i v_i \quad \text{for all} \quad \alpha^i \in V \right\}.$$

A linear combination is trivial if **all** the coefficients are zero, and otherwise, it is nontrivial. Since the linear combination is by definition a finite sum of vectors, we use the Einstein convention for the sum, and we denote

$$\alpha^1 v_1 + \alpha^2 v_2 + \cdots + \alpha^k v_k =$$

$$= \sum_{i=1}^{k} \alpha^i v_i =: (\alpha^i v_i)_k, = \alpha^i v_i \in V.$$

Usually, if there is no ambiguity, we drop the index k. We write $v = \alpha^i v_i$ and we mean that the vector v is a linear combination of the vectors $v_1, \ldots, v_k$. We can write, just as well, $v = v_i \alpha^i$ since the left or right action of $\mathbb{K}$ on V is the same.

Comment 3.1 The importance of linear combinations.

We consider linear combinations as the most important operation in linear algebra. It is very difficult to make a statement in linear algebra without using the term linear combination. For this reason, it might be useful to discuss various aspects of linear combinations. We first discuss linear combinations in $\mathbb{K}^n$ and then the analogous linear combinations in a given vector space V'.

We study a list of columns in $\mathbb{K}^n : A = (\vec{a}_1, \vec{a}_2, \ldots, \vec{a}_k)$ and the associated $n \times k$-matrix with the columns $\vec{a}_s \in \mathbb{K}^n$ with $s \in I(k)$ given by $[A] = [\vec{a}_1 \vec{a}_2 \ldots \vec{a}_k]$. We write $[A]$ for this matrix to distinguish it from the list $A = (\vec{a}_1, \vec{a}_2, \ldots, \vec{a}_k)$. Note that we use three kinds of brackets: "()" usually for a list, "[]" for a matrix, and the standard "{}" for a set. We identify of course $[A]$ with A automatically, but sometimes it is useful to make the difference. And we will do so also here when necessary.

To build formally a linear combination of k columns in $\mathbb{K}^n$, we may also use the map, ψ_A expressed as a matrix multiplication (see Remark 2.8) or as a linear combination:

$$\psi_A : \mathbb{K}^k \longrightarrow \mathbb{K}^n$$

$$\vec{\lambda} = \begin{bmatrix} \lambda^1 \\ \vdots \\ \lambda^k \end{bmatrix} \longmapsto \psi_A(\vec{\lambda}) := [A]\vec{\lambda} = [\vec{a}_1 \vec{a}_2 \ldots \vec{a}_k] \begin{bmatrix} \lambda^1 \\ \vdots \\ \lambda^k \end{bmatrix} = \vec{a}_s \lambda^s.$$

We therefore get for

$$\text{span } A = \text{span}(\vec{a}_1, \ldots, \vec{a}_k),$$

where

$$\text{span } A = \text{im } \psi_A = \psi_A(\mathbb{K}^k) := \{[A]\vec{\lambda} = \vec{a}_s \lambda^s : \text{ for all } \lambda^1, \ldots, \lambda^k \in \mathbb{K}\}.$$

The range im ψ_A is of course also the set of all possible linear combinations of the list A, and so we have im $\psi_A \leq \mathbb{K}^n$. We see immediately that im ψ_A is a subspace in $\mathbb{K}^n$ because it is the image of the linear map. Thus we have also another proof that span A is, as im ψ_A, a subspace of V, span $A \leq V$ (see Proposition 3.1).

In this context, the dimension of the subspace im ψ_A is also relevant in order to assign an important number to the list A. Of course, in our notation, the same is true for the matrix $[A]$). This leads to the next definition.

Definition 3.2 Rank of a list.
$\text{rank}(A) := \dim(\text{im } \psi_A) = \dim(\text{span } A).$

Usually, we use the map ψ_A when the list A is a basis. Then the map ψ_A is a basis isomorphism and is also called a parametrization. This is extensively discussed in Sect. 3.2.1. But here, A is taken as an arbitrary list, as it is not necessary to have a basis to obtain just a linear combination (see Comment 3.1). The same can be applied to a linear combination with $a_s \in V$ instead of $\vec{a}_s \in \mathbb{K}^n$:

$$\psi_A : \mathbb{K}^k \longrightarrow V,$$

$$\vec{\lambda} \longmapsto [a_1 a_2 \dots a_k] \begin{bmatrix} \lambda^1 \\ \vdots \\ \lambda^k \end{bmatrix} = a_s \lambda^s \in V.$$

So we again get $[A] = [a_1 \dots a_k]$ and im $\psi_A \leq V$ and, as above, we can also define for $A = (a_1, \dots, a_k)$

$$\text{rank } A = \text{rank}[A] := \dim(\text{im } \psi_A) = \dim(\text{span } A).$$

If we want, we can also identify ψ_A with $[A]$ and write

$$[A] : \mathbb{K}^k \longrightarrow V,$$

$$\vec{\lambda} \longmapsto [A]\vec{\lambda} = [a_1 \cdots a_k]\vec{\lambda} = [a_1 \cdots a_k] \begin{bmatrix} \lambda^1 \\ \vdots \\ \lambda^k \end{bmatrix} \in V.$$

Since span $A := \text{im } \psi_A$, we have already proved the following proposition, taking $A = (v_1, \dots, v_k)$ and $U = \text{span}(v_1, \dots, v_k)$.

Proposition 3.1 *The set* span(A) *is a vector space.*

$U = \text{span}(A)$ *with* $A = (v_1, \dots, v_k)$ *is a subspace of* V. ($U \leqslant V$).

Proof We present a second, different, proof.
This can easily be shown by using the criteria for a subspace given in Remark 2.2.

  i. The zero, $0 \in U$, since $0 = 0v_1 + \cdots + 0v_k \in U$;
 ii. The set, U, is closed under addition:

$$\alpha^i v_i + \beta^i v_i = (\alpha^i + \beta^i)v_i;$$

iii. The set, U, is closed under scalar multiplication:

$$\lambda(\alpha^i v_i) = (\lambda \alpha^i)v_i.$$

$\square$

Remark 3.1 The set $U = \text{span}(v_1, \ldots, v_k)$ is the smallest subspace of V containing the list $(v_1, \ldots, v_k)$.

Proof Indeed, a subspace W of V containing all v_i of the above list, contains also all linear combinations of them: $\text{span}(v_1, \ldots, v_k) \subset W$. This means that every such W is necessarily bigger than U. Hence $U \leqslant W$. $\qquad\qquad\qquad\qquad\qquad\qquad$ □

Comment 3.2 Lists and colists.

Here, we repeat some facts and add some notations (conventions). Since in this book we consider only finite-dimensional vector spaces, we have to use mostly finite families of elements belonging to a given set. We call such a finite family quite naturally a list. We may call the length or the size of the list, the number of elements, the cardinality of the list. Very often, the length of a list is either well-known or not relevant, in this case we do not specify the length. A list of length n is exactly what is usually called a n-tuple. But we consider the use of the name list as more elegant and more realistic (see also Example 3.1).

It is clear that in this book by definition the length of a list is always finite and as a family the order matters, and repetitions are allowed. As it is natural, we write a list horizontally. However, in many cases, and in particular in connection with matrices (e.g. for a list of rows). We may also need to write a list vertically. Therefore, in this book we introduce and we sometimes use the name "colist" for a vertically written list. So the word "colist" always refers to a vertically written list. But in some cases, very seldom, we allow ourselves to use the name "list" in a general way, meaning horizontally or vertically written, when it does not cause confusion.

The distinction between list and colist is especially useful in connection with our conventions about the position of indices (the tensor formalism convention).

We write a list of vectors always as a horizontal list and, correspondingly, use lower index notation. Accordingly, we write a list of covectors always as a vertical list (a colist) and, correspondingly, use the upper index notation. For scalars, depending on the situation, we write their indices again, as in every horizontal list, below, and in a colist of scalars we write the indices above as in every vertical list. The general rule is that if an element has its index below, it belongs to a list (written horizontally) and if an element has its index above, it belongs to a colist (written vertically).

Example 3.1 List, colist, and matrices.

The above considerations can easily be demonstrated if we take the vector space $\mathbb{R}^n$ and its dual $(\mathbb{R}^n)^*$:

$$(\mathbb{R}^n)^* = \mathrm{Hom}(\mathbb{R}^n, \mathbb{R}).$$

An element of $\mathbb{R}^n$ is a vector taken as column which is a $n \times 1$ matrix with scalar entries which are, here, real numbers. According to our given convention, this corresponds to a colist of numbers.

Until now, column and colist can be considered as synonyms, particularly if we consider the elements of $\mathbb{R}^n$ as points. But sometimes, as stated above, it is necessary and useful to draw the following distinction, for instance if we multiply matrices: A list and a colist are simply there as a set, as a given data, and if we want, we can later define some algebraic operations and it is usually uniquely clear from the context what we mean. Matrices are of course the well-known algebraic objects. In other words, a $1 \times n$-matrix with vector entries (columns) is more than simply a list. For a colist of length m with any element, we may write for example symbolically:

$$\begin{pmatrix} *^1 \\ *^2 \\ \vdots \\ *^m \end{pmatrix}.$$

For an $m \times n$-matrix, we may write symbolically ($m = 3, n \in \mathbb{N}$)

$$\begin{bmatrix} \sharp \cdots \sharp \cdots \sharp \\ \sharp \cdots \sharp \cdots \sharp \\ \sharp \cdots \sharp \cdots \sharp \end{bmatrix}.$$

We may add and multiply the elements $\sharp$ of the above matrix with each other. So we may now write for the $m \times 1$-matrix (column) with entries $\sharp$:

$$\begin{bmatrix} \sharp^1 \\ \vdots \\ \sharp^m \end{bmatrix}$$

and for a 1×1-matrix, we may have

$$[\sharp] \quad \text{or} \quad [\sharp_1 + \cdots + \sharp_n].$$

Taking $v, a \in \mathbb{R}^m$, we have for instance:

$$v = \begin{bmatrix} v^1 \\ \vdots \\ v^i \\ \vdots \\ v^m \end{bmatrix},$$

and

$$a = \begin{bmatrix} \alpha^1 \\ \vdots \\ \alpha^i \\ \vdots \\ \alpha^m \end{bmatrix},$$

$$v^i, \alpha^i \in \mathbb{R}, i \in I(m).$$

For the covector $\theta \in (\mathbb{R}^n)^*$, we may write the list

$$\theta = (\vartheta_1, \ldots, \vartheta_s, \ldots \vartheta_n), \quad \vartheta_s \in \mathbb{R}, \quad s \in I(n).$$

We may associate this with the row matrix $1 \times n$ and with the same symbol θ and without the comma:

$$\theta = [\vartheta_1 \vartheta_2 \cdots \vartheta_n].$$

For the sake of completeness, we write the associated colist of scalars, here numbers, to the vectors v and a above:

$$\begin{pmatrix} v^1 \\ \vdots \\ v^m \end{pmatrix}$$

and

$$\begin{pmatrix} \alpha^1 \\ \vdots \\ \alpha^m \end{pmatrix}$$

Example 3.2 List and colist in V.

We are going now to apply the above considerations to an abstract vector space V. For a list of vectors $v_1, \ldots, v_k \in V$, we write for a list

$$(v_1, \ldots, v_k)$$

and for the corresponding $1 \times k$-matrix with entries these vectors:

$$[v_1 \ldots v_k]$$

which is a row-matrix with entries vectors. Similarly, we write for the list of the covectors

$$\theta^1, \ldots, \theta^k \in V^*$$

the colist:

$$\begin{pmatrix} \theta^1 \\ \vdots \\ \theta^k \end{pmatrix}$$

and for the $k \times 1$-matrix with entries the above covectors

$$\begin{bmatrix} \theta^1 \\ \vdots \\ \theta^k \end{bmatrix}$$

which is a column-matrix with entries covectors.

Note that a vector in $\mathbb{K}^n$ corresponds to a colist (vertically written list) of scalars and a covector in $\mathbb{K}^n$ to a list of scalars.

We can now proceed with a few additional and important definitions to get to the notion of a basis in vector spaces which is a very special list of vectors with appropriate properties.

Definition 3.3 Spanning list, spanning set.

A list A or a set A in V spans V if every vector in V can be written as a linear combination of vectors in A, that is, if $V = \text{span } A$ holds. In this case, we say that A spans or generates V. In this book, A is typically finite and we can write $A = (a_1, \ldots, a_k), k \in \mathbb{N}$. So we get

$$V = \text{span}(a_1, \ldots, a_k)$$

and the vector space V is called finitely generated.

As usual in linear algebra, we consider mainly finitely generated vector spaces.

The notion of linearly independence is fundamental in linear algebra. It has a deeply geometrical character. But the usual definition is given in an algebraic form and seems quite abstract. For this reason, and to get a feeling of what is going on, we need some preparation and we therefore construct a model of what we mean by "linearly independent" and "linearly dependent", geometrically speaking.

We start with an informal definition. In some sense, it points indirectly to the geometric character of "linearly independent" and "linearly dependent". This definition is equivalent to the usual and more abstract algebraic formulation, as we shall show below. This sounds good, but it is difficult to check.

> **Definition 3.4** Linearly independent and linearly dependent.
> (Informal definition)
>   We say that the vectors $a_1, \ldots, a_m$ are linearly dependent, if there exists a vector in the list which is a linear combination of the others. Otherwise, they are linearly independent. Alternatively, we may define linear independence as follows: We say that the vectors $a_1, \ldots, a_m$ are linearly independent if no vector in this list is a linear combination of the others. Otherwise, this list is linearly dependent.

As we see, we have to check here a yes-no question. This, and all the equivalent definitions of linearly independent and linearly dependent, refer to an abstract vector space without any other structures, as, for example, volume and scalar product. Even more, it should be clear, that in this section, we do not know yet what a basis in a vector space is. This situation makes it difficult to recognize directly the geometric character of the above definition.

For our demonstration, we therefore choose our standard vector space $\mathbb{R}^n$, $n \in \mathbb{N}$. Here, we know of course the dimension, the volume, and the distance, and we have the canonical basis $E = (e_1, \ldots, e_n)$ in $\mathbb{R}^n$. In order to proceed, we have to remember what we mean by a k-volume in a fixed $\mathbb{R}^n$ where $(k \leqslant n)$. This is in itself interesting enough. We consider the k-vectors $(a_1, \ldots, a_k)$ which usually should define a nondegenerate k- parallelepiped $P_k := P_k(a_1, \ldots, a_k)$. We know its volume $vol_k(P_k) = vol_k(a_1, \ldots, a_k)$, a positive number (we do not need the orientation, here), which we may call k-volume. It is clear that in the case that P_k is degenerate, we have $vol_k(a_1, \ldots, a_k) = 0$. Similarly, if we consider $k < n$, in particular $(k \neq n)$, we have in an obvious notation $vol_n(a_1, \ldots, a_k) = 0$. For all that, we have of course our experience with our three-dimensional Euclidean space. The generalization to every fixed $n \in \mathbb{N}$ is quite obvious. In connection with this, it is useful to think of the following sequence of subspaces given by

$$\mathbb{R}^1 < \mathbb{R}^2 < \mathbb{R}^3 < \cdots < \mathbb{R}^k < \cdots < \mathbb{R}^n.$$

We may also see immediately the following results. The n-dimensional volume,

$$vol_n(P_k) = \begin{cases} \text{is positive} & \text{if } k = n \text{ and } P_k \text{ is nondegenerate} \\ = 0 & \text{if } k < n \\ & \text{if } n < k \text{ is not definite.} \end{cases}$$

It is not surprising that this can be expressed with the help of determinants. The parallelepiped $P_k(a_1, \ldots, a_k)$ corresponds to the matrix $A_k = [a_1 \cdots a_k]$ and the

Euclidean volume is given by $vol_k(P_k) = \det A_k$. Taking into account that every list $(a_1, \ldots, a_m)$ corresponds also to a parallelepiped, the above result may be stated differently:

If the list $A_m = (a_1, \ldots, a_m)$ "produces" enough dimension (enough "space"), which means that $\dim(\text{span } A_m) = m$, we have $vol_m(A_m)$ positive (nonzero). If this list "produces" not enough dimension, which means that $\dim(\text{span } A_m) < m$, we have $vol_m(A_m) = 0$. Furthermore, since we are only interested in values positive or zero, we may define an equivalence relation yes or no.

$$[A_m] = \text{yes} \quad \text{if } vol_m(A_m) \neq 0;$$
$$[A_m] = \text{no} \quad \text{if } vol_m(A_m) = 0.$$

This is our model for linearly independent and linearly dependent:

$A_m = (a_1, \ldots, a_m)$ is linearly independent if and only if $[A_m] = \text{yes};$

$A_m = (a_1, \ldots, a_m)$ is linearly dependent if and only if $[A_m] = \text{no}.$

In the above sense, we can say, to simplify, that linearly independent means "enough space" and linearly dependent "not enough space".

Example 3.3 Linearly dependent and linearly independent lists.
The following lists of vectors in $\mathbb{R}^n$ and in V, with $\dim V = n$, are linearly dependent.

(i) $(0, v_2, \ldots, v_m), (v_1, 0 \ldots, v_m), (v_1, \ldots, 0);$
(ii) $(v, v, v_3, \ldots, v_m), (v, v_2, v, \ldots, v_m), (v, \ldots, v);$
(iii) $(v_1, v_2, v_3, \ldots, v_m)$ with $v_3 = v_1 + v_2;$
(iv) $(v_1, v_2, v_3, \ldots, v_m)$ with $v_3 = v_1 \lambda^1 + v_2 \lambda^2, \lambda_1, \lambda_2 \in \mathbb{R}.$

The list $A_1 = (v)$ with $v \neq 0$ is linearly independent. The list $A_\emptyset = \emptyset$ is linearly independent since no vector is a linear combination of the rest. Now we are ready for the usual, more abstract, definition.

Definition 3.5 Linearly independent and linearly dependent.
A list of vectors $a_1, \ldots, a_m$ is called linearly independent if $a_i \xi^i = 0$ implies that the only possibility is $\xi^i = 0$ for all $i \in \{1, \ldots, m\}$. Otherwise, the list $a_1, \ldots, a_m$ is linearly dependent. This means that there exist $\xi^i \in \mathbb{K}$, not all of them zero, such that $a_i \xi^i = 0$.

The above definition can be formulated differently:

The list $(a_1, \ldots, a_m)$ is linearly independent if the equation $a_i \xi^i = 0$ has only the trivial solution: $\xi^i = 0$ for all i. Or else, the list $(a_1, \ldots, a_m)$ is linearly dependent which means that the equation $a_i \xi^i = 0$ has a nontrivial solution.

The next lemma shows a property of a linearly independent list which underlines the importance of being linearly independent. As we shall see, this property turns out to be an essential property of a basis in V.

Lemma 3.1 *Linear independence and uniqueness.*

Given a list $a_1, \ldots, a_m$ of vectors in V, the following statements are equivalent:

(i) The vectors $(a_1, \ldots, a_m)$ are linearly independent;
(ii) If $u = a_i \xi^i$ is a linear combination of the vector u ($u \in U = \mathrm{span}(a_1, \ldots, a_m)$) with the coefficients $\xi^i \in \mathbb{K}$, $i \in I(m)$, the coefficients ξ^i are uniquely determined.

Proof We start with (i). The list $(a_1, \ldots, a_m)$ is linearly independent.

If we write $u = a_i \xi^i$ and $u = a_i \eta^i$, for $\xi^i, \eta^i \in \mathbb{K}$, subtracting we have:

$$0 = a_i \xi^i - a_i \eta^i = a_i (\xi^i - \eta^i) \in U.$$

Since by assumption (i), $(a_1, \ldots, a_m)$ is linearly independent, we obtain

$$\xi^i - \eta^i = 0 \Leftrightarrow \xi^i = \eta^i$$

which shows that (ii) holds.

If we start with (ii), which states that every representation of $u \in U$ is unique and set $u = 0 \in U$, we have $0 = a_i \lambda^i$. One solution of this equation is $\lambda^i = 0$ for all i. By assumption of uniqueness, (ii), it follows that this is the only solution. By definition (see Definition 3.1), the list $(a_1, \ldots, a_m)$ is then linearly independent and so (i) holds. $\square$

The next lemma tells us essentially that the informal definition (see Definition 3.4) and Definition 3.5 are equivalent. It is closer to the geometric aspects of the definition. This means that in a linearly independent list, a vector loss leads to the loss of spanning space. For a linearly dependent list, there is always a redundant vector, the absence of which leaves the spanning space of the list invariant. For example, in the next lemma the vector a_j is redundant.

Lemma 3.2 *Linear dependence or a redundant vector.*
For a list of vectors $(a_1, \ldots, a_j, \ldots, a_m)$ in V, the following three statements are equivalent:

(i)   The vectors $a_1, \ldots, a_j, \ldots, a_m$ are linearly dependent;
(ii) There exists an index j with $a_j \in \operatorname{span}(a_1, \ldots, a_{j-1}, a_{j+1}, \ldots, a_m)$, that is, one of the vectors is a linear combination of the rest;
(iii) There exists an index j with:

$$\operatorname{span}(a_1, \ldots, a_m) = \operatorname{span}(a_1, \ldots, a_{j-1}, a_{j+1}, \ldots, a_m).$$

Proof We show that (i) $\Leftrightarrow$ (ii) and (i) $\Leftrightarrow$ (iii) which establishes the result. For this purpose, we define

$$A := (a_1, \ldots, a_m), \; \bar{A} := \bar{A}_j := (a_1, \ldots, a_{j-1}, a_{j+1}, \ldots, a_m).$$

We set $i \in I(m)$ and $s \in \{1, \ldots, j-1, j+1, \ldots, m\}$.

(i) $\Rightarrow$ (ii): We have to show that a_j is a linear combination of the rest: $a_j \in \operatorname{span}(\bar{A})$. From (i), it follows that the equation

$$a_i \xi^i = 0 \tag{3.1}$$

has a nontrivial solution. This means that, for example, there is some j for which $\mathcal{E}^j \neq 0$. Without loss of generality, by scaling if necessary, we may put $\xi^j = -1$ and we obtain from Eq. (3.1)

$$a_s \xi^s - 1 a_j = 0 \tag{3.2}$$

and $a_j = a_s \xi^s$, which proves (ii).

(ii) $\Rightarrow$ (i): (ii) means that we can choose $-a_j = a_s \xi^s$ which is $a_s \xi^s + a_j = 0$. This proves (i).

(i) $\Rightarrow$ (iii): From (i) we can conclude that if we take, without loss of generality, $\alpha^j = 1$:

$$a_j = a_s \alpha^s \quad \alpha^s \in \mathbb{K}. \tag{3.3}$$

If $v \in \operatorname{span} A$, we have

$$v = a_i v^i = a_s v^s + a_j v^j \quad \text{with} \quad j = j_0 \text{ fixed}, \; v^i \in \mathbb{K}. \tag{3.4}$$

We insert Eq. (3.3) into Eq. (3.4) with $j = j_0$ fixed ($v^i \in \mathbb{K}$) and we so obtain

$$v = a_s v^s + (a_s \alpha^s) v^j, \tag{3.5}$$

which is

$$v = a_s(v^s + \alpha^s v^j) \quad \text{with} \quad v^s + \alpha^s v^j \in \mathbb{K}. \tag{3.6}$$

Equation (3.6) shows that (iii) is proven.

"(iii) $\Rightarrow$ (i)": (iii) says that if $a_j \in$ span A, we have also $a_j \in$ span $\bar{A}$. This means that, as above,

$$a_j = a_s \alpha^s$$

and

$$-a_j + a_s \alpha^s = 0.$$

This proves (i) and altogether the above lemma.   □

The next lemma refers to a special feature of a spanning list. A spanning list is very "near" to a linearly dependent list.

> **Lemma 3.3** *Spanning and linearly dependent list.*
>
> $A_m = (a_1, \ldots, a_m)$ *is a spanning list in* V, $V = \text{span}(A_m)$. *If we add to* A_m *one more vector, we get* $A_{m+1} = (a_1, \ldots, a_m, v)$ *with* $v \in V$ *which is a linearly dependent list.*

Proof This is almost trivial: Since $V = \text{span } A_m$, we have $v \in \text{span } A_m$ which means that $-v = a_i \xi^i$, $i \in I(m)$. This is equivalent to $v + a_i \xi^i = 0$ which shows that A_{m+1} is linearly dependent.   □

The next proposition concerns a relationship between a linearly independent list and a spanning list. This relationship is completely evident in our above "model", following Definition 3.4, with $V = \mathbb{R}^n$ since whenever $k \leqslant n$, $\text{span}(e_1, \ldots, e_k) \leqslant \text{span}(e_1, \ldots, e_n) = \mathbb{R}^n$. Equivalently, if $A_k = (a_1, \ldots, a_k)$, like $(e_1, \ldots, e_k)$, is a linearly independent list, then we have again $\text{span}(A_k) \leqslant \mathbb{R}^n$. So the length (cardinality) of any linearly independent list is less or equal to the length of a spanning list.

But in our case, here in this section, we cannot use the above considerations. In this section, up to now with an abstract vector space V, we have not yet defined what a basis is. We neither know what a dimension of a vector space is. The following proposition will help us, among other things, to prove the existence of a basis in V.

> **Proposition 3.2** *The length of a linearly independent and a spanning list.*
>
> *The length of a linearly independent list is less than, or equal to the length of a spanning list in* V.

Proof Since V is an abstract vector space, we do not have enough structures to do calculations. We have to try very elementary steps. We try to exchange the vectors in the spanning list with those from the linearly independent list. This is the exchange procedure, which was very important in the older literature. We start with $A \equiv A_r :=$ $(a_1, \ldots, a_r)$, a linearly independent list and with $C \equiv C_m = (c_1, \ldots, c_m)$, a spanning list. We have to show that $r \leqslant m$ which is $\sharp(A_r) \leqslant \sharp(C_m)$. We add a_1 to C and we so obtain a new list (a_1, C) which is now, according to our preceding lemma (3.3), a linearly dependent list and, of course, again a spanning list. Using the linear dependence Lemma 3.2, we can throw out one of the vectors in C, and we obtain $C_1 = (a_1, c_2, \ldots, c_m)$, a spanning list again, with length m and eventually with a different numbering.

The next step leads quite similarly to $C_2 = (a_1, a_2, c_3 \ldots, c_m)$. Proceeding equally, we obtain $C_{r-1}^1 = (a_1, \ldots a_{r-1}, c_r, \ldots, c_m)$, a spanning list again.

The last step leads to $C_r = (a_1, \ldots a_r, c_{r+1}, \ldots, c_m)$, again a spanning list. As we see, $\sharp(C_r) = \sharp(C)$, of course. It is clear that we get $r \leqslant m$ ($\sharp(A_r) \leqslant \sharp(C_m)$). $\square$

So far, we discussed two important properties for a given fixed number of vectors $(a_1, \ldots, a_m)$ in V. Such a list of vectors can be linearly independent or not and spanning or not. The possibility of a list of vectors being both, linearly independent and spanning, seems more attractive than the other three possibilities. This leads to the definition of a basis for a finitely generated vector space V which we consider in this book.

Definition 3.6 Basis.
A list $B = (b_1, \ldots, b_n)$ of vectors in V is a basis for V if it is linearly independent and spans V.

We consider four equivalent definitions for a basis in V.

Proposition 3.3 *Equivalent definitions for a basis.*

The following four statements are equivalent.

(i) $B = (b_1, \ldots, b_n)$ *is a basis for V that is linearly independent and spanning V;*

(ii) *Every vector $v \in V$ is a unique linear combination of vectors in B;*

(iii) $B = (b_1, \ldots, b_n)$ *is a maximally linearly independent list in V;*

(iv) $B = (b_1, \ldots, b_n)$ *is a minimally spanning list in V.*

Proof We show that (i) ⇔ (ii), (i) ⇔ (iii), and (i) ⇔ (iv) which clearly establishes the result.

(i) ⇒ (ii): Given (i), every $v \in V$ is a linear combination of vectors in B, since B, according to (i), is also linearly independent. The above linear independence and uniqueness lemma states that this linear combination is unique. So we proved (ii).

(ii) ⇒ (i): Given (ii), every vector v is a linear combination of vectors in B, so B spans V. Since this linear combination is unique, the linear independence and uniqueness lemma tells that B is also linearly independent. This proves (i). So we proved the statement (i) ⇐ (ii).

(i) ⇒ (iii): Given (i), we have to show that the linearly independent list B is maximal. Since B spans V, if we add any vector $v \in V$ to B, we get $(B, v) = (b_1, \ldots, b_n, v)$, and according to the above remark, this list is now linearly dependent, and not linearly independent any more. This means that B is linearly independent and maximal. This proves (iii).

(iii) ⇒ (i): We have to show that B is linearly independent and spans V. Since B is already linearly independent and maximal, we have only to show that B spans V. B being maximally linearly independent, if we add any $v \in V$, we get (B, v) which is now linearly dependent. Therefore, v is a linear combination of B. So B spans V and (i) is proven. This is why the statement (i) ⇔ (iii) is proven too.

(i) ⇒ (iv): (i) means that $B = (b_1, \ldots, b_n)$ spans V and is linearly independent. According to the linearly dependent lemma above, if we delete a vector of this list and write, for example $B_0 = (b_1, \ldots, b_{n-1})$, then B_0 does not span V any more. So B spans V and is minimal. This proves (iv).

(iv) ⇒ (i): We start a spanning list with B minimal. This means for example that the list $B_0 = (b_1, \ldots, b_{n-1})$ does not span V any more: span$(B_0) \neq$ span(B). In this case, the linearly dependent lemma tells us that B is linearly independent. So B spans V and is linearly independent. This proves (i) and we proved the statement (i) ⇔ (iv). ☐

We considered all the above conditions in detail and will do so as well in what follows because a basis is our best friend in linear algebra!

The existence of a basis is given by the following proposition:

Proposition 3.4 *Basis existence.*

Every finitely generated vector space V possesses a basis.

Proof This can be seen as follows. Since V is finitely generated, we can start by a spanning list: Say span$(v_1, \ldots, v_m) = V$. If the list is not linearly independent, we throw out some vectors of it until we obtain a minimally spanning list. This is, according to the above proposition, a basis of V. ☐

The existence of bases does not mean a priori that every basis for V has the same number of vectors (the same length). But, as the next corollary shows, it does.

Corollary 3.1 *On the cardinality (length) of a basis.*

In a finitely generated vector space V in a basis, every basis has the same finite number of vectors.

Proof Here we can apply the Proposition 3.2. The length of a linearly independent list is less or equal to the length of a spanning list. We start with the two bases B and C. B is linearly independent and C spans V. So we have $\sharp(B) \leqslant \sharp(C)$. Similarly, we can say that C is linearly independent and B spans V so that we have $\sharp(B) \geqslant \sharp(C)$. It follows that $\sharp(B) = \sharp(C)$. $\square$

This means that the number of vectors in a basis, the length of a basis, is universal for all bases in a vector space V, and this is what we call the dimension of V.

Definition 3.7 Dimension.

The dimension of a finitely generated vector space is the length of any basis.

We denote the dimension by $\dim_{\mathbb{K}} V$. The dimension depends on the field $\mathbb{K}$. If the field for V is clear, we may write $\dim V$, but we have to know that, for example, $\dim_{\mathbb{R}} V \neq \dim_{\mathbb{C}} V$. It now becomes more apparent that the characteristic data of a vector space V are the field $\mathbb{K}$ and its dimension. This also justifies the isomorphism $V \cong \mathbb{K}^n$. But since $\mathbb{K}^n$ has much more structures than the abstract vector space V with $\dim_{\mathbb{K}} V = n$, the isomorphism refers to those structures in $\mathbb{K}^n$ which correspond only to the structure of V.

3.2   Basis Dependent Coordinate Free Representation

Now that we have a basis $B = (b_1, \ldots, b_n)$ for V, we may ask how many bases exist for V. As we already mentioned, Poincaré might have proposed this question to Einstein. We will discover that this goes very deeply into what relativity is (see also Chap. 4). Apart from this, to understand linear algebra, it is fundamental to have a good understanding of the space of bases. But here, the question initially arises what the individual bases are useful for.

It is therefore helpful to discuss what a given basis makes of an abstract vector space and, in particular, what a basis makes of a vector.

A given basis $B = (b_1, \ldots, b_n)$ determines for each abstract vector n numbers, its coordinates. This leads to the parameterization of the given abstract vector space, using the standard vector space $\mathbb{K}^n$ and, in particular, it allows to describe each abstract vector by n numbers. We also speak of the representation of a vector space and the representation of a vector by an $n \times 1$-matrix (a column).

3.2.1 Basic Isomorphism Between V and $\mathbb{R}^n$

To facilitate our discussion in the remaining part of this section, we consider a real vector space with $\dim V = n$. A basis B for V is given by a list of n linearly independent vectors.

$$B = (b_1, \ldots, b_n).$$

This allows to write a linear combination (unique representation) of every vector $v \in V$ with $\xi^i \in \mathbb{R}$ and $i \in I(n) := \{1, 2, \ldots, n\}$:

$$v = \sum_{I=1}^{n} \xi^i b_i = (\xi^i b_i)_n$$

We shall mostly use the notation for the scalars with small greek letters, vectors with small latin letters, and matrices with capital letters; covectors with small greek letters, taking care not to confuse them with the notation of scalars.

The scalars ξ^i are also called coefficients or components of v with respect to the basis B. We use the Einstein convention for the summation and in addition some obvious notations, as usual, setting, for example, $(\xi^i b_i)_n = \xi^i b_i$ whenever no confusion is possible. Further on, we consider the column vector or column-matrix $\vec{\xi}$:

$$\vec{\xi} = \begin{bmatrix} \xi^1 \\ \vdots \\ \xi^n \end{bmatrix} = (\xi^i)_n = (\xi^i)$$

as an element of $\mathbb{R}^n$ identifying $\mathbb{R}^n$ with $\mathbb{R}^{n \times 1}$ (the column-matrices) and using again an obvious notation. What follows corresponds to Sect. 3.1 and to the notation presented there. For a fixed B, this leads to a bijection between the elements of $\mathbb{R}^n$ and V, as $\vec{\xi} \leftrightarrow v$, or more precisely to a linear bijection or isomorphism ψ_B (basis isomorphism):

$$\psi_B : \mathbb{R}^n \longrightarrow V,$$

$$\vec{\xi} \longmapsto \psi_B(\vec{\xi}) := \xi^i b_i = [B]\vec{\xi} = [b_1 \ldots b_n] \begin{bmatrix} \xi^1 \\ \vdots \\ \xi^n \end{bmatrix}.$$

If we think in terms of manifolds, a basis B induces here a (global) linear parametrization for the abstract vector space V and equivalently a (global) linear chart or coordinate map $\psi_B^{-1} =: \phi_B$ from V to $\mathbb{R}^n$:

$$\phi_B : V \longrightarrow \mathbb{R}^n,$$

$$v \longmapsto \phi_B(v) = v_B = \vec{\xi} \in \mathbb{R}^n.$$

The linear map ϕ_B, given by the basis B, is also called a representation. It is, as well as ψ_B, a linear bijection, an isomorphism, given by the basis B and therefore also called basis isomorphism.

We might want to identify ψ_B with B and $\psi_B^{-1} = \phi_B$ with B^{-1}, and $[B]$ with B and write

$$B : \quad \mathbb{R}^n \ \overset{\cong}{\to} \ V \qquad \text{and}$$
$$B^{-1} : \quad V \ \overset{\cong}{\to} \ \mathbb{R}^n. \tag{3.7}$$

3.2.2 The Space of Bases in V and the Group Gl(n)

As already stated, it is fundamental in mathematics, as well as in physics, to have a good understanding of the space of bases in a given vector space. Therefore, our aim is now to determine and discuss the space $B(V) := \{B, \dots\}$ of all bases of V. We therefore use the identification $\psi_B \overset{!}{\equiv} [B] \equiv [b_1 \dots b_n] \overset{!}{\equiv} B$ and we have to consider the space of basis isomorphisms

$$B(V) = Iso(\mathbb{R}^n, V) = \{\psi_B\}.$$

This point of view allows determining the spaces $B(V)$ and finding the correct behavior under bases and coordinate changes. It also allows us to determine precisely what a coordinate-free notion means in a formalism, mainly when this formalism depends explicitly on coordinates. This is also the case with tensor calculus.

What follows is a very interesting and important example and application of Sect. 1.3 about the group action and the definitions there. The key observation is that the group $Gl(n)$ of linear transformations ("transformation" in this book is a synonym for "bijection") in $\mathbb{R}^n$, surprisingly acts also on the space $B(V)$, even if V is an abstract vector space where the dimension n is not visible as with $\mathbb{R}^n$. On the other hand, it is clear that we have the following $Gl(n)$ actions on $\mathbb{R}^n$:

$$Gl(n) \times \mathbb{R}^n \quad \longrightarrow \quad \mathbb{R}^n,$$
$$(g, \vec{\xi}) \quad \longmapsto \quad g\vec{\xi} \tag{3.8}$$

and also (with $[B] \equiv \psi_B$)

$$B(V) \times Gl(n) \quad \longrightarrow \quad B(V),$$
$$(B, g) \quad \longmapsto \quad \psi_B \circ g \equiv [B] \circ g \equiv Bg. \tag{3.9}$$

This action is naturally given by the diagrams

$$g : \mathbb{R}^n \longrightarrow \mathbb{R}^n \quad [B] : \mathbb{R}^n \overset{\sim}{\cong} V \quad \text{and}$$

$$\mathbb{R}^n \overset{g}{\longrightarrow} \mathbb{R}^n \overset{[B]}{\longrightarrow} V.$$

$$[B] \circ g$$

The following proposition concerning the $Gl(n)$ action on $B(V)$ which we give without proof, answers our question about the space of bases in V.

Proposition 3.5 $B(V) \underset{bij}{\cong} Gl(n)$.

The group $Gl(n)$ acts on $B(V)$ free and transitively. So we get

$$Gl(n) \underset{bij}{\longmapsto} B(V),$$

$$g \longmapsto \Psi(g) := B_0 g, \quad B_0 \in B(V).$$

This means that there is a bijection between the elements of $B(V)$ and the elements of $Gl(n)$ (i.e., $B \leftrightarrow g$). Because of transitivity, for a fixed basis B_0, for each B, we have $\exists! g \in G$ with $B = B_0 g$.

$$B = B_0 g \quad \text{or} \quad B(V) = B_0 Gl(n).$$

$B(V)$ is an orbit of $Gl(n)$ relative to B_0. Note that $B(V)$ is the so-called $Gl(n)$ torsor.

A space which is an orbit of a group G is called a homogeneous space (see Remark 1.2). So $B(V)$ is a homogeneous space of the group $Gl(n)$. A free action means that for every B_1 and $B_2 \in B(V)$ there exists a unique $g_{12} \in Gl(n)$ so that $B_2 = B_1 g_{12}$. This is analogous to the connection between a vector space V and its associated affine space. The homogeneous space $B(V)$ corresponds to the affine space $X = (V, T(V), \tau)$ as discussed in Sect. 2.5 and the group $Gl(n)$ corresponds to the abelian group V.

The group $Gl(n)$ is also called the structure group of the vector space $V \cong \mathbb{R}^n$. Relativity here means that the geometric object $v \in V$ can be represented by an element of $\mathbb{R}^n$ relative to the coordinate system B as $\psi_B^{-1}(v) = v_B \in \mathbb{R}^n$.

In addition, every different coordinate system $C \equiv \psi_C \in B(V)$ is good enough to represent V and can be obtained from B by a transformation: $\bar{g} \in Gl(n)$. So we obtain $C = B\bar{g}$ and $\psi_C^{-1}(v) = v_C \in \mathbb{R}^n$. Obviously, the vector space V is characterized by the $Gl(n)$ relativity and in connection with this, the $Gl(n)$ group is also called the structure group, and usually in physics, the symmetry group of the theory.

3.2.3   The Equivariant Vector Space of V

Our next step is to construct a new vector space $\tilde{V}$ which contains in a precise way all the representations of the vectors v in V and can be identified with our original vector space V. As we shall see, the result is a coordinate-free formulation of the $Gl(n)$-relativity of V. Coordinate-free here means, by the explicit use of coordinate systems, that we work with all coordinate systems simultaneously. In other words, the tensor calculus, as applied in physics and engineering, which depends explicitly on coordinates, can be formulated in a precise coordinate independent way. In this sense, it is equivalent to any coordinate-free formulation if done right in a consistent notation.

The vector space $\tilde{V}$ is given as a set $\tilde{V} = \{\tilde{z}, \tilde{y}, \tilde{x} \ldots\}$ of $Gl(n)$-equivariant maps from $B(V)$ to $\mathbb{R}^n$ (see Definition 1.9). As we saw, $B(V)$ is a right $Gl(n)$ space (see Definition 1.4) and we consider $\mathbb{R}^n$ as a left $Gl(n)$ space. As we saw in Eqs. (3.8) and (3.9), both actions are canonically given. This justifies the equivariance property we demand, so we have for $\tilde{z} \in \tilde{V}$

$$\tilde{z} : B(V) \longrightarrow \mathbb{R}^n,$$
$$B \longmapsto \tilde{z}(B),$$

with

$$\tilde{z}(Bg) = g^{-1}\tilde{z}(B). \tag{3.10}$$

See also Comment 1.1 on the meaning of the right action. This may also be shown by the commutative diagram

$$
\begin{array}{ccc}
B(V) & \xrightarrow{\tilde{z}} & \mathbb{R}^n \\
g \downarrow & & \downarrow g^{-1} \\
\mathcal{B}(V) & \xrightarrow{\tilde{z}} & \mathbb{R}^n.
\end{array}
\tag{3.11}
$$

In Eq. (3.11), we interpret the $Gl(n)$ action on $\mathbb{R}^n$ also as a right action:

$$
\begin{array}{ccc}
\mathbb{R}^n \times Gl(n) & \longrightarrow & \mathbb{R}^n, \\
(\vec{\xi}, g) & \longmapsto & g^{-1}\vec{\xi}.
\end{array}
\tag{3.12}
$$

Definition 3.8 The equivariant vector space $\widetilde{V}$.

Taking into account Eqs. (3.10), (3.11), and (3.12), the vector space $\widetilde{V}$ can be written, with $\mathcal{B} := B(V)$, as

$$\widetilde{V} := \widetilde{Map}(\mathcal{B}, \mathbb{R}^n) = Map_{equ}\big(B(V), \mathbb{R}^n\big). \tag{3.13}$$

For good reasons, we may call $\widetilde{V}$ the equivariant vector space of V. We have here an example of a "complicated" vector space (see Comment 2.9 in Sect. 2.6)! $\widetilde{V}$ is a vector space since it is a vector valued map ($\mathbb{R}^n$—valued). Furthermore, $\dim \widetilde{V} = \dim V$ holds since $\tilde{z} \in \widetilde{V}$ is an equivariant and the group $Gl(n)$ acts transitively on $B(V)$. So if we define $\tilde{z}$ at one given $B_0 \in B(V)$, then its value is also given by the equivariance property in every other basis B, for example, $B = B_0 g$. This leads to

$$\tilde{z}(B) = \tilde{z}(B_0 g) = g^{-1} \tilde{z}(B_0). \tag{3.14}$$

The vector space $\widetilde{V}$ has the same dimension: $\dim \widetilde{V} = \dim \mathbb{R}^n = \dim V = n$. The rest is shown by the following proposition:

Proposition 3.6 *The equivariant vector space of V.*

The vector space $\widetilde{V} = \widetilde{Map}(\mathcal{B}, \mathbb{R}^n)$ is canonically isomorphic to V, so we have $V \underset{k}{\cong} \widetilde{V}$.

Proof We already know that $\widetilde{V}$ is a vector space and that its dimension is $\dim \tilde{V} = n$. Addition and scalar multiplication are given as follows:

For $\tilde{z}, \tilde{y} \in \widetilde{V} : \tilde{z}, \tilde{y} : B(V) \longrightarrow \mathbb{R}^n$,

we have $(\tilde{z} + \tilde{y})(B) := \tilde{z}(B) + \tilde{y}(B)$,

and $(\lambda \tilde{z})(B) := \lambda \tilde{z}(B)$.

The canonical isomorphism $V \underset{k}{\overset{\cong}{\to}} \widetilde{V}$ is defined by

$$k : V \longrightarrow \widetilde{V} = \widetilde{Map}(\mathcal{B}, \mathbb{R}^n),$$
$$v \longmapsto k(v) =: \tilde{v}$$

with

$$\tilde{v}(B) := B^{-1}(v) \in \mathbb{R}^n \tag{3.15}$$

and applying various identifications, we have

$$\tilde{v}(B) = [B]^{-1}(v) = B^{-1}(v) = \psi_B^{-1}(v) = \phi_B(v) = v_B = \vec{v}_B \in \mathbb{R}^n. \quad (3.16)$$

We use the notation $\tilde{v}(B) = \vec{v}_B \in \mathbb{R}^n$ which might be more familiar.

It is evident that $k(v) = \tilde{v}$ depends only on v in this canonical way, independently of a specific B or, more precisely, $\tilde{v}$ depends on all B simultaneously in an equivariant way, as in Eqs. (3.10) and (3.11) We could also say that this property is consistent with the group $Gl(n)$ action. What is now left is to show that the explicitly defined map $\tilde{v}(B) := B^{-1}(v)$ is indeed an equivariant map. So we have altogether:

$$B : \quad \mathbb{R}^n \longrightarrow V,$$

$$B \circ g : \mathbb{R}^n \underset{g}{\longrightarrow} \mathbb{R}^n \underset{B}{\longrightarrow} V,$$

$$\mathbb{R}^n \underset{g^{-1}}{\longleftarrow} \mathbb{R}^n \underset{B^{-1}}{\longleftarrow} V : (B \circ g)^{-1},$$

$$\tilde{v}(Bg) = (Bg)^{-1}(v) = g^{-1} \circ B^{-1}(v) = g^{-1}\tilde{v}(B). \quad (3.17)$$

This represents simultaneously the effect of any change of basis and shows that $k(v) \equiv \tilde{v}$ is indeed equivariant. The proposition is proven, and the identification between V and $\tilde{V}$ is established. □

This means that geometrically we can work with V or with $\tilde{V}$, it is completely equivalent. The identification of $\tilde{v} \equiv v$ should now be clear. For every basis B, $\tilde{v}$ gives its representation with this basis (coordinate system). $\tilde{v} \equiv v$ is the quintessence of all the representations of the given abstract vector $v \in V$ relative to all authorized (here $Gl(n)$) coordinate systems simultaneously. In this sense, $\tilde{v} \equiv v$ is coordinate-free!

In order to reformulate our results in a more familiar formalism, we have to necessarily consider, once again, some of the various notations which also appear in the literature. So we have for example for the basis B in V

$$\psi_B \equiv B \equiv (b_1, \ldots, b_1) \equiv [b_1 \ldots b_n] \equiv [B].$$

The expression $[b_1 \ldots b_n]$ which we sometimes also abbreviate by $[b_i]$, is a $1 \times n$-matrix, that is, a row-matrix with vector entries. It could also represent (as the list $(b_1, \ldots, b_n)$) the isomorphism ψ_B. This was also presented in the previous section.

What we are doing here is to identify the linear map $\psi_B : \mathbb{R}^n \to V$ (which is a parametrization of the abstract vector space V with the standard vector space $\mathbb{R}^n$) with the list of the basis vectors $(b_1, \ldots, b_b)$, and with the $1 \times n$-matrix $[b_1 \ldots b_n]$ with basis vector entries. Then we denote all these by the symbol B which we call simply a basis. For the equivariant map $\tilde{v}$ we may write, as above, similarly:

$$\tilde{v}(B) \equiv \vec{v}_B \equiv (v_B^i)_n \equiv [v]_B \equiv v_B \in \mathbb{R}^n.$$

For example, $[v]_B$ is a $n \times 1$ column-matrix with entries scalars. The diagram (3.11) and Eq. (3.12) express a change of basis via equivariance: $B \xmapsto{g} B' = Bg$ and we get

$$\vec{v}_{B'} = \tilde{v}(B') = \tilde{v}(Bg) = g^{-1}\tilde{v}(B) = g^{-1}\vec{v}_B. \tag{3.18}$$

If we set $g^{-1} = h$, we have

$$\vec{v}_{B'} = \tilde{v}(B') = \tilde{v}(Bh^{-1}) = h\tilde{v}(B) = h\vec{v}_B. \tag{3.19}$$

The last equation is the usual form for a change of basis for the coefficient vectors. In what follows, we recall change of basis in the standard form, as usually done in physics.

Taking a second basis $C = (c_1, \ldots, c_n)$ for V, we have analogously $\tilde{v}(C) = v_C = \vec{v}_C = [v^i]_C \in \mathbb{R}^n$. So there exists a matrix $T \in Gl(n)$ with scalar entries $\tau^i_s \in \mathbb{R}$:

$$i, s \in I(n) := \{1, \cdots, n\},$$

$$T = (\tau^i_s) \equiv [\tau^i_s],$$

so that

$$\psi_B = \psi_C \circ T \text{ or equivalently } B = CT \Leftrightarrow C = BT^{-1}. \tag{3.20}$$

So we have for $v \in V$

$$v = \psi_B(\vec{v}_B) = \psi_C(\vec{x}_C) \tag{3.21}$$

in various notations:

$$v = \psi_B(\vec{v}_B) \equiv [B][v^i]_B = B\vec{v}_B = C\vec{v}_C = \psi_C(\vec{v}_C). \tag{3.22}$$

The result of the map $\tilde{v}$ is again given by $\tilde{v}(C) = \vec{v}_C$. Using Eqs. (3.20), (3.21), and (3.22), we can write:

$$\vec{v}_C = \tilde{v}(C) = \tilde{v}(BT^{-1}) = T\tilde{v}(B) = T\vec{v}_B. \tag{3.23}$$

The result of Eq. (3.23), $\vec{v}_C = T\vec{v}_B$, is exactly the result of the equivariant property of $\tilde{v}$.

The appearance here of $v \in V$ as the map $\tilde{v}$ which is explicitly coordinate (basis) dependent, legitimizes the formalism of coordinates of linear algebra and the tensor calculus to be as rigorous as any coordinate-free formulation.

3.2.4   The Associated Vector Space of V

There is, in addition, a different formalism which shows another aspect of coordinate independence. There is a second canonical isomorphism of vector spaces:

$$\bar{V} \cong V,$$

where $\bar{V}$ is defined in analogy to the vector bundle formalism and by using again the $Gl(n)$ action on $\mathcal{B}(V)$ and $\mathbb{R}^n$. We will discuss this here shortly since it offers a different but equivalent point of view. Besides that, it is a very interesting and important example of Sect. 1.2 dealing with quotient spaces.

We consider the set $M := \mathcal{B}(V) \times \mathbb{R}^n = \{(B, \vec{x})\}$ which is the space of pairs (basis, coordinate vector). M is canonically a $Gl(n)$ space. Setting $G = Gl(n)$ we have an action defined as:

$$M \times G \quad \longrightarrow \quad M,$$
$$\big((B, \vec{x}), g\big) \longmapsto (Bg, g^{-1}\vec{x}) =: (B, \vec{x})g.$$

For every pair $(C, \vec{y})$ we consider its G orbit:

$$(C, \vec{y})G := \{(C, \vec{y})\, g = (Cg, g^{-1}\vec{y}) : g \in G\}$$

and we define the equivalent class

$$[C, \vec{y}] := (C, \vec{y})G.$$

It is not difficult to recognize that this (basis, coordinate vector) class corresponds bijectively to a unique vector in V. We expect for example $[B, \vec{x}] \leftrightarrow \psi_B(\vec{x})$. This leads to the definition ($G := Gl(n)$):

$$\bar{V} := (\mathcal{B}(V) \times \mathbb{R}^n) \,/\, G = \{[B, \vec{x}]\}.$$

We see that $\bar{V}$ is a G orbit space since every element $[B, \vec{x}] = (B, \vec{x})G$ is a G orbit; see also Remark 1.2 on homogeneous spaces. Then the following proposition is valid.

> **Proposition 3.7** $\bar{V}$, *the associated vector space to* V.
>
> $\bar{V} = (\mathcal{B}(V) \times \mathbb{R}^m) \,/\, Gl(n)$ *is a vector space and is canonically isomorphic to* V.

$$\lambda : \bar{V} \longrightarrow V,$$
$$[B, \vec{x}] \longmapsto \psi_B(\vec{x}) = x \in V.$$

We may call $\bar{V}$ the associated vector space to V and λ the isomorphism of the structure since the vector space $\bar{V}$ can also be considered as a model for the abstract vector space.

Proof The vector space structure of V is not trivial to reveal:
Addition: $[B, \vec{x}] + [C, \vec{y}] :=?$
We set $C = Bg$ and we have

$$[C, \vec{y}] = [Bg, \vec{y}] = [B, g^{-1}\vec{y}] = [B, \vec{z}]$$

with $\vec{z} = g^{-1}\vec{y}$, then

$$[B, \vec{x}] + [C, \vec{y}] = [B, \vec{x}] + [B, \vec{z}] := [B, \vec{x} + \vec{z}].$$

Scalar multiplication:
$$\alpha[B, \vec{x}] := [B, \alpha\vec{x}], \quad \alpha \in \mathbb{R}.$$

□

So we have for the isomorphic $\bar{V} \cong V$.

Remark 3.2 The four isomorphic vector spaces.

The following isomorphisms are valid.

$$\mathbb{R}^n \cong V \cong \widetilde{V} \cong \bar{V}!$$

The isomorphism $\mathbb{R}^n \cong V$ is not canonical, that is, it depends on the specific $B \in \mathcal{B}(V)$ we chose. However, $V \cong \widetilde{V} \cong \bar{V}$ are, as stated above, canonical isomorphisms.

3.3 The Importance of Being a Basis *Hymn to Bases*

We have already demonstrated the usefulness of a basis for V. This makes it possible to express an abstract vector $v \in V$ simply by a list of numbers (scalars). This allows us to communicate to everybody this special vector just by numbers. The price for this achievement is for example that, in the end, one basis is not enough and that we have to consider all the bases $B(V)$ altogether. This is demonstrated in Sect. 3.2 by the equivariant map $\tilde{v} : B(V) \to \mathbb{R}^n$. Stated differently, the price is that instead of a single element v, we have to know a special function $\tilde{v}$ or the equivalence class $[v]$. That is, we believe, a fair price!

It gives us even more: with given bases in V and V', we can, in addition, describe a linear map $f \in \mathrm{Hom}(V, V')$ with a finite amount of numbers. This list of numbers is organized by a matrix, as is well-known, and there is a linear bijectivity or an isomorphism between linear maps and matrices. In addition, many properties and many proofs within the category of vector spaces can be easily formulated by explicitly using bases. This is what we are going to demonstrate in what follows. Even more, we are going to realize in this book that bases are our best friends in linear algebra.

The proposition below shows that a linear map is uniquely determined by the values of the basis vectors for the domain space:

Proposition 3.8 *Basis and linear map.*

Given a basis $B = (b_1, \ldots, b_n)$ of V, then the linear map $f \in \mathrm{Hom}(V, V')$ is given uniquely by the values $f(b_i) = w_i \in V'$ $i \in I(n) = \{1, 2, \ldots, n\}$.

Proof For every v the basis B delivers a unique expression $v = \xi^i b_i$.
The coefficients $\xi^i \in \mathbb{K}$ are uniquely defined. This follows from the fact that B is also a linearly independent list and from Lemma 3.1 on linear independence and uniqueness. We define by linearity:

$$f(v) := f(\xi^i b_i) := \xi^i w_i. \tag{3.24}$$

This shows that the value $f(v)$ is given uniquely: the coefficients ξ^i are uniquely defined and the values $w_i = f(b_1)$ are uniquely given. Therefore, there exists at most one such map.   □

The following proposition shows that we can choose a tailor-made basis B_0, leading to further essential conclusions, as for example Theorem 3.1 below about the normal form of linear maps which reveals its geometric character and many very important corollaries.

Proposition 3.9 *Tailor-made bases and linear map.*
Let $f : V \to V'$ be a linear map, $(w_1, \ldots, w_r)$ be a basis of $\mathrm{im}\, f$, and $(z_1, \ldots, z_k)$ be a basis of $\ker f$. We choose arbitrary vectors $b_i \in f^{-1}(w_i)$, $i \in I(r)$. Then the list $B_0 := (b_1, \ldots, b_r, z_1, \ldots, z_k)$ is a basis for V.

Proof We show first that span $B_0 = V$.
For $v \in V$ we have $f(v) = (\eta^i w_i)_r$, $\eta^i \in \mathbb{K}$.
Define $u \in V$ by $u := \eta^i b_i$.
So we have

$$f(u) = f(\eta^i b_i) = \eta^i f(b_i) = (\eta^i w_i)_r$$

and we obtain

$$f(v) = f(u) \implies f(v - u) = 0 \implies v - u \in \ker f \implies$$
$$v - u = (\xi^\mu z_\mu)_k, \ \mu \in I(k) \implies$$
$$v = u + (\xi^\mu z_\mu)_k = (\eta^i b_i)_r + (\xi^\mu z_\mu)_k \implies v \in \text{span } B_0.$$

We show that B_0 is linearly independent: We choose $0 = (\lambda^i b_i)_r + (\rho^\mu z_\mu)_k$. Applying f to this equation, we obtain, as $\{w_i\}$ is a basis for im f,

$$0 = \left(\lambda^i f(b_i)\right) + 0 \implies \lambda^i w_i = 0 \implies \lambda^i = 0 \ \forall i \in I(r)$$

so $0 = (\rho^\mu z_\mu)_k$ is left.

Since $(z_\mu)_k$ is a basis for ker f, it follows that $\rho^\mu = 0$. This shows that B_0 is also linearly independent. The list B_0 spans V and is linearly independent. We so managed to find B_0, a tailor-made basis for V! $\qquad\square$

It is clear that this proposition determines the numbers $\dim(\ker f) = k$ and $\dim(\text{im } f) = r$. Combined with $\dim V = n$ and $\dim V' = m$, it shows important geometric aspects of the map f.

From this proposition and the proof, we obtain the following corollaries directly. These characterize substantially the structure of the vector space homomorphisms $\text{Hom}(V, V')$ with $V' \neq V$ and subsequently also the structure of linear algebra itself. Even more, we could say that these corollaries summarize the entire representation theory of linear maps with $V' \neq V$ or essentially also the endomorphisms $\text{Hom}(V, V)$ if we take two different bases (see the singular value decomposition, SVD, in Sect. 12.2). On the other hand, if we use for the description of the endomorphisms $\text{Hom}(V, V)$ only one basis, then the problem is more challenging and leads to more advanced linear algebra (see Chaps. 9 up to 13). We consider, as usual in linear algebra, finite-dimensional vector spaces.

Corollary 3.2 : *Rank-nullity theorem.*
$\dim(\ker f) + \dim(\text{im } f) = \dim V$.

Proof From the length of the bases and for the basis-independent subvector spaces ker f, im f and V, $k = \dim(\ker f)$, $r = \dim(\text{im } f)$ and $n = \dim V$, and the basis above B_0, we see: $k + r = n$ $\qquad\square$

Corollary 3.3 *Dimension of an affine space $f^{-1}(w)$.*

For $w \in V'$, then $\dim f^{-1}(w) = \dim(\ker f)$ *holds.*

Proof Taking into account Definition 2.18 and Exercise 2.38 about affine spaces and linear maps, and $f^{-1}(w) = A(v) = v + \ker f$ from Fig. 2.4, we get $\dim f^{-1}(w) = \dim A(v) = \dim \ker f$. We so obtain directly from the rank-nullity theorem, $\dim f^{-1}(w) = \dim V - \dim(\operatorname{im} f)$ and $\dim f^{-1}(w) = \dim(\ker f)$. $\qquad\square$

Corollary 3.4 *Equivalence for equal dimensions.*

For $f : V \to V'$ linear and $\dim V = \dim V'$*, the following conditions are equivalent:*

(i) f is injective,
(ii) f is surjective,
(iii) f is bijective. $\qquad\square$

See Exercise 3.22.

Corollary 3.5 *Criterion for injectivity.*

Let $f : V \to V'$ be linear, $B = (b_1, \ldots, b_n)$ be a basis for V and $f(b_i) = w_i$ for $i \in I(n)$. Then f is injective if and only if the list $(w_1, \ldots, w_n)$ is linearly independent.

Proof In Proposition 3.9, we have for $r = n$ and from $B_0 = (b_1, \ldots, b_r, z_1 \ldots, z_{k=0})$ since span $B_0 = V$. Thus, $\dim(\ker f) = 0$ and so $\ker f = 0$. This shows the injectivity. $\qquad\square$

Corollary 3.6 *Criterion for isomorphism.*

The map $f \in \operatorname{Hom}(V, V')$ is an isomorphism if and only if for a basis $B = (b_1, \ldots, b_n)$ in V and a basis $B' = (b'_1, \ldots, b'_n)$ in V', $f(b_s) = b'_s, s \in I(n)$ holds.

Corollary 3.7 *Canonical basis and basis isomorphism.*

For V a vector space and $B = (b_1, \ldots, b_n)$ a basis for V, there exists one canonical isomorphism ψ_B (basis-isomorphism)

$$\psi_B : \mathbb{K}^n \longrightarrow V \text{ with } \psi_B(e_i) = b_i, \quad i \in I(n).$$

Let $E = (e_1, \ldots, e_n)$ be the canonical basis for $\mathbb{K}^n$. ψ_B may be considered as a parametrization of V. With the inverse map $\phi_B := \psi_B^{-1}$, the pair (V, ϕ_B) is a global linear coordinate chart on V.

This was used many a time so far!

Corollary 3.8 *Identification* $\text{Hom}(\mathbb{K}^n, \mathbb{K}^m)$ *with* $\mathbb{K}^{m \times n}$.

For every linear map

$$f : \mathbb{K}^n \longrightarrow \mathbb{K}^m,$$

there exists precisely one matrix $F \in \mathbb{K}^{m \times n}$ such that $f(\vec{x}) = F\vec{x}$.

This shows that in this case we do not have to distinguish between linear maps and matrices, the above $F\vec{x}$ being of course a matrix multiplication.

Proof Notice that $f(e_i) \equiv Fe_i := f_i$ are the columns of the matrix F (see Example 2.23 and Sect. 2.4). So we have $F = [f_1 \ldots f_n]$. □

Corollary 3.9 *Representation of linear maps by matrices.*

Given two vector spaces V with basis $B = (v_1, \ldots, v_n)$ and V' with basis $C = (w_1, \ldots, w_m)$.

Then for every linear map $f : V \to V'$, there exists precisely one matrix $F = (\varphi_r^i) \in \mathbb{K}^{m \times n}$ such that $f(v_r) = w_i \varphi_r^i$ for $r \in I(n)$ and $i \in I(m)$. The map

$$M_{CB} : \text{Hom}(V, V') \longrightarrow \mathbb{K}^{m \times n},$$
$$f \longmapsto F := M_{CB}(f)$$

is an isomorphism: $\text{Hom}(V, V') \cong \mathbb{K}^{m \times n}$.

> For the given bases B and C, $F = M_{CB}(f)$ is a representation of the linear map f by the matrix F.

Proof We use the Einstein convention. The position of the indices i and r upstairs and downstairs respectively refers also to the basis transformation properties $(Gl(n), Gl(m))$. As C is a basis for V', the linear combinations $w_i \varphi_r^i$ are uniquely determined and with the index r fixed,

$$f_r = \begin{bmatrix} \varphi_r^1 \\ \vdots \\ \varphi_r^i \\ \vdots \\ \varphi_r^m \end{bmatrix}$$

is the rth column of the matrix F.

We now show that M_{CB} is linear: For a second map g with matrix $G = (\gamma_r^i)$ we have

$$(f + g)(v_r) = f(v_r) + g(v_r) = w_i \varphi_r^i + w_i \gamma_r^i = w_i(\varphi_r^i + \gamma_r^i)$$

and

$$(\lambda f)(v_r) = \lambda w_i \varphi_r^i = w_i(\lambda \varphi_r^i).$$

So we have with $M := M_{CB}$

$$M(f + g) = M(f) + M(g)$$
$$M(\lambda f) = \lambda M(f).$$

Since B is a basis for V, f is, by Proposition 3.8, uniquely defined by the condition $f(v_s) := w_i \varphi_s^i$. Therefore, F determines uniquely the values of $f : F = [f_1 \ldots f_n]$, $f(B) = [C]F$ and $M_{CB}(f) = F$. So M_{CB} is bijective. $\qquad\square$

The most impressive consequence of the usefulness of tailor-made bases is that they can be used to obtain the normal form for linear maps. The following theorem reveals the geometric character of a linear map. It relates directly to the vector spaces involved and their dimensions. Therefore, it could also be considered as the fundamental theorem of linear maps.

> **Theorem 3.1** *Normal form of linear maps.*
> *Given $f : V \to V'$ linear, $n = \dim V$, $m = \dim V'$. There exist bases B_0 for V and C_0 for V' so that*

$$M_{C_0 B_0}(f) = \begin{bmatrix} \mathbb{1}_r & 0 \\ 0 & 0 \end{bmatrix} \quad where \quad \mathbb{1}_r = \begin{bmatrix} 1 & & 0 \\ & \ddots & \\ 0 & & 1 \end{bmatrix} \in \mathbb{R}^{r \times r}.$$

Proof From Proposition 3.9 we have essentially the tailor-made bases B_0 and C_0 for V and V'

$$B_0 = (b_1, \ldots, b_r, z_1, \ldots, z_k) \in B(V).$$

We extend the basis in im f, $(w_1, \ldots, w_r) \in B(\text{im } f)$, to a basis C_0 of V':

$$C_0 = (w_1, \ldots, w_r, w_{r+1}, \ldots, w_m) \in B(V').$$

Then

$$f(b_i) = w_i \text{ for } i \in I(r) \text{ and}$$
$$f(z_j) = 0 \text{ for } j \in I(k).$$

$\square$

Remark 3.3 Normal form.

The name "normal form" is historical. Behind it, however, are the notions of equivalence, relation and quotient space as introduced in Sect. 1.2. Theorem 3.1 is a very prominent example. It corresponds to the simplest possible representation in every equivalence class.

Similarly, for example, the normal form $M_0 = \begin{bmatrix} \mathbb{1}_r & 0 \\ 0 & 0 \end{bmatrix}$ for an $m \times n$-matrix M, is a special representative of the corresponding matrices which are equivalent to the given matrix M. This corresponds to an improved form of the reduced row echelon form (rref). Thus, the set of all these "normal" matrices $\{M_0\}$ is bijective to the corresponding quotient space, as discussed in Sect. 1.2. If we take $m \leq n$ without loss of generality, we have:

$$\mathbb{K}^{m \times n}/\sim \ : \{[M] = M \in \mathbb{K}^{m \times n}, \text{rank}(M) = r, \ r \in I(m)\}.$$

So we get:

$$\mathbb{K}^{m \times n}/\sim \ \overset{\cong}{\underset{bij}{=}} \ \{\begin{bmatrix} \mathbb{1}_r & 0 \\ 0 & 0 \end{bmatrix} \mid \quad r \in I(m)\}.$$

We observe that the set $\mathbb{K}^{m \times n}$ with infinite cardinality has the quotient space $\mathbb{K}^{m \times n}/\sim$ which is finite:

$$\mathbb{K}^{m \times n}/\sim \ \overset{\cong}{\underset{bij}{=}} \ I_0(n) = \{0, 1, 2, \ldots, n\}.$$

The relevant equivalence relation $\sim$ here, is given by the following definition:

Definition 3.9 Equivalent linear maps and equivalent matrices.
The linear maps f and g, f, $g \in \text{Hom}(V, V')$ are equivalent: $f \sim g$ if and only if there are automorphisms Φ and Φ' (bijective linear maps in V and V') and a commutative diagram:

$$
\begin{array}{ccc}
V & \xrightarrow{f} & V' \\
\Phi \uparrow & & \uparrow \Phi' \\
V & \xrightarrow{g} & V'
\end{array}
$$

so that $f \circ \Phi = \Phi' \circ g$ or equivalently $g = \Phi'^{-1} \circ f \circ \Phi$.
    Similarly, the matrices A and B, A, B, $\in \mathbb{K}^{m \times n}$ are equivalent ($A \sim B$) if and only if F and F', invertible matrices, exist so that $B = F'^{-1} A F$ holds.

Definition 3.10 Similar operators and similar matrices.
In Definition 3.9, if $V = V'$, $\Phi = \Phi'$ and $F = F'$, then $f, g \in \text{Hom}(V, V)$ are similar if and only if $f \circ \Phi = \Phi \circ g$ or, equivalently, if $g = \Phi^{-1} \circ f \circ \Phi$ holds. A, $B \in \mathbb{K}^{n \times n}$ are similar if and only if $B = F^{-1} A F$ holds.

Remark 3.4 On the normal form of endomorphisms.

    There is a simple question: Does the same simple normal form (see Theorem 3.1 for a linear map $f \in \text{Hom}(V, V')$ also apply to the case where $V' = V$ and $C_0 = B_0$, that is, for an endomorphism $f \in \text{Hom}(V, V)$? The answer here is, no. This is a difficult problem and it leads to the Jordan form. It is the question of diagonalization or non-diagonalization of endomorphisms (operators) and square matrices (see Chap. 9, Sect. 9.5).
    But what would such a simple normal form mean? The operator f, for example, would in any case have a representation with a diagonal matrix, that is, a direct decomposition of the space V, a discrete list of $n = \dim(V)$ scalars $(\lambda_1, \lambda_2, \ldots \lambda_n)$ and

$$
f(u_i) = \lambda_i u_i, \quad i \in I(n),
$$

into one-dimensional subspaces U_i, with $u_i \in U_i$,

$$V = U_1 \oplus U_2 \oplus \cdots \oplus U_n,$$

and correspondingly into a decomposition of f of the form

$$f_{|U_i} = \lambda_i id_{U_i}.$$

We could simply represent each such one-dimensional space U_i as a null space of $f - \lambda_i id_V$:

$$U_i = \ker(f - \lambda_i id_V).$$

It is clear that if for a given index, for instance $i = 1$, the scalar λ_1 is zero, then we have $U_1 = \ker f$ and, in addition,

$$im\ f = U_2 \oplus U_3 \oplus \cdots \oplus U_n.$$

We thus obtain also a direct $\ker f - im\ f$ decomposition

$$\ker f \oplus im\ f = V.$$

It may be plausible that the search for such a simple decomposition of the vector space V cannot be straightforward. See also Proposition 3.13.

Comment 3.3 Notations for matrices of linear maps.

For the matrix $M_{CB}(f)$, there exist many different notations in the literature, and we are going to add a few more! The symbol "$\equiv$" here means that a different notation is used for the same object.

$$M_{CB}(f) \equiv M_{BC}(f) \equiv M_B^C(f) \equiv f_{CB} \equiv [f]_{CB}$$
$$\equiv F_{CB} \equiv [f(b_1)_C \ldots f(b_n)_C] \equiv [F(B)]_C.$$

For fixed B and C we can define a basis for $\mathrm{Hom}(V, V') : f_{ir} \equiv f_r^i : V \to V'$:

$$f_{ir} \equiv f_r^i : V \to V' : \quad f_{ir}(v_s) \begin{cases} w_i & \text{for } s = r \\ 0, & s \neq r \end{cases}.$$

Then $M_{CB}(f_{ir}) = E_{ir}$ where $E_{ir} = (0, \ldots, 0, e_i, 0, \ldots 0)$ with e_i in the rth position:

$$E_{ir} \equiv E_r^i := \begin{bmatrix} & & 0 & & \\ & & \vdots & & \\ 0 \cdots & 0 & \begin{matrix}i\\0\\0\end{matrix} & 0 \cdots & 0 \\ & & \vdots & & \\ & & 0 & & \end{bmatrix}.$$

The entry 1 is in the rth column and in the ith row. All the entries are zero.

This is another proof of Corollary 3.8. $\{f_{ir}\}$ is a basis of $\mathrm{Hom}(V, V')$ and E_{ir} is a basis of $\mathbb{K}^{m \times n}$. It should be evident that an isomorphism is a map that sends a basis to a basis.

3.4 Sum and Direct Sum Revisited

It is instructive to start with the direct product or Cartesian product we know very well because the comparison with the direct sum provides interesting insights for both. We first briefly recall its definition:

Definition 3.11 Direct product (Cartesian product) of vector spaces.
The direct product of $U_1 \times \cdots \times U_m$ is given by

$$U_1 \times \cdots \times U_m := \{(u_1, \ldots, u_m) : u_1 \in U_1, \ldots, u_m \in U_m\}$$

with addition $(u_1, \ldots, u_m) + (w_1, \ldots, w_m) := (u_1 + w_1, \ldots, u_m + w_m)$ and scalar multiplication by

$$\lambda(u_1, \ldots, u_m) := (\lambda u_1, \ldots, \lambda u_m).$$

Remark 3.5 The dimension of the direct product.
The dimension of $U_1 \times \cdots \times U_m$ is given by

$$\dim(U_1 \times \cdots \times U_m) = \dim U_1 + \cdots + \dim U_m.$$

A characteristic property of the direct product is directly related to the notion of linear independence and we could call it a (block) linear independence. This leads to the following definition:

> **Definition 3.12** Linear independence of a list of vector spaces.
> We call a list of vector spaces $(U_1, \ldots, U_m)$ linearly independent if the following holds:
> Any list of the form $A = (a_1, \ldots, a_m)$ with $a_1 \in U_1, \ldots, a_m \in U_m$ is linearly independent.

Using the definition in Sect. 2.7, the following results for given subspaces $U_1, \ldots, U_m$ of V are directly obtained.

(i) $U_1 + \cdots + U_m \leqslant V$;
(ii) $U_1 + \cdots + U_m = \text{span}(U_1 \cup \cdots \cup U_m)$;
(iii) $\dim(U_1 + \cdots + U_m) \leqslant \dim U_1 + \cdots + \dim U_m$.

For the sum of two vector spaces, U_1 and U_2, particularly $U_1 \cap U_2 \neq \{0\}$ not being excluded, we have $\dim(U_1 + U_2) \leq \dim(U_1) + \dim(U_2)$.

The exact relation is given by the following statement.

> **Proposition 3.10** *Dimension of a sum of two vector spaces.*
>
> $$\dim(U_1 + U_2) = \dim U_1 + \dim U_2 - \dim(U_1 \cap U_2).$$

Proof In an obvious notation taking

$$B_1 = (a_1, \ldots, a_k), \quad B_2 = (b_1, \ldots, b_l), \quad B_3 = (c_1, \ldots, c_m),$$

bases of U_1, U_2, and $U_3 = U_1 \cap U_2$, we may obtain new bases B_1' and B_2' for U_1 and U_2.

$$B_1' = (c_1, \ldots, c_m, a_{m+1}, \ldots, a_k) \text{ and } B_2' = (c_1, \ldots, c_m, b_{m+1}, \ldots b_l).$$

Then we obtain a basis B_4 for $U_1 + U_2$ given by

$$B_4' = (c_1, \ldots, c_m, a_{m+1}, \ldots, a_k, b_{m+1}, \ldots b_l).$$

So we can read immediately $\dim(U_1 + U_2) = m + (k - m) + (l - m) = k + l - m$. This is

$$\dim(U_1 + U_2) = \dim U_1 + \dim U_2 - \dim U_1 \cap U_2.$$

$\square$

Corollary 3.10 *Basis of the direct sum of two vector spaces.*
A basis B of the direct sum $U_1 \oplus U_2$ is given by the disjoint union of the two bases B_1 and B_2 of U_1 and U_2:

$$B = B_1 \sqcup B_2 \quad hence \quad \dim(U_1 \oplus U_2) = \dim U_1 + \dim U_2.$$

Note: The symbol "$\sqcup$" in $B_1 \sqcup B_2$ here means disjoint union. If we consider B_1, B_2 as lists, we may write B as a new list $B = (B_1, B_2)$.

Proof This follows directly from the above proof. Here we have $m = 0$ so that $B_1 = (a_1, \ldots, a_k)$, $B_2 = (b_1, \ldots, b_l)$ which is $B_1 \cap B_2 = \varnothing$ so that $B = (a_1, \ldots, a_k, b_1, \ldots, b_l)$ and of course $B = B_1 \sqcup B_2$ which means also $\dim(U_1 \oplus U_2) = \dim U_1 + \dim U_2$.   □

Another result that is very important and almost evident is given below.

Proposition 3.11 *A complementary subspace to U.*
If U is a subspace of V, there is always a complementary subspace W such that $V = U \oplus W$.

Proof Using appropriate bases in V and U, as in the proof of Proposition 3.10, we see immediately the result: $B_1 = (a_1, \ldots, a_k)$ is a basis of U and $B = (a_1, \ldots, a_k, c_n, \ldots, c_l)$ a basis of V. We set $W = \operatorname{span}(c_1, \ldots, c_l)$. We so have $V = U + W$ and since $U \cap W = \{0\}$, we obtain

$$V = U \oplus W.$$

□

It is evident that the choice of W is not unique. So we may have, for example, another subspace Y, such that again $V = U \oplus Y$.

Remark 3.6 Complement in set theory.
The set-theoretic complement U^c of U in V is different: $U^c = V \setminus U \neq W$.

Remark 3.7 On $\ker f - \mathrm{im}\, f$ decomposition of an operator.
The notion of an f-invariant subspace of V is central here. A subspace of V is f-invariant if $F(U) \subseteq U$ holds. Remark 3.4 and Theorem 3.1 can lead to the following question. Can an operator $f \in \mathrm{Hom}(V, V)$ lead to an f-invariant decomposition of V such that

$$V = \ker f \oplus \mathrm{im}\, f$$

holds? For this problem, the spaces $\ker f, \ker f^2, \mathrm{im}\, f^2, \mathrm{im}\, f$ are relevant, as is their behavior. All these spaces are f-invariant subspaces of V and in particular the relation

$$\ker f \leq \ker f^2 \tag{3.25}$$

holds.
This follows from

$$x_0 \in \ker f \Rightarrow f x_0 = 0 \Rightarrow f^2 x_0 = f(f x_0) = f(0) = 0$$

and so

$$x_0 \in \ker f^2.$$

The following proposition provides the answer to the above question.

Proposition 3.12 $\ker f - \mathrm{im}\, f$ *decomposition.*
Let $f \in \mathrm{Hom}(V, V)$. Then the following assertions are equivalent.

(i) $V = \ker f \oplus \mathrm{im}\, f$,
(ii) $\ker f \cap \mathrm{im}\, f = \{o\}$,
(iii) $\ker f^2 = \ker f$,
(iv) $\ker f^2 \leqslant \ker f$.

Proof We already saw that $\ker f \leq \ker f^2$ (3.25). So assertion (iii) is equivalent to condition (iv). Therefore, it is enough to show that (i) and (ii) are equivalent to (iv). We now show that (iv) $\Leftrightarrow$ (ii) and (ii) $\Leftrightarrow$ (i) which establishes the result.

– (iv) $\Rightarrow$ (ii)
  Given $\ker f^2 \leq \ker f$, we have to show that $\ker f \cap \mathrm{im}\, f = \{0\}$: Let $z \in \ker f \cap \mathrm{im}\, f$ which means $z \in \mathrm{im}\, f$ or $z = f(x)$, and also $z \in \ker f$ which means $0 = f(z) = f(f(x)) = f^2(x)$ or $x \in \ker f^2$. Assertion (iv) leads to $x \in \ker f$ which means $f(x) = 0$, and to $z = f(x) = 0$, which proves $\ker f \cap \mathrm{im}\, f = \{0\}$ which is assertion (ii).

– (ii) $\Rightarrow$ (iv)

Given $\ker f \cap \operatorname{im} f = \{0\}$, we have to show $\ker f^2 \leq \ker f$: Let $x \in \ker f^2$. Then $f^2(x) = 0$ which means $f(f(x)) = 0$ and $f(x) \in \ker f$. Since $f(x) \in \operatorname{im} f$, we have $f(x) \in \ker f \cap \operatorname{im} f = \{0\}$ such that $f(x) = 0$, $x \in \ker f$ and $\ker f^2 \leq \ker f$ which proves (iv).

– (ii) $\Leftrightarrow$ (iv)

The implication (i) $\Rightarrow$ (ii) is clear by Proposition 2.3 since the direct sum $U_1 \oplus U_2 = V$ means that $U_1 \cap U_2 = \{0\}$.

– (ii) $\Rightarrow$ (i)

Given $\ker f \cap \operatorname{im} f = \{0\}$, we have to show $\ker f + \operatorname{im} f = V$. According to Proposition 2.3, $\ker f \cap \operatorname{im} f = \{0\}$ means direct sum:

$$\ker f + \operatorname{im} f = \ker f \oplus \operatorname{im} f.$$

The rank-nullity theorem (Corollary 3.2) states that $\dim(\ker f) + \dim(\operatorname{im} f) = \dim V$. Since $\ker f \oplus \operatorname{im} f$ is a subspace of V, we obtain $\ker f \oplus \operatorname{im} f = V$ which proves (i) and completes the prove of the proposition.

$\square$

Proposition 3.13 *Direct product and direct sum.*

The linear map $\Phi : U_1 \times \cdots \times U_m \longrightarrow U_1 \oplus \cdots \oplus U_m$ *given by* $(u_1, \ldots, u_m) \longmapsto \Phi(u_1, \ldots, u_m) = u_1 + \cdots + u_m$ *is an isomorphism:* $(U_1 \times \cdots \times U_m) \cong U_1 \oplus \cdots \oplus U_m.$

Proof We first show that Φ is injective, that is, $\ker \Phi = \{0\}$: If $\Phi(z_1, \ldots, z_m) = 0$, we have $z_1 + \cdots + z_m = 0$. Since the sum on the right hand side above is direct, the uniqueness of the decomposition of 0 leads to $z_1 = 0, \ldots, z_m = 0$ which shows that $\ker \Phi = \{0\}$ and Φ are injective. Further more, the rank-nullity theorem,

$$\dim(\ker \Phi) + \dim(\operatorname{im} \Phi) = \dim(U_1 \times \ldots \times U_m),$$

with $\dim(\ker \Phi) = 0$ gives

$$\dim(\operatorname{im} \Phi) = \dim(U_1 \times \ldots \times U_m),$$

which shows that Φ is also surjective. So we proved that Φ is an isomorphism. $\square$

This means that the dimension of $U_1 \oplus \ldots \oplus U_m$ is also given by

Corollary 3.11 $\dim(U_1 \oplus \cdots \oplus U_m) = \dim(U_1) + \cdots \dim(U_m)$.

3.5 The Origin of Tensors

As we saw, there are various possibilities to construct a new vector space out of the two vector spaces U_1 and U_2. The role of the two bases B_1 and B_2 is particularly important in this construction. As we saw (Corollary 3.10), for the case of the direct sum $U_1 \oplus U_2$, we have $U_1 \oplus U_2 = \text{span}(B_1 \amalg B_2)$ with $\dim(U_1 \oplus U_2) = \dim U_1 + \dim U_2$. Note that we may also write $B_1 \amalg B_2 = (B_1, B_2)$ and likewise $U_1 \oplus U_2 = \text{span}(B_1, B_2)$.

Now we may ask the provocative question: If we take the Cartesian product $B_1 \times B_2$ instead of the disjoint union $B_1 \amalg B_2$, what can we say about the corresponding vector space $W = \text{span}(B_1 \times B_2)$?

Using the same notation as in the corollary above, the basis B_W of W is given by

$$B_W = \{(a_s, b_i) : a_s \in B_1, b_i \in B_2\} \quad s \in I(k), i \in I(l).$$

Let us now for simplification reasons consider real vector spaces. As we know, an abstract vector space is completely determined by its dimension, so we have: $\dim W = \dim U_1 \dim U_2$. The vector space W is what is called a tensor product of U_1 and U_2 and we write $W = U_1 \otimes U_2$.

In addition, it is clear that for W nothing changes if we write for the basis vectors $(a_s, b_i) \equiv a_s b_i \equiv a_s \otimes b_i$ and for good reasons they may be called product or even tensor product of the basis vectors a_s and b_i. So we may write $B_W = B_1 \times B_2 = \{a_s \otimes b_i\}$ and we have for $w \in U_1 \otimes U_2$

$$w = w^{si} a_s \otimes b_i \quad \text{with} \quad w^{si} \in \mathbb{R}.$$

It may also be clear that the new vector space W, $W = U_1 \otimes U_2$ is not a subspace of V. This justifies our characterization of the above question as provocative.

The tensor space $U_1 \otimes U_2$ depends only on U_1 and U_2, regardless of where these come from.

As one may already realize, we can hardly find a subject that does not use tensors in physics. In the Chaps. 8 (First Look at Tensors) and 14 (Tensor Formalism), we are going to learn much more about tensor products.

3.6   From Newtonian to Lagrangian Equations

As we shall see, the transition from Newtonian to Lagrangian equations is essentially the transition from a linearly dependent to a linearly independent system. In Newtonian mechanics, the Newton Axioms tell us for example the equations of motion for a point particle with n degrees of freedom in $\mathbb{R}^n$. The motion of this point particle takes place in $\mathbb{R}^n$ without further conditions, except of course the physical forces that act on this particle. We use Cartesian coordinates for Newton's equation, these correspond to a chosen inertial frame. Usually, the space Q in which the motion takes place, is called configuration space, and we have here $Q = \mathbb{R}^n$. If the configuration space Q is not $\mathbb{R}^n$ but rather a manifold or equivalently a n-dimensional surface in $\mathbb{R}^m$, that is, $Q \subseteq \mathbb{R}^m$, we have to derive the Lagrangian equations, starting from the Newtonian equations. In this sense, the Lagrangian equation may be considered as the appropriate Newtonian equation for the motion in a manifold (Fig. 3.1).

An instrument to derive the Lagrangian equation starting from the Newtonian equation, is traditionally D'Alambert's principle. The crucial step here has to do with linear algebra. It is the transition from a linearly dependent system, usually expressed in a Cartesian coordinate system in $\mathbb{R}^m$, to a linearly independent system. This linearly independent system corresponds to an appropriate coordinate system in the submanifold Q with dim $Q = n < m$. This fact is often underestimated. Here, we do the opposite and we take the point of view of linear algebra in order to perform the same derivation, starting from the Newtonian equation in $\mathbb{R}^m$.

The motion is constrained to be on the surface Q which we may parametrize with $\psi = (\psi^i)$, using a simplified notation as follows:

$$i \in I(m), \quad s \in I(n)$$

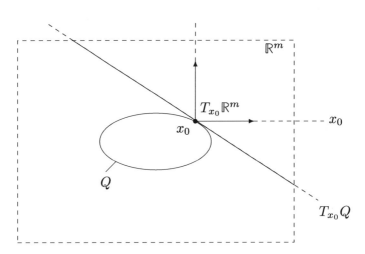

Fig. 3.1   From Newtonian to Lagrangian equations

and the functions

$$\psi^i : \mathbb{R}^m \longrightarrow \mathbb{R},$$
$$(q^1, \ldots q^n) \longmapsto \psi^i(q^1, \ldots, q^n) \equiv x^i(q^s),$$

describing the configuration space Q of dimension n, see Fig. 3.1. The variables q^s are the usual generalized coordinates used in classical mechanics. For our demonstration, the time dependence is not relevant and is therefore left out. For the mass m of the particles, we take, without loss of generality, $m_i = m = 1$ for all $i \in I(m)$. It is clear that we cannot solve the Newtonian equation in the usual form (with $m = 1$ and the force F):

$$\ddot{x}^i(t) = F^i, \tag{3.26}$$

since for the Cartesian coordinate x^i we have the infinitesimal constrains

$$dx^i = \frac{\partial x^i}{\partial q^s} dq^s. \tag{3.27}$$

For this construction, in order to use linear algebra, we direct our attention to the point $p_0 \in Q$. We consider the vector space $V = T_{p_0}\mathbb{R}^m \cong \mathbb{R}^m$ and its subspace $W := T_{p_0}Q \cong \mathbb{R}^n$. Heuristically, we have to "project" Eq. (3.24, 3.26) onto the configuration space Q and particularly onto the corresponding vector space W at the position $p_0 \in Q$. This leads to the work done by the force F with respect to the displacement dx in W. Using the dot product $<|>$ in $\mathbb{R}^m$ and the Newtonian equation (3.24, 3.26) we obtain

$$< \ddot{x} \mid dx > = < F \mid dx > \tag{3.28}$$

and the corresponding tensor equation ($F_i = F^i$):

$$\ddot{x}_i dx^i = F_i dx^i. \tag{3.29}$$

In this representation, we may consider $\ddot{x}_i$ and F_i as scalars and $dx^i(p_0)$ as element of the dual of W:

$$dx^i(x_0) \in W^* = (T_{p_0}Q)^*. \tag{3.30}$$

It is clear that we cannot drop the dx^i in Eq. (3.28, 3.29). The covectors dx^i are linearly dependent as we saw in Eq. (3.26, 3.27). The covectors $dq^s(p_0)$ are by definition linearly independent. At the same time it is also clear that we have to use Eq. (3.26, 3.27) in order to express Eq. (3.29) with the covectors $dq^i(p_0)$. The latter are linearly independent and so we obtain:

$$\ddot{x}_i \frac{\partial x^i}{\partial q^s} dq^s = F_i \frac{\partial x^i}{\partial q^s} dq^s. \tag{3.31}$$

Since dq^s are linearly independent, we may write for all $s \in I(n)$:

$$\ddot{x}_i \frac{\partial x^i}{\partial q^s} = F_i \frac{\partial x^i}{\partial q^s}. \tag{3.32}$$

From now on, we can proceed exactly the same way as in the physical literature. For the sake of completeness, we continue with our simplified prerequisites ($m_i = 1$ and time independence) and we obtain the Lagrangian equation: For the right hand-side of the Eq. (3.32), we may write:

$$\mathcal{F}_s := F_i \frac{\partial x^i}{\partial q^s}. \tag{3.33}$$

The quantifiers $\mathcal{F}_s$ are called generalized forces. In the case of the existence of a potential V, we may write

$$\mathcal{F}_s = -\frac{\partial V}{\partial q^s} + \frac{d}{dt} \frac{\partial V}{\partial q^s}. \tag{3.34}$$

For the left-hand side of Eq. (3.32), using essentially the product rule of differentiation, we obtain:

$$\begin{aligned}
\ddot{x}_i \frac{\partial x^i}{\partial q^s} &= \frac{d}{dt} \left(\dot{x}_i \frac{\partial x^i}{\partial q^s} \right) - \dot{x}_i \frac{d}{dt} \left(\frac{\partial x^i}{\partial q^s} \right) \\
&= \frac{d}{dt} \left(\dot{x}_i \frac{\partial x^i}{\partial q^s} \right) - \dot{x}_i \frac{\partial \dot{x}^i}{\partial q^s}.
\end{aligned} \tag{3.35}$$

We write $v^i = \dot{x}^i$ for the velocity coefficients. Taking into account that the kinetic energy T of the system is given by $T = \sum_{(i=1)}^{m} \frac{1}{2} v_i v^i$ where $v_i = v^i$, and using Eq. (3.27), we get:

$$\frac{dx^i}{dt} \equiv \dot{x}^i = v^i(q^s, \dot{q}^s) = \frac{\partial x^i}{\partial q^s} \dot{q}^s, \tag{3.36}$$

$$\frac{\partial \dot{x}^i}{\partial \dot{q}^s} = \frac{\partial x^i}{\partial q^s} \tag{3.37}$$

and

$$\frac{\partial x^i}{\partial q^s} = \frac{\partial \dot{x}^i}{\partial \dot{q}^s} = \frac{\partial v^i}{\partial \dot{q}^s}. \tag{3.38}$$

This way, Eq. (3.35) takes the form

$$\ddot{x}_i \frac{\partial x^i}{\partial q^s} = \frac{d}{dt}\left(v_i \frac{\partial v^i}{\partial \dot{q}^s}\right) \qquad -\left(v_i \frac{\partial v^i}{\partial q^s}\right)$$

$$= \frac{d}{dt}\left(\frac{\partial}{\partial \dot{q}^s}\left(\tfrac{1}{2}v_i v^i\right)\right) - \frac{\partial}{\partial q^s}\left(\tfrac{1}{2}v_i v^i\right),$$

so $\qquad \ddot{x}_i \dfrac{\partial x^i}{\partial q^s} = \dfrac{d}{dt}\left(\dfrac{\partial T}{\partial \dot{q}^s}\right) \qquad - \dfrac{\partial T}{\partial q}.$ $\qquad\qquad$ (3.39)

From Eqs (3.32) and (3.34), we have:

$$\ddot{x}_i \frac{\partial x^i}{\partial q^s} = \mathcal{F}_s = -\frac{\partial V}{\partial q^s} + \frac{d}{dt}\left(\frac{\partial V}{\partial \dot{q}^s}\right). \qquad\qquad (3.40)$$

The equations (3.39) and (3.40) give

$$\frac{d}{dt}\left(\frac{\partial T}{\partial \dot{q}^s}\right) - \frac{\partial T}{\partial q} = \frac{d}{dt}\left(\frac{\partial V}{\partial \dot{q}^s}\right) - \frac{\partial V}{\partial q^s}, \qquad\qquad (3.41)$$

and

$$\frac{d}{dt}\left(\frac{\partial}{\partial \dot{q}^s}(T - V)\right) - \frac{\partial}{\partial q^s}(T - V) = 0. \qquad\qquad (3.42)$$

$$L := T - V \qquad\qquad (3.43)$$

is the Lagrangian function, and as we know from Eq. (3.42),

$$\frac{d}{dt}\left(\frac{\partial L}{\partial \dot{q}^s}\right) - \frac{\partial L}{\partial q^s} = 0 \qquad\qquad (3.44)$$

are the Lagrangian equations. As we see, the essential part for the derivation of the Lagrangian equation up to Eq. (3.32), is just an application of linear algebra.

Summary

We have examined the role of bases from all angles. This role is extremely positive, particularly concerning physics. We have also pointed out certain potential drawbacks when expressing certain basis dependent statements. But for these drawbacks, a satisfactory response was provided. We showed that one can look at all possible bases simultaneously to a avoid coordinate dependence.

In order to define what a basis of a vector space is, it was necessary to introduce several elementary concepts, such as the concept of a generating system of a vector space and linear dependence or independence. When the number of elements in a generating list is finite, we call such a vector space "finitely generated". These are precisely the vector spaces we discuss in this book. The associated dimension of a vector space was then simply defined by the number of elements in a basis.

In the remaining chapter, some of the most important advantages resulting from the use of bases were discussed. Perhaps the most significant is that through bases, abstract vectors and abstract maps can be expressed by a finite number of scalars. Thus, bases enable concrete calculations of a theory to be performed and compared with experiments whose results are essentially numerical.

Bases allow us, in particular, to maintain the geometric character of a linear map. By choosing suitable tailored bases, one can find the simplest possible representation matrix of the corresponding map, which in most cases only has entries on the diagonal that are nonzero. This is the so-called normal form of a linear map, it may be considered essentially as the fundamental theorem of linear maps.

With the help of bases, this chapter presented the first and probably the easiest access to tensors. At the end of the chapter, a perhaps surprising application of linear algebra to classical mechanics was discussed.

Exercises with Hints

Exercise 3.1 *The span of a list in a vector space is the smallest subspace containing this list.*
Let V be a vector space and $A = (a_1, \ldots, a_k)$ a list of vectors in V. Show that span A is the smallest subspace of V containing all the vectors of the list A.

Exercise 3.2 *A linearly independent sublist in a linearly dependent list.*
If in the linearly dependent list $(a_1, \ldots, a_r, v)$ of vectors in V, the sublist $(a_1, \ldots, a_r)$ is linearly independent, show that the vector v is a linear combination of the list $(a_1, \ldots, a_r)$.

The following exercise is a variation of Exercise 3.2.

Exercise 3.3 If the list $(a_1, \ldots, a_r)$ in a vector space V is linearly independent and $v \in V$, then show that the list $(a_1, \ldots, a_r, v)$ is linearly independent if and only if $v \notin \text{span}(a_1, \ldots, a_r)$.

Exercise 3.4 *This exercise shows that the length of a spanning list plus one vector more, is always linearly dependent.*
Suppose that the list $A_m = (a_1, \ldots, a_m)$ of m vectors spans the vector space V (span $A_m = V$). Show that any list A_{m+1} with $m + 1$ vectors (not necessarily containing A_m) is always linearly dependent.

The following exercise proves the existence of a basis in a finitely generated vector space by extending a linearly independent list, using the results of the previous Exercise 3.3.

Exercise 3.5 Suppose that the vectors $(a_1, \ldots, a_r)$ are linearly independent. Show that either $(a_1, \ldots, a_r)$ is a basis in V or that there are vectors $(a_{r+1}, \ldots, a_n)$ such that $(a_1, \ldots, a_r, a_{r+1}, \ldots, a_n)$ is a basis of V.

Exercise 3.6 *Using the extension of a linearly independent list to a basis as in the previous Exercise 3.5, we obtain the following result.*
Show that all bases of a vector space have the same length. This means that if $B(V)$ is the set of bases in V and $B_1, B_2 \in B(V)$, then $l(B_1) = l(B_2)$.

Now we give another definition for the dimension of a vector space which does not rely on a property of a basis, as usual in literature. Only the notion of linearly dependent or linearly independent will be used.

Definition: Dimension of a vector space V:
   We consider the following set of integers given by

$$N(V) := \{m \in \mathbb{N} : \text{any } m + 1 \text{ vectors of } V \text{ are linearly dependent}\}$$

and
$$\dim V := \min \mathbb{N}(V).$$

Note that for finitely generated vector spaces, as we consider them in this book, the set $\mathbb{N}(V)$ is nonempty, $\mathbb{N}(V) \neq \emptyset$ and if $V \neq \{0\}$, then $\dim V \geq 1$.

Exercise 3.7 Given the above definition, show that if $\dim V = \min \mathbb{N}(V) = n$, then the length $l(B)$ of any basis B of V is given by $l(B) = n$.

The next two exercises are almost trivial. Using the definition of dimension by $\mathbb{N}(V)$ might make them even easier to prove.

Exercise 3.8 *Subspace and its dimension.*
If U is a subspace of a vector space V, then show that

$$\dim U \leq \dim V.$$

Exercise 3.9 *Subspaces with the maximal dimension.*
If U is a subspace of a vector space V and $\dim U = \dim V$, then show that

$$U = V.$$

In the next two exercises, if we know the dimension of a vector space, we consider one additional condition for a list to become a basis.

Exercise 3.10 *Linearly independent list in a vector space V with length equal to* $\dim V$.
Let $A = (a_1, \ldots, a_k)$ be a linearly independent list of vectors in V with $\dim V = n$. Show that if $k = n$, then A is a basis of V.

Exercise 3.11 *Spanning list in a vector space V with length equal* $\dim V$.
Let $A = (a_1 \ldots, a_k)$ be a spanning list of vectors in V with $\dim V = n$. Show that if $k = n$, then A is a basis of V.

Exercise 3.12 *Linear combinations and the basis map.*
Let $A = (a_1, \ldots, a_k)$ be a list in the vector space V and a linear combination

$$a_s \lambda^s = A\vec{\lambda} \quad \text{with} \quad \lambda^s \in \mathbb{K}, \vec{\lambda} \in \mathbb{K}^n$$

and the basis map Ψ_A given by

$$\Psi_A : \quad \mathbb{K}^n \longrightarrow V$$
$$e_s \longmapsto a_s.$$

Show the following assertions:

(i) A is a basis in V if and only if Ψ_A is an isomorphism;
(ii) A is a linearly independent list if and only if Ψ_A is an injection;
(iii) A is a spanning list if and only if Ψ_A is a subjection.

The following two exercises correspond to simple examples of linear maps.

Exercise 3.13 *Linear maps on a one-dimensional vector space.*
If $f \in \text{Hom}(V, V)$ with $\dim V = 1$, show that f is a scalar multiplication, that is, there is a $\lambda \in \mathbb{K}$ such that :

$$f(v) = \lambda v \quad \text{for all} \quad v \in V.$$

Exercise 3.14 If f is a linear map, $f \in \text{Hom}(\mathbb{K}^n, \mathbb{K}^m)$, show that there exist scalars $\varphi_s^i \in \mathbb{K}$ with $s \in I(n)$ and $i \in I(m)$ such that for every $\vec{v} = e_s v^s \in \mathbb{K}^n$, $v^s \in \mathbb{K}$ and $(e_s)_n$ the standard basis in $\mathbb{K}^n$

$$F : \begin{bmatrix} v^1 \\ \vdots \\ v^n \end{bmatrix} \longrightarrow \begin{bmatrix} \varphi_s^1 v^s \\ \vdots \\ \varphi_s^m v^s \end{bmatrix} \quad \text{holds.}$$

The following exercises concern properties of linear maps.

Exercise 3.15 *The image of a basis already determines a linear map. This ensures the existence of a linear map as was required in Proposition 3.8.*
Let $B = (b_1, \ldots, b_n)$ be a basis of a vector space V and $(w_1, \ldots, w_n)$ any list of vectors in a second vector space W. Show that there exists a unique linear map $f : V \to W$ such that $f(b_s) = w_s \ \forall \, s \in I(n)$.

Exercise 3.16 *Any linear map preserves linear dependence.*
Let f be a linear map $f : V \to V'$. If the list $(v_1, \ldots, v_k)$ in $\mathbb{K}$ is linearly dependent, show that the list $(f(x_1), \ldots, f(v_k))$ in V' is also linearly dependent.

Exercise 3.17 *The preimage of a linear map preserves the linear independence in the following sense.*
Let f be a linear map $f : V \to V'$. Show that the list $(v_1, \ldots, v_r)$ in V is linearly independent if the list $(f(v_1), \ldots, f(v_r))$ in V' is linearly independent.

Exercise 3.18 *If a linear map is injective, it preserves the linear independence.*
Let f be an injective linear map $f : V \to V'$. If the list $(v_1, \ldots, v_r)$ in V is linearly independent, show that the list $(f(v_1), \ldots, f(v_r))$ in V' is also linearly independent.

Exercise 3.19 *The inverse map of a bijective linear map is also linear.*
If the map $f : V \to V'$ is an isomorphism, show that the inverse map

$$f^{-1} : V' \longrightarrow V$$

is also an isomorphism (bijective and linear).

Exercise 3.20 *All isomorphic vector spaces have the same dimension.*
Show that vector spaces (finite-dimensional) are isomorphic if and only if they have
the same dimension.

Exercise 3.21 *Criterion for isomorphism. (Corollary 3.6)*
Show that the map $f \in \mathrm{Hom}(V, V')$ is an isomorphism if and only if for any basis
$B = (b_1, \ldots, b_n)$ of V

$$f(B) = \{f(b_1), \ldots, f(b_n)\}$$

is a basis of V'.

Exercise 3.22 *Equivalence for equal dimensions. (Corollary 3.4)*
Show that when $f : V \to V'$ is linear and $\dim V = \dim V'$, then the following conditions are equivalent.

   (i) f is injective ;
  (ii) f is surjective ;
 (iii) f is bijective .

Exercise 3.23 *Injectivity and dimensions.*
Let V and V' be vector spaces with $\dim V > \dim V'$. Show that any $f \in \mathrm{Hom}(V, V')$
is not injective.

Exercise 3.24 *Surjectivity and dimensions.*
Let V and V' be vector spaces with $\dim V < \dim V'$. Show that $f \in \mathrm{Hom}(V, V')$ is
not surjective.

> The following three exercises concern sums and direct sums of a vector space.

Exercise 3.25 Let $U_1, \ldots, U_m$ be subspaces of a vector space V. Verify the following results (see Definition 3.12 and the pages thereafter):

   (i) $U_1 + \cdots + U_m \leq V$;
  (ii) $U_1 + \cdots + U_m = \mathrm{span}(U_1 \cup \ldots \cup U_m)$;
 (iii) $\dim(U_1 + \cdots + U_m) \leq \dim U_1 + \cdots + \dim U_m$.

Exercise 3.26 *Equivalent conditions for a direct sum of subspaces of a vector space.*
Let $U_1, \ldots, U_m$ be subspaces of a vector space V and $U = U_1 + \cdots + U_m$. Show
that the following conditions for a direct sum

$$U = U_1 \oplus \cdots \oplus U_m$$

are equivalent.

(i) Every $u \in U$ has a unique representation $u = u_1 + \cdots + u_m$ with $u_j \in U_j$ for each $j \in I(m)$;
(ii) Whenever $u_1 + \cdots + u_m = 0$ and $u_j \in U_j$, for each $j \in I(m)$ we have $u_j := 0$ for all $U_j \in I(m)$;
(iii) For every $j \in I(m)$, $U_j \cap (U_1 + \cdots + U_{j-1} + U_{j+1} + \cdots + U_m) = 0$.

Exercise 3.27 *Equivalent conditions for a direct sum decomposition of a vector space.*
Let $U_1, \ldots, U_m$ be subspaces of a vector space V. Show that the following conditions are equivalent:

(i) $V = U_1 \oplus \cdots \oplus U_m$;
(ii) For every $j \in I(m)$ and every basis of U_j, $B^j = (b_1^j, \ldots, b_{n_j}^j)$, $B = (B^1, \ldots, B^m)$ is a basis of V;
(iii) $V = U_1 + \cdots + U_m$ and $\dim V = \dim U_1 + \cdots + \dim U_m$.

The next exercises concern another point of view concerning the origin of tensors (Sect. 3.5) and needs some preparation. We have to compare the Cartesian product with the tensor product. The role of the scalar field is different. For this comparison we consider the following two exercises.

Exercise 3.28 *Cartesian product.*
Let U and V be two vectors spaces with $\dim U = k$ and $\dim V = l$. We denote their Cartesian product by

$$U \times V \equiv (U \times V, \cdot)$$

and the scalar action of the field $\mathbb{K}$ explicitly by the dot $\cdot$. So we have, as usual, for $\lambda \in \mathbb{K}$ and $(u, v) \in U \times V$:

$$\lambda \cdot (u, v) \equiv \lambda(u, v) := (\lambda u, \lambda v).$$

Show that $(U \times V, \cdot)$ is a vector space and that its dimension is $\dim(U \times V) = k + l$.

Exercise 3.29 *Tensor product.*
Let U and V be two vector spaces with $\dim U = k$ and $\dim V = l$. We denote their tensor product by $U \otimes V = (U \times V, \odot)$ and the scalar action on $U \otimes V$ by $\odot$. The scalar action of $\mathbb{K}$ is given by

$$\lambda \odot (u, v) \equiv \lambda(u, v) = (\lambda u, v) = (u, \lambda v),$$

and we have the bilinearity conditions

$$(\lambda_1 u_1 + \lambda_2 u_2, v) = \lambda_1(u_1, v) + \lambda_2(u_2, v) \quad \text{and}$$
$$(u_1, \lambda_1 v_1 + \lambda_2 v_2, v) = \lambda_1(u, v_1) + \lambda_2(u, v_2).$$

To distinguish from $U \times V$ in Exercise 3.28, we write (u, v) as $u \otimes v$ from now on. The elements of $U \otimes V$ are a linear combination of the type $u \otimes v$. Show that $U \otimes V$ is a vector space and that its dimension is dim $U \otimes V = k \cdot l$.

References and Further Reading

1. S. Axler, *Linear Algebra Done Right* (Springer Nature, 2024)
2. S. Bosch, *Lineare Algebra* (Springer, 2008)
3. G. Fischer, B. Springborn, *Lineare Algebra. Eine Einführung für Studienanfänger*. Grundkurs Mathematik (Springer, 2020)
4. S.H. Friedberg, A.J. Insel, L.E. Spence, *Linear Algebra* (Pearson, 2013)
5. S. Hassani, *Mathematical Physics: A Modern Introduction to its Foundations* (Springer, 2013)
6. K. Jänich, *Mathematik 1. Geschrieben für Physiker* (Springer, 2006)
7. N. Jeevanjee, *An Introduction to Tensors and Group Theory for Physicists* (Springer, 2011)
8. N. Johnston, *Introduction to Linear and Matrix Algebra* (Springer, 2021)
9. M. Koecher, *Lineare Algebra und analytische Geometrie* (Springer, 2013)
10. G. Landi, A. Zampini, *Linear Algebra and Analytic Geometry for Physical Sciences* (Springer, 2018)
11. J.M. Lee, *Introduction to Smooth Manifolds*. Graduate Texts in Mathematics (Springer, 2013)
12. J. Liesen, V. Mehrmann, *Linear Algebra* (Springer, 2015)
13. P. Petersen, *Linear Algebra* (Springer, 2012)
14. S. Roman, *Advanced Linear Algebra* (Springer, 2005)
15. B. Said-Houari, *Linear Algebra* (Birkhäuser, 2017)
16. A.J. Schramm, *Mathematical Methods and Physical Insights: An Integrated Approach* (Cambridge University Press, 2022)
17. G. Strang, *Introduction to Linear Algebra* (SIAM, 2022)
18. R.J. Valenza, *Linear Algebra. An Introduction to Abstract Mathematics* (Springer, 2012)

Chapter 4
Spacetime and Linear Algebra

It is well known that Newtonian mechanics and electrodynamics are fundamental theories of physics and also theories of our physical spacetime. Any theory of spacetime is, of course, a geometrical theory. As we already saw, linear algebra is, in many aspects, also geometric. The surprise is that linear algebra, along with some group theory, allows us to describe spacetime geometry in Newtonian mechanics and electrodynamics. Consequently, to describe this surprise, we will show the relation of linear algebra with the two central principles of Newtonian mechanics and electrodynamics, in particular with the law of inertia and the relativity principle, with linear algebra.

For thousands of years, it was thought to be an absolute convention that space and time are a priori given and physics itself had to be formulated in this context. This, by itself a very plausible position, was also taken in science for more than two thousand years. It has its roots in the fascination and power of the Euclidean axioms. But since Gauss and Riemann we had suspected, and since Einstein we have known that physics is the one that determines the structure of spacetime. This is demonstrated in this chapter.

4.1 Newtonian Mechanics and Linear Algebra

In Newtonian mechanics, it turns out that we only need the first Newtonian law to determine the structure of spacetime.

First Newtonian law: Every body continues in its state of rest or of uniform rectilinear motion, except if it is compelled by forces acting on it to change that state.

This law refers particularly to a trajectory $\vec{x}(t)$ of a mass point. It corresponds to the well-known equation of motion without the presence of a force:

© The Author(s), under exclusive license to Springer Nature Switzerland AG 2024
N. A. Papadopoulos and F. Scheck, *Linear Algebra for Physics*,
https://doi.org/10.1007/978-3-031-64908-0_4

$$\frac{d^2}{dt^2}\vec{x}(t) = 0. \tag{4.1}$$

This means that we here postulate the existence of a special reference frame in which the solutions of the above equation are straight lines with constant velocity (vanishing acceleration). Without going into details, we realize intuitively that the space where the movement takes place must be a manifold which contains straight lines. For instance, vector spaces and affine spaces (see Sect. 2.5) are such manifolds which may contain straight lines as a subspace. On the other hand, there is not enough room for straight lines in a sphere or a cube.

Definition 4.1 Reference frames (choice of coordinates) in which the first Newtonian law above has indeed the analytic form $\frac{d^2}{dt^2}\vec{x}(t) = 0$ are called inertial frames.

In other words, Newton postulates by the above equation of motion, as Galilei, the existence of an inertial frame. This is the principle of the law of inertia:

A force-free body remains at rest or in a state of rectilinear and uniform motion if the spatial reference system and the time scale are chosen appropriately.

The law of inertia leads first to a four-dimensional affine spacetime, and it is valid for both Newtonian mechanics and electrodynamics (special relativity) with additional different geometric structures.

This may seem a pretty harmless formulation at first, but it is a massive step at once for physics. There are not many manifolds where straight lines have enough space. As already stated above, there has to be an affine space with additional structures.

For Newton, presumably, all the above discussion was only a question of consistency since he assumed, according to the spirit of that time, that a priori the motion takes place in a Euclidean space E^3, which is a three-dimensional affine space with an inner product (dot product). His equation of motion was consistent with the mathematically given physical space or spacetime. Where time is concerned, Newton also assumes a priori an affine one-dimensional Euclidean space E^1.

Although the notion of spacetime was not current in those times, we may assume that his point of view of spacetime would be a four-dimensional manifold M. If M is not precisely equal to the Cartesian product $E^1 \times E^3$, it is at least isomorphic to it:

$$M \cong E^1 \times E^3.$$

Our intention is not to discuss the physical relevance of the law of inertia but to compare the physical situation with mathematics, especially with linear algebra. We hope that this helps appreciate the different roles of mathematics and physics and their connection. This relation can be demonstrated within linear algebra in a very transparent way. As we saw, in physics, we have to postulate the existence of an inertial frame. This is a huge step in understanding our world. Its validity has to be examined and tested by experiments, of course. It is understood that here is not the place for such discussions. But still, turning back to linear algebra, we may first assume, for the moment, as a model for the comparison, that our spacetime within linear algebra corresponds to a vector space. In this case, in mathematics, the existence of an inertial frame corresponds to the existence of a basis in a vector space which has only to be proven. As we see, the situation in mathematics is obvious. It is also clear that we cannot do the same in physics. The existence of a frame of inertia has to be postulated.

After having postulated in physics the existence of a frame of inertial and having proven in linear algebra the existence of a basis in a vector space (see Proposition 3.4), we may now ask naively how many frames of inertia do exist and in analogy how many bases exist in a vector space. Stated differently, we look for the set of Newtonian inertial frames (which we denote by $I F(M)$ with M the Newtonian spacetime), and similarly, for the set of bases $B(V)$ in a vector space V. This leads to the principle of relativity.

The equations describing the laws of physics have the same form in all admissible frames of reference.

Bearing this in mind, we have to determine those admissible transformations which transform one inertial frame into the other. We expect that the set of all these transformations builds a group called the Galilean group or Galilean transformations for good reasons.

In linear algebra we already answered this question analogously: The set of all bases $B(V)$ of an n-dimensional vector space V is given by the group of automorphism $\mathrm{Aut}(V)$ in V, the linear transformations in V, which is isomorphic to the real group $Gl(n)$. So we have the isomorphism between groups:

$$\mathrm{Aut}(V) = Gl(n).$$

The group $\mathrm{Aut}(V)$ consists precisely of those transformations which respect the linear structure of V. We here consider first an abstract vector space without further structure. This can be described by the action of the group $Gl(n)$ on $B(V)$. With the basis $B = (b_1, \ldots, b_n)$ and $g = (\gamma_s^i) \in Gl(n); \gamma_s^i \in \mathbb{R}; i, s \in I(n) = \{1, \ldots, n\};$ we have:

$$B(V) \times Gl(n) \longmapsto B(V)$$
$$(B, g) \longmapsto B' := Bg = [b_i \gamma_1^i, \ldots, b_i \gamma_n^i].$$

It is well-known that $Gl(n)$ acts on $B(V)$ freely and transitively from the right, from which follows that the two sets, even having quite different structures, are still bijective (see Proposition 3.5):

$$B(V) \underset{bij}{\cong} Gl(n).$$

For this reason, we may call $Gl(n)$ the structure group of V. Indeed, the above action of $Gl(n)$ on $B(V)$ completely characterizes the linear structure on the set V via the set of the basis $B(V)$ of V. This is the deeper meaning of the isomorphism between groups

$$\text{Aut}(V) \cong Gl(n).$$

This discussion within linear algebra is the model that significantly clarifies the corresponding discussion within the Newtonian mechanics and spacetime. So we can apply the above procedure also in Newtonian mechanics.

As already stated, the transformations which are implied by the law of inertia and the relativity principle are given by the Galilean group. To simplify, we here consider the part of the Galilean group which is connected with the identity $G(a.k.a.G \equiv G_1 \equiv \text{Gal} \overset{(+)}{\uparrow})$. This is the so-called proper orthochronous Galilean group $\text{Gal} \overset{(+)}{\uparrow}$. In an obvious notation, the element $g \in G$ that maps inertial frames to inertial frames is given by the following expression:

$$t \longmapsto t + s$$
$$\vec{x} \longmapsto \vec{x}\,' := R\vec{x} + \vec{w}t + \vec{a} \qquad (4.2)$$

with $R \in SO(3)$, $\vec{w}, \vec{a} \in \mathbb{R}^3$, $s \in \mathbb{R}$. R corresponds to a rotation and $\vec{a}$ to a translation in space. $\vec{w}$ corresponds to a velocity transformation and s to a time translation. Velocity transformation $\vec{w}$ means that for a given frame of inertial IF, any other frame F' which moves with constant velocity $\vec{w}$ relatively to IF, is also a frame of inertia ($F' = IF'$). It is clear that all the above expressions are mathematics which belong to linear algebra. But linear algebra is in addition useful as it gives an analogy, a simple model, for the situation in physics.

It also helps to clarify the role of the Galilean group in determining the structure of spacetime in Newtonian mechanics. So we may now proceed similarly as in linear algebra. We already know the Galilean group in physics, which was determined for the law of inertia and the relativity principle. We consider the set of inertial frames $IF(M)$ in the spacetime of Newtonian mechanics. The Galilean group G acts freely and transitively from the right on $IF(M)$:

$$IF(M) \times G \longrightarrow IF(M)$$
$$(IF, g) \longmapsto IF' := IFg.$$

So we have as in linear algebra before the bijection

$$IF(M) \underset{bij}{\cong} G.$$

The Galilean group is the structure group of spacetime; this means that we have the isomorphism of groups:

$$\text{Aut}(M) \cong G.$$

That means that the Galilean group ultimately determines the Newtonian spacetime M structure: We first see from the above expression for the Galilean transformations that M is an affine space with additional structure. We have now to determine and describe the additional structure in this affine space. It is helpful to return to our linear algebra model. We learned from the equation of motion in Newtonian mechanics that our model of spacetime as a vector space V is not realistic and should be an affine space. So we have indeed to take the affine space corresponding to the vector space V, which we denote by A given by the triple (A, V, ψ) as discussed in Sect. 2.5 with ψ the action of V on A.

$$\psi : V \times A \longrightarrow A$$
$$(v, p) \longrightarrow p + v.$$

This action is by definition free and transitive. We therefore have the bijection

$$V \underset{bij}{\cong} A.$$

This is still not enough. We need additional structures in the space A. A realistic and typical example would be to introduce a Euclidean structure to A, for example, our three-dimensional Euclidean space. This transforms A into a Euclidean space E. This is done by definition in the corresponding vector space V (also called the difference space of A), and we obtain a Euclidean vector space. So we have $V \equiv (V, < | >)$. We have to distinguish between the notion of a Euclidean space which is always an affine space (affine Euclidean space) and a Euclidean vector space. An inner product or, more generally, a scalar product (symmetric nondegenerate bilinear form) to an affine space, is defined directly on the difference space of an affine space which is exactly the corresponding vector space.

A simplified but quite realistic linear algebra model of spacetime inspired by Newtonian mechanics would be a Euclidean space E as discussed above.

We come back to Newtonian mechanics. We found that the spacetime M in Newtonian mechanics is an affine space given by the triplet (A^4, V^4, ψ) with V^4 the four-dimensional difference space of A^4 with possibly additional structure. If we denote by x a (global) coordinate chart on M, we have for this coordinate system in an obvious notation:

$$x(A^4) = \{(t, \vec{x}) \in \mathbb{R}^4\}$$
$$x(V^4) = \{(t_2 - t_1, \vec{x}_2 - \vec{x}_1) =: (\tau, \vec{\xi}) \in \mathbb{R}^4\}. \tag{4.3}$$

The Galilean transformation (Eq. 4.2) in the coordinate of V^4 is given by

$$\tau' = \tau$$
$$\vec{\xi}' = \vec{w}\tau + R\vec{\xi}. \tag{4.4}$$

This leads immediately to the two invariants:

$$(1) \quad \tau$$
$$(2) \quad \| \vec{\xi} \| \text{ if } \tau = 0 \quad (\| \vec{\xi} \| = \sqrt{< \xi | \xi >}). \tag{4.5}$$

Summarizing, we can say that we found that the Newtonian spacetime M is an affine space A^4 with an additional structure. This can be described by the two invariants in Eq. (4.5) τ and $\| \vec{\xi} \|$, the latter being the length or norm defined only if $\tau = 0$. So we may write

$$M = (A^4, \Delta t, \| \Delta \vec{x} \| \text{ if } \Delta t = 0).$$

The invariant $\tau = \Delta t$ is the duration between two events that characterize the absolute time as assumed by Newton. The second invariant $\| \vec{\xi} \| = \| \Delta \vec{x} \|$ is the Euclidean distance between two *simultaneous events*. Therefore, it is clear that the spacetime M of the Newtonian mechanics is not a Euclidean or semi-Euclidean or an affine space with certain scalar products. The reason for this "complication" is the velocity transformation in Galilean transformations. (Eq. (4.2)).

Despite this, time and space are each separately regarded as Euclidean spaces E^1 and E^3, as considered by Newton. We believe, and hopefully, the reader can also see, that a good understanding of linear algebra is necessary to clarify the structure of spacetime completely.

4.2 Electrodynamics and Linear Algebra

In electrodynamics, both the law of inertia and the relativity principle hold. As in the previous section, the spacetime of electrodynamics that we denote now by $\mathcal{M}$ is an affine space A^4 as in Newtonian mechanics but with a different additional structure. To determine the additional structure, we have to use the laws of electrodynamics. Following Einstein, we take from electrodynamics only the existence of a photon. This turns out to be entirely sufficient to determine the structure of spacetime in electrodynamics. From our experience with Newtonian mechanics, we learned that we have to search for the relevant invariants. Therefore, it is quite reasonable to use the velocity of light c. For this reason, we consider the speed of light (of a photon) in two different frames of inertia, and we have:

$$\text{In } IF \text{ with coordinates } (t, \vec{x}) : c^2 = \frac{\| \Delta\vec{x} \|^2}{\Delta t^2}. \tag{4.6}$$

$$\text{In } IF' \text{ with coordinates } (t', \vec{x}') : c'^2 = \frac{\| \Delta\vec{x}' \|^2}{\Delta t'^2}. \tag{4.7}$$

The invariance of velocity of light given by

$$c' = c \tag{4.8}$$

leads to the equation

$$\frac{\| \Delta\vec{x}' \|^2}{\Delta t'^2} = \frac{\| \Delta\vec{x} \|^2}{\Delta t^2} = c^2 \tag{4.9}$$

or equivalently to the equation

$$c^2 \Delta t'^2 - \| \Delta\vec{x}' \|^2 = c^2 \Delta t^2 - \| \Delta\vec{x} \|^2 = 0. \tag{4.10}$$

To proceed, we make an assumption that is, in principle, not necessary but which simplifies our derivation significantly in a very transparent way. We assume that the Eq. (4.10) is valid also in the form:

$$c^2 \Delta t'^2 - \| \Delta\vec{x}' \|^2 = c^2 \Delta t^2 - \| \Delta\vec{x} \|^2 = K \tag{4.11}$$

with $K \in \mathbb{R}$. This means that the invariant K could also be different from zero.

Defining $\Delta x^0 := c\Delta t$ which has the dimension of length as Δx^i with $i \in \{1, 2, 3\}$ and $\mu, \nu \in \{0, 1, 2, 3\}$, Eq. (4.11) takes first the form

$$(\Delta x^{0'})^2 - \| \Delta\vec{x}' \|^2 = (\Delta x^0)^2 - \| \Delta\vec{x}' \|^2 \tag{4.12}$$

and we may define

$$\Delta s^2 := \sigma_{\mu\nu} \Delta x^\mu \Delta x^\nu \tag{4.13}$$

with

$$S = (\sigma_{\mu\nu}) = \begin{bmatrix} 1 & 0 & 0 & 0 \\ 0 & -1 & 0 & 0 \\ 0 & 0 & -1 & 0 \\ 0 & 0 & 0 & -1 \end{bmatrix}. \tag{4.14}$$

The expressions in Eqs. (4.13) and (4.14) correspond to the relativistic scalar product (symmetric nondegenerate bilinear form) which is nondegenerate and nonpositive definite. Δs^2 is invariant and we have, for example, for the two different frames of inertia IF and IF':

$$\Delta s^2(IF') = \Delta s^2(IF) \tag{4.15}$$

which shows that Δs^2 is universal. This means that the spacetime $\mathcal{M}$ of electrodynamics is an affine space A^4 with a scalar product given by the matrix $S = (\sigma_{\mu\nu})$, a symmetric covariant tensor called also metric tensor or Minkowski Metric tensor. After diagonalization, this tensor has the canonical form given by Eq. (4.14). Concluding, we may state that the spacetime $\mathcal{M}$ of electrodynamics is given by the pair

$$\mathcal{M} = (A^4, S). \tag{4.16}$$

Since here the scalar product S is not positive definite as in the case of a Euclidean space, $\mathcal{M}$ is here known as a semi-Euclidean or pseudo-Euclidean space or Minkowski spacetime.

It is interesting to notice that the space $\mathcal{M}$ in electrodynamics is mathematically much simpler than the space M in Newtonian mechanics. The space $\mathcal{M}$ is formally almost a Euclidean space, whereas in M, as we saw, the invariants are mathematically not as simple as a scalar product. On the other hand, the physics of $\mathcal{M}$, the spacetime of electrodynamics (special relativity), is much more complicated and complex than the physics of M, the spacetime of Newtonian mechanics because the duration τ of the two events is not any more an invariant. At the same time, $\mathcal{M}$, the semi-Euclidean or Minkowski spacetime, is the spacetime of elementary particle physics or simply the spacetime of physics without gravity. This causes all the well-known difficulties which enter into relativistic physics.

Now having found the structure of spacetime $\mathcal{M}$, it is equally interesting to determine its structure group G.

$$\text{Aut}(\mathcal{M}) = G. \tag{4.17}$$

Since we know that the space $\mathcal{M}$ is a semi-Euclidean space, we expect that the structure group G consists of semi-Euclidean transformations. So G is isomorphic to the well-known Poincaré (Poin) or inhomogeneous Lorentz group. This is a special affine group similar to a Euclidean group (affine Euclidean group). We may write $G = Poin$ and we have in an obvious notation:

$$Poin = \{(a, \Lambda) : a = (\alpha^\mu) \in \mathbb{R}^4, \Lambda = (\lambda^\mu_\nu) \in O(1, 3)\}. \tag{4.18}$$

The Poincare transformations are given by

$$x^\mu \longrightarrow x^{\mu'} := \lambda^\mu_\nu x^\nu + \alpha^\mu \tag{4.19}$$

or in a matrix form

$$x \mapsto x' = \Lambda x + a. \tag{4.20}$$

Λ corresponds to a special linear transformation (semiorthogonal transformation) given by:

$$O(1, 3) = \{\Lambda, \Lambda^\mathsf{T}\sigma\Lambda = \sigma\}. \tag{4.21}$$

As we see, all the mathematics we used in this chapter belong formally to linear algebra. After this experience, we may expect that all the mathematics we need for symmetries in physics also belongs to linear algebra.

Summary

In this chapter, we discussed one of the most important applications of linear algebra to physics. Starting from two fundamental theories of physics, Newtonian mechanics and electrodynamics, essentially using linear algebra alone, we described the structure of spacetime.

Using Newton's axioms, specifically employing the principles of inertia and relativity, we derived the spacetime structure of Newtonian mechanics, which is famously associated with the Galilean group.

For the description of the spacetime of electrodynamics, which simultaneously represents the spacetime of elementary particle physics and essentially the spacetime of all physics if one wants to exclude gravitational interaction, we followed Einstein's path: from electrodynamics, we only adopted the properties of the photon, the elementary particle associated closely with the electromagnetic force. With the photon, the principle of relativity, and linear algebra, we described the spacetime of physics without gravity. This spacetime is famously also closely connected with a group of transformation, the Poincaré group.

Reference and Further Reading

1. F. Scheck, *Mechanics. From Newton's Laws to Deterministic Chaos* (Springer, 2010)

Chapter 5
The Role of Matrices

What is a matrix? We consider matrices as elements of $\mathbb{K}^{m \times n}$. We have first to clarify a potential source of confusion. Matrices by themselves are not tensors in the sense of tensor calculus. They do not have a specific transformation rule, so, for example, the transformation property of a matrix representing a linear map is different to the transformation property of a matrix representing a given scalar product. Therefore we should always clarify how to use matrices in a given situation. Matrices seem to be very flexible objects, and we use this feature in mathematics and physics. We do not consider matrices only with scalars, but we also use matrices with entries other than scalars, for example, entries with vectors or covectors (linear forms), or even matrices. The most prominent property of all matrices is that we can add and, under certain conditions, multiply matrices. The multiplication rule seems at first sight quite complicated, but it turns out to be a reasonable and practical approach. In most cases, matrices are used as a representation of linear maps. In addition, the multiplication rule is justified by the composition of linear maps.

5.1 Matrix Multiplication and Linear Maps

One of the most important applications of matrices, is the ability to add, and, under certain conditions, to multiply matrices. This makes matrices look like numbers, perhaps something like super numbers. This possibility opens up when you consider matrices as linear maps. More precisely, it turns out that the composition of maps induces the product of matrices:

We consider the linear maps $f : U \to V$ and $g : V \to W$.

Let $X = (u_1, \ldots, u_n) \equiv (u_r)_n$ be a basis of U, $Y = (v_1, \ldots, v_p) \equiv (v_\mu)_p$ a basis of V, and $Z = (w_1, \ldots, w_m) \equiv (w_i)_m$ a basis of W. For the indices we choose $r \in I(n)$, $\mu \in I(p)$, and $i \in I(m)$. The composition of f and g is given by $h = g \circ f$ with $h : U \to W$.

© The Author(s), under exclusive license to Springer Nature Switzerland AG 2024
N. A. Papadopoulos and F. Scheck, *Linear Algebra for Physics*,
https://doi.org/10.1007/978-3-031-64908-0_5

Suppose now that we do not know anything about matrix multiplication. We want to define the matrix multiplication to be compatible with the composition of linear maps, that is, obtain a homomorphism between linear maps and matrices.

The values of f, g, h at basis vectors are given:

$$f(u_r) = v_\mu \varphi_r^\mu, \quad g(v_\mu) = w_i \psi_\mu^i, \quad h(u_r) = w_i \chi_r^i, \tag{5.1}$$

$$\text{for} \quad \varphi_r^\mu, \quad \psi_\mu^i, \quad \chi_r^i \in \mathbb{K}.$$

The composition $h = g \circ f$ leads to

$$h(u_r) = (g \circ f)(u_r) = g\big(f(u_r)\big) = g(v_\mu \varphi_r^\mu) = g(v_\mu)\varphi_r^\mu =$$
$$= w_i \psi_\mu^i \varphi_r^\mu \quad \text{and}$$
$$h(u_r) = w_i \chi_r^i = w_i \psi_\mu^i \varphi_r^\mu. \tag{5.2}$$

So, comparing (5.1) with (5.2), we obtain

$$\psi_\mu^i \varphi_r^\mu = \chi_r^i \tag{5.3}$$

which is the standard matrix multiplication in tensor notation. The corresponding matrices are given by

$$F = (\varphi_r^\mu) \qquad G = (\psi_\mu^i) \qquad H = (\chi_r^i)$$

and we have $GF = H$. More precisely, we may write

$$F \equiv f_{YX} \qquad G \equiv g_{ZY} \qquad H = h_{ZX}.$$

Thus we see (Eqs. 5.1, 5.2, 5.3) the homomorphism between linear maps and matrices underlining the role of bases:

$$h_{ZX} = g_{ZY} f_{YX}. \tag{5.4}$$

Remark 5.1 Products of linear maps and matrix multiplications have the same algebraic properties, so that we have

associativity : $\quad l \circ (g \circ f) = (l \circ g) \circ f$, $L(GF) = (LG)F$,

distributivity : $\quad g \circ (f_1 + f_2) = g \circ f_1 + g \circ f_2$, $G(F_1 + F_2) = GF_1 + GF_2$.

To make the analysis of the above results in algebraic terminology easier, we change the notation of matrices to

$$A := G \quad B := F \quad C := H, \tag{5.5}$$

so we have

$$A = (\alpha^i_\mu) \quad B = (\beta^\mu_r) \quad C := (\gamma^i_r). \tag{5.6}$$

We demonstrate various aspects to present a matrix, taking as example the matrix A in Eqs. (5.6) and (5.7), $A \in \mathbb{R}^{m \times p}$. So we write

$$A = (a_\mu)_p = [a_1 \dots a_p] : \text{ a row } (1 \times p\text{-matrix}) \text{ with columns (vectors) entries },$$

$$A = (\alpha^i)_m = \begin{bmatrix} \alpha^1 \\ \vdots \\ \alpha^m \end{bmatrix} : \text{ a column } (m \times 1\text{-matrix}) \text{ with rows (covectors) entries }.$$

Then for the columns of A, we have $a_\mu \in \mathbb{R}^m$ and for the rows (covectors) of A, we have $\alpha^i \in (\mathbb{R}^p)^*$. We have similar expressions for B and C in Eq. (5.6) and in Eqs. (5.8) and (5.9) below. In Eqs. (5.7), (5.8) and (5.9), we also see the block matrix form of A, B and C, with blocks columns or rows. The matrix multiplication is given by the following map, written as juxtaposition:

$$\mathbb{K}^{m \times p} \times \mathbb{K}^{p \times n} \longrightarrow \mathbb{K}^{m \times n},$$

$$(A, B) \longmapsto C := AB.$$

This also leads to various aspects of matrix multiplication: Summarizing and using an obvious notation, we write

$$A = (\alpha^i_\mu) = (a_\mu)_p = (\alpha^i)_m, \tag{5.7}$$

$$B = (\beta^\mu_r) = (b_r)_n = (\beta^\mu)_p, \tag{5.8}$$

$$C = (\gamma^i_r) = (c_r)_n = (\gamma^i)_m. \tag{5.9}$$

For the various components of the product matrix C, we have, using Eqs. (5.7), 5.8, and (5.9), the following very compact and transparent expressions:

$$\alpha^i_\mu \beta^\mu_r = \gamma^i_r, \tag{5.10}$$

$$a_\mu \beta^\mu_r = c_r \quad Ab_r = c_r, \tag{5.11}$$

$$\alpha^i_\mu \beta^\mu = \gamma^i \quad \alpha^i B = \gamma^i, \tag{5.12}$$

$$a_\mu \beta^\mu = C. \tag{5.13}$$

Equation (5.10) is the standard form of multiplication. This is actually the well-known low level multiplication.

Equation (5.11) means that the linear combination of the columns of the first matrix A with the coefficients of the rth column of the second matrix B gives the rth column of the product matrix C or it means simply that all columns of $C = AB$ are linear combinations of the columns of A. In Eq. (5.11) we see explicitly that the action of the matrix A on the column b_r gives the column c_r of the product.

Equation (5.12) is analog to rows: the linear combination of the rows of the second matrix B with the ith row coefficients of the first matrix A gives the ith row of the matrix C. Equivalently, the right action of the matrix B on the ith row of the matrix A, gives the ith row of the product.

In Eq. (5.13), for fixed μ, the product $\alpha_\mu \beta^\mu$ is the matrix product between an $m \times 1$ and a $1 \times n$-matrix.

The identification between linear maps and matrices provides the notion of rank for matrices as well. If we denote by f_A the linear map related to the matrix A, then we can also define a rank for matrices:

Definition 5.1 Rank of a matrix.
The rank of an $m \times n$-matrix A is given by the rank of the linear map:

$$\mathbb{K}^n \longrightarrow \mathbb{K}^m$$
$$\vec{\xi} \longmapsto f_A(\vec{\xi}) := A\vec{\xi} = a_s \xi^s \in \mathbb{K}^m.$$

So we have

$$\operatorname{rank}(A) := \operatorname{rank}(f_A) := \dim(\operatorname{im}(f_A)) \equiv \dim(\operatorname{im} A).$$

This definition corresponds to the column rank ($c\operatorname{rank}(A) \equiv \operatorname{rank}(A)$), as defined in the next section, Sect. 5.2.

Linear maps not only provide the notion of rank for matrices, they also provide the corresponding estimate for the rank of the product of two matrices. In an obvious notation, we define the following vector space and linear maps:
Let

$$V := \mathbb{K}^n, \qquad V' := \mathbb{K}^p, \qquad V'' := \mathbb{K}^m,$$
$$f := f_B, \qquad g := f_A, \qquad h := F_C := g \circ f,$$
$$\bar{g} := g|\operatorname{im} f$$

and suppose we have maps:

$$V \xrightarrow{f} V' \xrightarrow{g} V'',$$
$$V \xrightarrow{f} \operatorname{im} f \xrightarrow{\bar{g}} \operatorname{im} \bar{g}.$$

This leads to the following proposition.

Proposition 5.1 *Rank inequalities for the composition $g \circ f$.*

For the vector spaces V, V', V'' and the linear maps $f : V \to V'$ and $g : V' \to V''$, the following inequalities hold:

$$\text{rank}(f) + \text{rank}(g) - \dim V' \leq \text{rank}(g \circ f) \leq min\{\text{rank } f, \text{rank}(g)\}.$$

Proof Define $\bar{g} := g|_{\text{im } f}$. We consider the image of $\bar{g}$. Then we have:

$$\text{im } \bar{g} = \text{im}(g \circ f) \tag{5.14}$$

and the following inequalities:

$$\text{im } \bar{g} \subseteq \text{im } g \tag{5.15}$$

and

$$\dim(\text{im } \bar{g}) \leq \dim(\text{im } f) \quad \text{and} \quad \text{rank}(\bar{g}) \leq \text{rank}(f). \tag{5.16}$$

Equations (5.15) and (5.16) lead to

$$\text{rank}(g \circ f) \leq \min\{\text{rank}(f), \text{rank}(g). \tag{5.17}$$

To get the other inequality, consider the kernel of $\bar{g}$. Then

$$\ker \bar{g} = \ker g \cap \text{im } f. \tag{5.18}$$

The rank-nullity theorem gives

$$\dim(\ker g) = \dim V' - \text{rank}(g) \tag{5.19}$$

and the following inequalities:

$$\ker \bar{g} \leq \ker g \quad \text{and} \quad \dim(\ker \bar{g}) \leq \dim(\ker g). \tag{5.20}$$

Using the rank-nullity theorem and Eq. (5.16), this leads for the rank$(g \circ f)$ to

$$\text{rank}(g \circ f) = \text{rank}(\bar{g}) = \text{rank}(f) - \dim(\ker \bar{g}) \tag{5.21}$$

and to

$$\text{rank}(g \circ f) \geq \text{rank}(f) - \dim(\ker g). \tag{5.22}$$

Inserting Eq. (5.18), we obtain

$$\mathrm{rank}(g \circ f) \geq \mathrm{rank}(f) - \{\dim V' - \mathrm{rank}(g)\} \tag{5.23}$$

and

$$\mathrm{rank}(g \circ f) \geq \mathrm{rank}(f) + \mathrm{rank}(g) - \dim V'. \tag{5.24}$$

The inequalities of Eqs. (5.17) and (5.24) give

$$\mathrm{rank}(f) + \mathrm{rank}(g) - \dim V'' \leq \mathrm{rank}(g \circ f) \leq \min\{\mathrm{rank}(f), \mathrm{rank}(g)\}.$$

$\square$

This corresponds for matrices, using the above notation, to the corollary:

Corollary 5.1 *Rank inequalities.*

For matrices $A \in \mathbb{K}^{m \times p}$ and $B \in \mathbb{K}^{p \times n}$, and $AB \in \mathbb{K}^{m \times n}$, it holds that

$$\mathrm{rank}(A) + \mathrm{rank}(B) - p \leq \mathrm{rank}(AB) \leq \min\{\mathrm{rank}(A), \mathrm{rank}(B)\}.$$

$\square$

Comment 5.1 Rank inequality from matrix multiplication.

A direct application of the matrix multiplication, as in Eq. (5.11), leads also to the inequality:

$$\mathrm{rank}\, AB \leq \mathrm{rank}\, A. \tag{5.25}$$

Since we have

$$c_r = a_\mu \beta_r^{l\mu}, \tag{5.26}$$

every column of AB is a linear combination of the columns of A. This means that

$$\mathrm{span}\, AB \leq \mathrm{span}\, A$$

and

$$c\,\mathrm{rank}\, AB \leq c\,\mathrm{rank}\, A. \tag{5.27}$$

which is by definition $\mathrm{rank}\, AB \leq \mathrm{rank}\, A$. $\square$
See Definitions 5.3 and 5.4 in Sect. 5.2 below.

5.2 The Rank of a Matrix Revisited

Perhaps the most important parameter of a matrix is its rank. We would like to remember that if we consider a $m \times n$-matrix $A = (\alpha_s^i)$, it is, as we know, just a rectangular array with scalar (number) entries. We may think that A is a list of n vectors (columns),

$$A = (a_1, \ldots, a_s, \ldots, a_n), \ a_s \in \mathbb{R}^m, \ s \in I(n),$$

or that A is a list of m rows

$$(\alpha^1, \ldots, \alpha^i, \ldots, \alpha^m), \ \alpha^i \in (\mathbb{R}^n)^*, \ i \in I(m),$$

which we write vertically. We may call it a colist, this was already discussed in Sect. 3.1.

$$B = \begin{pmatrix} \alpha^1 \\ \cdot \\ \cdot \\ \alpha^i \\ \cdot \\ \cdot \\ \alpha^m \end{pmatrix}.$$

If we want to stress that by A we mean the matrix face of A (the matrix A, $A \equiv [A]$) or analogously for B, we write

$$A = [A] = [a_1 \cdots a_s \cdots a_m] \text{ or } [B] = \begin{bmatrix} \alpha^1 \\ \cdot \\ \cdot \\ \alpha^i \\ \cdot \\ \cdot \\ \alpha^m \end{bmatrix}.$$

Usually, we identify A with $[A]$ and $[B]$. But the list A is different from the colist B, $(A \neq B)$. The elements of the list A belong to $\mathbb{R}^m$ whereas the elements of the list B (list $B = \alpha^1, \ldots, \alpha^m$) belong to $(\mathbb{R}^n)^*$.

In what follows, we do not initially consider the matrix A as a map. We stay at a very elementary level and use only the terms linearly independent and span. These were introduced in Sect. 3.1.

We are only interested in the rank of the above two lists A and B. More precisely, we only want to compare the rank of the above different lists. The rank of a list is given by the dimension of the spanned subspace or, equivalently, by the size of a *maximal linearly independent sublist*. Therefore, we consider the two relevant subspaces which are related to the matrix A:

Definition 5.2 Column space and row space of the matrix A.

The column space $C(A) := \text{span}(a_1, \ldots, a_n) \le \mathbb{R}^m$;

The row space $R(A) := \text{span}(\alpha^1, \ldots, \alpha^m) \le (\mathbb{R}^n)^*$.

We have $R(A) \cong C(A^\mathsf{T}) \le (\mathbb{R}^n)$. Note that A^T is the column face of rows of A. This leads to the following definition:

Definition 5.3 Row and column rank.

The row rank of the matrix A is given by $r\,\text{rank}(A) := \dim C(A^\mathsf{T})$.
The column rank of the matrix A is given by $c\,\text{rank}(A) := \dim C(A)$.

It is further on clear that if we set $t := r\,\text{rank}$ and $c := c\,\text{rank}$, t is also the number of linearly independent rows and c is also the number of linearly independent columns.

We are now able to formulate a main theorem of elementary matrix theory.

Theorem 5.1 *Row and column rank, the first fundamental theorem of linear algebra.*

The row rank of a matrix is equal to the column rank: $r\,\text{rank}(A) = c\,\text{rank}(A)$, so that $t = c$.

Proof The goal is to try to reduce A to a specific $[t \times c]$-matrix $\tilde{A}$, with t linearly independent rows and r linearly independent columns if possible. Without loss of generality, we choose the first t rows to be linearly independent and we call the remaining, the linearly dependent rows, superfluous. Similarly, we choose the first r columns to be linearly independent and we call the rest (that is, the linearly dependent columns) superfluous. For the rows, in order to express the linear dependence, we split the index i in j and μ. Taking $\sigma \in I(n)$, $i \in I(m)$, $j \in I(t)$ and $\mu \in \{t+1, \ldots, m\}$ with $\rho_j^\mu \in \mathbb{K}$, we have:

$$\alpha^\mu = \rho_j^\mu \alpha^j \text{ and } \alpha_\sigma^\mu = \rho_j^\mu \alpha_\sigma^j. \tag{5.28}$$

For the columns, we consider the columns and their column rank. According to our choice above, we need to rearrange and write the r linearly independent columns first in the list and we have

$$c\,\text{rank}(A) = c\,\text{rank}(a_1, \ldots, a_r, a_{r+1}, \ldots a_n) = c\,\text{rank}(a_1, \ldots, a_r). \tag{5.29}$$

We may also write for the column, $\alpha_\sigma = \begin{pmatrix} \alpha_\sigma^j \\ \alpha_\sigma^\mu \end{pmatrix}$. The linear independence of the columns $(a_1, \ldots, a_s, \ldots, a_r)$, $(s \in I(r)$, and i, j, μ as above), means that: If

$$\alpha_s \lambda^s = 0 \quad \text{or equivalently}$$
$$\alpha_s^i \lambda^s = 0, \tag{5.30}$$
$$(\alpha_s^j \lambda^s = 0 \quad \text{and} \quad \alpha_s^\mu \lambda^s = 0)$$
$$\text{for all} \quad i \in I(n),$$

then it follows that

$$\lambda^s = 0 \quad \text{for all} \quad s \in I(r). \tag{5.31}$$

Now we throw out the $m - t$ superfluous rows and we are so left with the shortened columns which we denote by $\bar{A}_s$, the shortened matrix or list $\bar{A} = (\bar{a}_1, \ldots, \bar{a}_s, \ldots, \bar{a}_n)$.

The point is that the row operation above does not affect the column rank and we get the equality:

$$c \operatorname{rank}(\bar{a}_1, \ldots, \bar{a}_r) = c \operatorname{rank}(a_1, \ldots, a_r). \tag{5.32}$$

This is equivalent to the statement that also the shortened column list $(\bar{a}_1, \ldots, \bar{a}_r)$ is linearly independent like the given list $(a_1, \ldots, a_r)$. This means that we have to show the assertion:
If

$$\bar{a}_s \lambda^s = 0 \quad (\alpha_s^j \lambda^s = 0, \forall j \in I(t)), \tag{5.33}$$

it follows that

$$\lambda^s = 0 \quad \forall s \in I(r). \tag{5.34}$$

Given that the equation

$$a_s \lambda^s = 0 \quad \text{(see Eq. 5.30)}, \tag{5.35}$$

leads to the equation

$$\lambda^s = 0 \quad \text{(see Eq. 5.31)}, \tag{5.36}$$

it is sufficient to show that equation

$$\bar{a}_s \lambda^s = 0 \quad \text{(see Eq. 5.30)}, \tag{5.37}$$

leads to the equation

$$a_s \lambda^s = 0. \tag{5.38}$$

This can be shown as follows: The equations $a_s \lambda^s = 0$ or $\alpha_s^i \lambda^s = 0$ contain the equations $\bar{a}_s \lambda^s = 0$ or $\alpha_s^j \lambda^s = 0$, $j \in I(t), i \in I(m)$, $t < m$. So what is left, is to check the equations

$$\alpha_s^\mu \lambda^s = 0, \ \mu \in \{t+1, \ldots, m\}. \tag{5.39}$$

The next calculation shows that Eq. (5.39) is indeed valid.

From Eq. (5.28), we have $\alpha_s^\mu = \rho_j^\mu \alpha_s^j$. This gives us, with Eq. (5.39) and using Eq. (5.30),

$$\alpha_s^\mu \lambda^s = \rho_j^\mu \alpha_s^j \lambda^s = \varrho_j^\mu (\alpha_s^i \lambda^s) = 0. \tag{5.40}$$

So it is proven that $c \operatorname{rank}(\bar{A}) = r$ and $c \operatorname{rank}(\bar{A}) = c \operatorname{rank}(A)$.

This means that by row operations, the column rank stays invariant. The row rank stays in any case invariant by construction.

The result is that $\bar{A}$ is a $t \times n$-matrix with $c \operatorname{rank}(\bar{A}) = r$ and $r \operatorname{rank}(\bar{A}) = t$.

Using now the same procedure as before but interchanging the role of rows and columns, we can get rid of the superfluous columns to create a new matrix with the same row rank as originally.

This new matrix is now the $t \times r$-matrix $\bar{\bar{A}}$. In this matrix, both, the rows and the columns should be by construction linearly independent.

This is only possible in this case if the equation $t = r$ holds and so we proved the equation $r \operatorname{rank}(A) = c \operatorname{rank}(A)$.   □

The assertion of this theorem is highly nontrivial.

With this result, we can set the usual definition for the rank of a matrix.

Definition 5.4   Rank of a matrix.
$\operatorname{rank}(A) := c \operatorname{rank}(A)$.

We may think that the number of rank (A) is the quintessence of the matrix A and indicates the "true" size of A.

The following sections will justify this point of view.

5.3   A Matrix as a Linear Map

In this section, we will discuss the geometric properties of a $m \times n$-matrix F considered as a linear map and the role of the rank. We will then compare the general situation of the two abstract vector spaces V and V'. We consider for the sake of simplicity, $\mathbb{R}$ vector spaces. We will take into consideration that in $\mathbb{R}^n$ (and $\mathbb{R}^m$), we have the canonical Euclidean structure given by the dot product and the canonical basis $E = (e_1, \ldots, e_n)$.

The scalar product allows an enjoyable result: a unique decomposition where coim f, a subspace of $\mathbb{R}^n$ (coim $f \le \mathbb{R}^n$), is the (orthogonal) complement of ker f and coker f, a subspace of $\mathbb{R}^m$ (coker $f \le \mathbb{R}^m$, the (orthogonal) complement of im f given by the matrix F, acting as a linear map f. We identify f with F and, slightly misusing the notation (see Comment 5.2 below), we have:

$$\mathbb{R}^n = \ker f \oplus \operatorname{coim} f \xrightarrow{f} \operatorname{im} f \oplus \operatorname{coim} f = \mathbb{R}^m \quad \text{with}$$

$$\operatorname{coim} f \cong \operatorname{im} f \quad \text{or, equivalently,}$$

$$\dim \left(\operatorname{coim} f \right) = \dim(\operatorname{im} f) = \operatorname{rank} f.$$

So we have, more precisely:

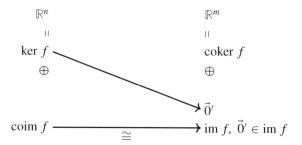

But coker $f \not\cong \ker f$ or equivalently dim(coker f) $\neq$ dim(ker f) since we have $n \neq m$ in general.

In an abstract vector space there is no such decomposition. In order to obtain a similar behavior, we have to consider, as we will see in Sect. 6.2, the dual spaces V^* and V'^* or to introduce explicitly a scalar product in V and V' (see below and in Sect. 6.3). Here, we have $V = \mathbb{R}^n$ and $V' = \mathbb{R}^m$. The linear map f is given in our smart notation with $r, s \in I(n)$ and $i \in I(m)$ by the matrix $F = (\varphi_s^i) = (f_s)_n = (\phi^i)_m$: $\varphi_s^i \in \mathbb{R}, \ f_r \in \mathbb{R}^m, \phi^i \in (\mathbb{R}^n)^*$:

$$f : \mathbb{R}^n \longrightarrow \mathbb{R}^m$$

$$x \longmapsto f(x) := Fx.$$

Fx is the matrix multiplication. If $E := (e_1, \ldots, e_n)$ and $E' := (e_1', \ldots, e_m')$ are the canonical bases in $\mathbb{R}^n$ and $\mathbb{R}^m$, we may also think the matrix F as a representation of the map f with respect to the bases E and E' and in our notation $F = f_{E'E} \in \mathbb{R}^{m \times n}$. In components (coefficients, coordinates), the map f is given by the following expressions:

$$\text{Setting} \quad y = f(x) \quad \text{with} \quad x = e_s \xi^s \equiv (\xi^s)_n \equiv \vec{\xi}$$

$$\text{and} \quad y = e_i' \eta^i \equiv (\eta^i)_m \equiv \vec{\eta}, \quad s \in I(n), \quad i \in I(m) \quad \text{and} \quad \xi^s, \eta^i \in \mathbb{R}.$$

$f(x)$ is given by $\eta^i = \varphi_s^i \xi^s$ or equivalently by $f(e_s) = e_i' \varphi_s^i, \ e_i' \in \mathbb{R}^m$. This is apparent in the following elementary steps:

$$y = e_i' \eta^i = f(x) = f(e_s \xi^s) = f(e_s) \xi^s = e_i' \varphi_s^i \xi^s = e_i' (\varphi_s^i \xi^s) \Rightarrow \eta^i = \varphi_s^i \xi^s.$$

Remark 5.2 *F* as a row-matrix with entries columns.

We have $f(e_s) = f_s$ and of course $F = [f_1 \ldots f_n]$, a $1 \times n$-matrix with entries columns.

Comment 5.2 The role of a rank (rank $f = r$).

To reveal the role of rank f, we would like to point out that given the map $f \in \mathrm{Hom}(V, V')$, with dim $V = n$ and dim $V' = m$, the subspaces ker f in V and im f in V' are always uniquely defined. If we want to consider two abstract vector spaces V and V', that is, without a scalar product and without using any specific bases, this leads to a nonunique decomposition of V and V'. So we may have for example:

$$V \cong \ker f \oplus U_1 \xrightarrow{f} \operatorname{im} f \oplus \Omega_1 \cong V' \quad \text{with}$$
$$V \cong \ker f \oplus U_2 \xrightarrow{f} \operatorname{im} f \oplus \Omega_2 \cong V' \quad \text{with}$$
$$U_1 \cong U_2 \cong \operatorname{im} f \qquad \text{or, equivalently,}$$
$$\dim U_1 = \dim U_2 = \dim(\operatorname{im} f) = \operatorname{rank} f = r.$$

In general, we have $n \neq m$ and

$$\Omega_1 \cong \Omega_2 \qquad \ncong \ker f \quad \text{or, equivalently}$$
$$m - r = \dim \Omega_1 = \dim \Omega_2 \neq \dim(\ker f) = n - r.$$

We see this directly if we take a tailor-made basis for f as in Proposition 3.9. It is usual to call the equivalence class $U_1 \cong U_2 \cong \ldots =: co \operatorname{im} f$ and the isomorphism class $\Omega_1 \cong \Omega_2 \ldots =: \operatorname{coker} f$. It is clear that coim f and coker f are uniquely defined since coim f is the quotient to the above equivalence relation. The same holds of ker f. So

$$V \cong \ker f \oplus \operatorname{coim} f \xrightarrow{f} \operatorname{im} f \oplus \operatorname{coker} f \cong V'.$$

Remark 5.3 The role of the transpose.

In the present case with $(\mathbb{R}^n, E)$ and $(\mathbb{R}^m, E')$, with E and E' the corresponding canonical (standard) basis in $\mathbb{R}^n$ and $\mathbb{R}^m$, the situation is now significantly better than in Comment 5.2. We have the transpose matrix $F^{\mathsf{T}} \in \mathbb{R}^{n \times m}$ and concretely this leads also to a unique decomposition. In our case, here, we have im $f = C(F) := \operatorname{span}(f_1, \ldots, f_n) \leqslant \mathbb{R}^m$ and ker $f = N(F) := \{x \in V : Fx = 0\} \leqslant \mathbb{R}^n$ with $C(F)$ the column space and $N(F)$ the null space of F.

F^{T} is closely connected with the dot products in $\mathbb{R}^n$ and $\mathbb{R}^m$. This gives

$$\mathbb{R}^m \xrightarrow{g} \mathbb{R}^n,$$

$$w \longmapsto g(w) := F^{\mathsf{T}} w.$$

So we get

$$\mathbb{R}^n \xrightarrow{f \equiv F} \mathbb{R}^m,$$

$$\mathbb{R}^n \xleftarrow{g \equiv F^{\mathsf{T}}} \mathbb{R}^m.$$

Now we have im $g = C(F^{\mathsf{T}})$. This is the row space of F in form of columns and it is a subspace of $\mathbb{R}^n$. The null space of $F^{\mathsf{T}} N(F^{\mathsf{T}}) = \ker g$ is a subspace of $\mathbb{R}^m$. As we already showed in Theorem 5.1, $\dim C(F^{\mathsf{T}}) = \dim C(F) = \operatorname{rank}(F) = r$. So we also obtain

$$\dim(\operatorname{im} f) = \dim(\operatorname{im} g) = \operatorname{rank} f = r. \tag{5.41}$$

In addition, we can show that im $g = C(F^{\mathsf{T}})$ and ker $f = N(F)$ are not only complementary but also orthogonal:

Given $z \in N(F)$ and $v \in C(F^{\mathsf{T}})$, we have with $w \in \mathbb{R}^n$:
$Fz = 0$ and $F^T w = v$. The transpose of the last equation is given by

$$v^{\mathsf{T}} = w^{\mathsf{T}} F. \tag{5.42}$$

Using Eq. (5.42) and $Fz = 0$ in $v^{\mathsf{T}} z$, we obtain $v^{\mathsf{T}} z = w^{\mathsf{T}} F z = 0$.

The equation $v^{\mathsf{T}} z = 0$ indicates the orthogonality $v \perp z$. So now the subspaces in $\mathbb{R}^n$, $N(F)$ and $C(F^{\mathsf{T}})$, are orthogonal and we may write

$$N(F) \perp C(F^{\mathsf{T}}) \tag{5.43}$$

or

$$\ker f \perp \operatorname{im} g. \tag{5.44}$$

Similarly, we obtain

$$C(F) \perp N(F^{\mathsf{T}}) \tag{5.45}$$

for $C(F)$ and $N(F^{\mathsf{T}}) \leq \mathbb{R}^m$.

It is clear that now coim $f = C(F^\mathsf{T})$ and coker $f = N(F^\mathsf{T})$ are uniquely determined. It is interesting that in this case we obtain with $\mathbb{R}^n$ and $\mathbb{R}^m$ not only uniquely the decomposition

$$\mathbb{R}^n = \ker f \oplus \operatorname{coim} f \xrightarrow{f} \operatorname{im} f \oplus \operatorname{coker} f = \mathbb{R}^m, \qquad (5.46)$$

but even more: the unique orthogonal decomposition denoted by the symbol $\ominus$:

$$\mathbb{R}^n = \ker f \ominus \operatorname{coim} f \xrightarrow{f} \operatorname{im} f \ominus \operatorname{coker} f. \qquad (5.47)$$

Since we have $f \equiv F$, we can also write this equivalently as

$$\mathbb{R}^n = \ker F \ominus \operatorname{im} F^\mathsf{T} \xrightarrow{F} \operatorname{im} F \ominus \ker F^\mathsf{T} = \mathbb{R}^m. \qquad (5.48)$$

So the following theorem was proven.

Theorem 5.2 *The fundamental theorem of linear maps.*
Given a matrix $F \in \operatorname{Hom}(\mathbb{R}^n, \mathbb{R}^m)$, the following orthogonal decomposition holds:
$$\mathbb{R}^n = \ker F \ominus \operatorname{im} F^\mathsf{T} \to \operatorname{im} F \ominus \ker F^\mathsf{T} = \mathbb{R}^m.$$

For good reasons, this theorem may be considered as the second fundamental theorem of elementary linear algebra. It states more precisely the situation at the beginning of this section, written as:

$$\mathbb{R}^n = \ker F \oplus \operatorname{coker} F \to \operatorname{im} F \oplus \operatorname{coker} F = \mathbb{R}^m.$$

The result of the above theorem may be represented symbolically by Fig. 5.1.

5.4 Vector Spaces and Matrix Representations

If we want to perform concrete calculations with abstract vectors, we have to use matrices. For vectors, it is standard to use columns and, for maps, rectangular matrices. In physics, we often start with concrete calculations, that is to say, we use matrices of the appropriate type or size from the beginning. Nevertheless, it is often useful and necessary to also take an overview of the abstract situation. So firstly, we are going to present an abstract vector space with columns, and the presentation of linear maps will be treated in the next section. Although we already covered most of the necessary background knowledge concerning vector spaces, especially in Sect. 3.2, we are now going to review and summarize briefly some of the results as this

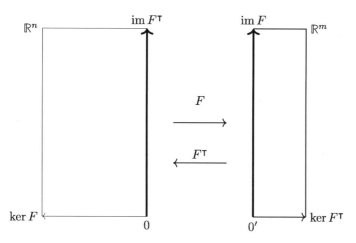

Fig. 5.1 The fundamental theorem of linear maps symbolically

will be a useful preparation for the presentation of linear maps too. Section 3.2 gives a in-depth knowledge, but what follows here is self-sufficient.

We start with an abstract vector space V with $\dim(V) = n$, and, as we already know, we need a basis $B = (b_1, \ldots, b_s, \ldots, b_n)$ in order to construct the appropriate representation. This provides the following basis isomorphism:

$$\phi_B : V \longrightarrow \mathbb{K}^n,$$
$$b_s \longmapsto \phi_B(b_s) := e_s \quad s \in I(n), \tag{5.49}$$

$$\text{or} \quad v \longmapsto \phi_B(v) := \begin{bmatrix} v^1 \\ \cdot \\ \cdot \\ v^n \end{bmatrix} = v_B = \vec{v}_B. \tag{5.50}$$

The elements of the list $(v^1, \ldots, v^s, \ldots, v^n)$, $v^s \in \mathbb{K}$ are called coordinates, coefficients, or components. With $E = (e_1, \ldots, e_s, \ldots, e_n)$, we denote the canonical basis in $\mathbb{K}^n$. In Eq. (5.50), we also use the notation $\vec{v}_B$ instead of v_B when we want to emphasize that v_B belongs to $\mathbb{K}^n$. The index B has to be used if there is also another basis involved. In Eq. (5.49), the basis isomorphism ϕ_B is given by the transformation from the basis list B to the basis list E (see Proposition 3.8):

$$(\phi_B(b_1), \ldots, \phi_B(b_s) \ldots, \phi_B(b_n)) := (e_1, \ldots, e_s, \ldots, e_n)$$

or

$$\phi_B(B) := E.$$

Comment 5.3 Comparison of the use of bases with the theory of manifolds.

In Sect. 3.2 we used exclusively the isomorphism ψ_B with $\psi_B^{-1} = \phi_B$, in linear algebra both are called basis isomorphisms. In the theory of manifolds, we use the notations (V, ϕ_B), a (global) chart and (V, ψ_B), a (global) parametrization.

We can now see the four faces of a basis:

- $B = (b_1, \ldots, b_n)$, a list,
- $[B] = [b_1 \cdots b_n]$, a matrix,
- ϕ_B, a chart,
- ψ_B, a parametrization,

and we write $v = \psi_B(\vec{v}) = [B]\vec{v} = [b_1 \cdots b_n] \begin{bmatrix} v^1 \\ \vdots \\ v^n \end{bmatrix}$.

We might now ask ourselves what the main purpose of a basis in practice is. With a given basis, we can replace an abstract vector space by a concrete vector space, and since this consists of number lists, we can calculate not only with vectors, but we can also send these vectors elsewhere, for example, from Nicosia to Berlin. This is possible if the observers at different positions previously agreed on the basis to be used. So we can think that we gained a lot. But what is the price for this gain?

We lose uniqueness: any other basis yields quite different values, coordinates for v. So we actually need to know how to go from a basis B to any other basis C.

In Sect. 3.2, we learned that the best way to think about the abstract vector v is to present v with all its representations simultaneously. This means to consider all the bases, $B, C, D, \ldots$, at the same time. But concretely, it is actually enough to use only one more basis, for example C, and to determine the transition from B to C. This means, we use the set of all bases $B(V)$ (think of relativity!). Thus, we can think that we can present this v with all its representations simultaneously and we know that we can reach and use all the bases in V, as shown and discussed in Sect. 3.2. We can describe this, using the bases B and C, in the commutative diagram in Fig. 5.2:

We choose a new basis $C = (c_1, \ldots, c_n)$ and the corresponding basis isomorphism ϕ_C. The transition map is given by $T_{CB} = \phi_C \circ \phi_B^{-1}$ which we identify immediately with the matrix $T := T_{CB} = (\tau_s^\mu)$, $\tau_s^\mu \in \mathbb{K}$. In what follows, we take $\mu, \nu, s, r \in I(n)$. The matrix T is invertible (so that $T \in Gl(n)$), so we have $TT^{-1} = \mathbb{1}_n$ and $T^{-1}T = \mathbb{1}_n$, with $T^{-1} = (\bar{\tau}_\mu^s)$, or equivalently

$$\tau_\mu^r \bar{\tau}_s^\mu = \delta_s^r \quad \text{and}$$
$$\bar{\tau}_s^\mu \tau_\nu^s = \delta_\nu^\mu.$$

Fig. 5.2 Vector space and representations B and C

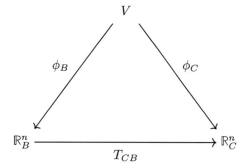

Since, with our smart indices, we distinguish clearly μ, ν, from s, r, we may write

$$\bar{\tau}^s_\mu \equiv \tau^s_\mu.$$

In this sense, $\bar{\tau}^s_\mu$ is a pleonasm and is used only if we want to emphasize that τ^s_μ belongs to T^{-1}. It is unfortunate that the notation can be confused with the transpose $T^\top$, and one should notice that generally, $T^{-1} \neq T^\top$.

Now, we can describe the change of basis also by this definition:

Definition 5.5 Change of basis.
  For a given basis $B = (b_1, \ldots, b_s, \ldots, b_n)$ and $C = (c_1, \ldots, c_\mu, \ldots, c_n)$ in V, we write $b_s = c_\mu \tau^\mu_s$ or equivalently $[B] = [C]T$:

$$[b_1 \cdots b_n] = [c_1 \cdots c_n] \begin{bmatrix} \tau^1_1 & \cdots & \tau^1_n \\ & \cdot & \\ \tau^m_1 & \cdots & \tau^m_n \end{bmatrix}.$$

The following lemma is commonly used in physics.

Lemma 5.1 *Change of basis.*
  The assertions (i) and (ii) are equivalent.

  (i) $b_s = c_\mu \tau^\mu_s$,
  (ii) $v^\mu_C = \tau^\mu_s v^s_B$,

with $v^\mu_C, v^s_B \in \mathbb{K}$ *or in matrix form* $\vec{v}_C = T \vec{v}_B$.

Proof (i) $\Rightarrow$ (ii)
Given $v = b_s v^s_B$ and $v = c_\mu v^\mu_C$, we have

$$v = b_s v_B^s \overset{(i)}{=\!=\!=} c_\mu T_s^\mu v_B^s \overset{!}{=\!=\!=} c_\mu v_C^\mu$$
$$\Longrightarrow v_C^\mu = T_s^\mu v_B^s$$

which proves (ii). $\qquad\qquad\qquad\qquad\qquad\qquad\qquad\qquad\qquad\qquad\qquad\square$

Proof (ii) $\Rightarrow$ (i)
Given

$$v = c_\mu v_C^\mu \overset{(ii)}{=\!=\!=} c_\mu T_s^\mu v_B^s,$$
$$\text{or} \quad v = b_s v_B^s \quad \forall \quad v_B \in \mathbb{K}^n,$$

we obtain

$$b_s = c_\mu T_s^\mu$$

which proves (i). $\qquad\qquad\qquad\qquad\qquad\qquad\qquad\qquad\qquad\qquad\qquad\square$

Comment 5.4 Right and left action.

It is interesting to realize that the matrix T acts on the basis from the right and on the coefficient vectors from the left:

$$[B] = [C]T \Leftrightarrow [C] = [B]T^{-1} \quad \text{and}$$
$$\vec{v}_C = T\vec{v}_B \Leftrightarrow \vec{v}_B = T^{-1}\vec{v}_C.$$

This is also an example for the discussion in Sect. 1.3, particularly for Remark 1.3.

Comment 5.5 $T_{CB} = id_{CB}$.

If we want we can use an equivalent diagram with a different notation from the one given in Fig. 5.2, for the same situation.

$$
\begin{array}{ccc}
V & \xrightarrow{\ id_V\ } & V \\
\phi_B \downarrow & & \downarrow \phi_C \\
\mathbb{R}_B^n & \xrightarrow[\ id_{CB}\ ]{} & \mathbb{R}_C
\end{array}
$$

and we have

$$id_{CB} \circ \phi_B = \phi_C \circ id_V.$$

The meaning of T_{CB} is $T_{CB} \equiv id_{CB}$. This is what we call in physics a passive symmetry transformation. It corresponds to a coordinate change while the physical system remains fixed.

In the theory of manifolds as well as in linear algebra, we have a cycle relation:

Lemma 5.2 *Composition of change of basis.*
If we have $v_C = T_{CB} v_B$ and $v_D = T_{DC} v_C$, $B, C, D \in B(V)$, the cycle relation $T_{DB} = T_{DC} T_{CB}$, holds.

This is explained in the following commutative diagram (Fig. 5.3).

Remark 5.4 The set of all representations of V.

It is clear that the set of all representations of V is given by the set of bases in V, $B(V) = \{B\}$. If we consider the set of all bases in V, every basis B leads to a basis isomorphism ϕ_B or to a linear chart (V, ϕ_B). It is also called a representation of the vector space V by $n \times 1$-matrices or by columns of length n, or by the coefficient vectors in $\mathbb{K}^n$.

As discussed in Eqs. (5.49) and (5.50) at the beginning of this section, if we want, we can also use the same notation for all these representations with the letter M (like matrix) and so get for every $B \in B(V)$ the isomorphism

$$M : V \longrightarrow \mathbb{K}^n \equiv \mathbb{K}^{n \times 1},$$

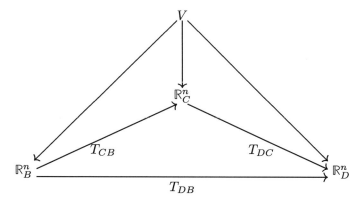

Fig. 5.3 Composition of change of basis: $T_{DB} = T_{DC} \circ T_{CB}$

$$v \longmapsto M(v) = \vec{v}.$$

In this sense, we here use a "universal" notation. In the next section, we shall see that we may use the same universal notation for the representation of linear maps too.

5.5   Linear Maps and Matrix Representations

In this section, we use a similar notation and apply the results of the previous sections.

For the representation of linear maps, we return now to the general case where V and V' are vector spaces without further structure. In order to describe a given linear map

$$f : V \longrightarrow V', \, f \subset \mathrm{Hom}(V, V'),$$

we have to choose a basis $B = (b_1, \ldots, b_n)$ and $B' = (b'_1, \ldots, b'_m)$ corresponding to V and V'. The connection with the matrix $F \equiv f_{B'B} \in \mathbb{K}^{m \times n}$ which represents the map, is shown in the following cumulative diagram:

$$
\begin{array}{ccc}
V & \xrightarrow{\ f\ } & V' \\
{\scriptstyle \phi_B}\big\downarrow & & \big\downarrow{\scriptstyle \phi_{B'}} \\
\mathbb{R}^n_B & \xrightarrow[\ F\ ]{} & \mathbb{R}^m_{B'}
\end{array}
$$

So we have $\phi_B(b_s) = e_s$ and $\phi'_B(b'_i) = e'_i$ with $s \in I(n)$ and $i \in I(m)$. The map f is, according to Proposition 3.8 in Sect. 3.3, determined uniquely by the values of the basis B:

$$f(b_s) = b'_i \varphi^i_s, \quad \varphi^i_s \in \mathbb{K}. \tag{5.51}$$

This also determines uniquely the matrix

$$f_{B'B} \equiv F := (\varphi^i_s). \tag{5.52}$$

Equation (5.51) can also be written as follows:

$$f(b_1, \ldots, b_n) = (f(b_1), \ldots, f(b_n)) \tag{5.53}$$

$$\text{or}$$

$$f[b_1 \ldots b_n] = [f(b_1) \ldots f(b_n)],$$
$$f[b_1 \ldots b_n] = [b'_s \varphi^s_1 \ldots b'_s \varphi^s_n]$$

$$\text{and}$$

$$f[b_1 \ldots b_n] = [(b'_1) \ldots (b'_m)] F \tag{5.54}$$

$$\text{or}$$

$$f[B] = [B']F. \tag{5.55}$$

Equations (5.54) and (5.55) correspond to the above diagram:

$$\phi_{B'} \circ f = F_{B'B} \circ \phi_B \quad \text{or} \tag{5.56}$$
$$\phi_{B'} \circ f = F \circ \phi_B. \tag{5.57}$$

This gives also

$$F = \phi_{B'} \circ f \circ \phi_B^{-1}. \tag{5.58}$$

As mentioned in the previous Sect. 5.4, (ϕ_B, V) and $(\phi_{B'}, V')$ are what is called in the theory of manifolds (local) charts. In linear algebra, these are of course global charts. Equations (5.51) and (5.52) exposed in the form of a diagram as above, give:

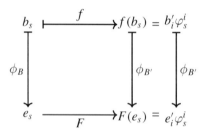

So we have from $\phi_B(b_s) = e_s$, $\phi_{B'}(b'_i) = e'_i$, using linearity and Eqs. (5.51) and (5.52),

$$\phi_{B'}(f(b_s)) = \phi_B(b'_i \varphi^i_s) = \phi_{B'}(b'_i)\varphi^i_s \quad \text{and} \tag{5.59}$$
$$\phi_{B'}(b'_i \varphi^i_s) = e'_i \varphi^i_s. \tag{5.60}$$

This justifies the equations in the above diagram and the following correspondence:

$$\begin{array}{ccc} b_s & f & b_i \\ B \updownarrow & \updownarrow & \updownarrow B' \\ e_s & F & e'_i \end{array}$$

This shows, stated in simple words, that we can perfectly describe what happens at the "top" level of $V \xrightarrow{f} V'$, at the "bottom" level of $\mathbb{R}^n \xrightarrow{F} \mathbb{R}^m$. This is the essence of the representation theory of linear maps. Using here in addition the universal notation with the symbol M, we can write

$$v \xmapsto{F} w \quad = F(v),$$
$$M(v) \xmapsto{M(f)} M(w) = M(f)M(v)$$
$$\text{or} \quad \vec{v}_B \longmapsto \vec{w}_{B'} \quad = F\vec{v}_B. \tag{5.61}$$

We obtain Eq. (5.62) using again Eqs. (5.51) and (5.52):

$$f(b_s v_B^s) = f(b_s)v_B^s = b_i'\varphi_s^i v_B^s \overset{!}{=} b_i' w_{B'}^i.$$ (5.62)

So we have

$$w_{B'}' = \varphi_s^i v_{B'}^s, \ w_B^i, \ v_B^s \in \mathbb{K}$$ (5.63)

which is

$$\vec{w}_{B'}' = F\vec{v}_B$$ (5.64)

with $\vec{v}_B \in \mathbb{K}^n$ and $\vec{w}_{B'} \in \mathbb{K}^m$. If we consider the various faces of the representation matrix $F \in \mathbb{K}^{m \times n}$, we can write:

$$F = [\varphi_s^i] = [f_1 \cdots f_s \cdots f_n] = \begin{bmatrix} \phi^1 \\ \cdot \\ \cdot \\ \phi^i \\ \cdot \\ \cdot \\ \phi^m \end{bmatrix}.$$ (5.65)

Since in the above diagram, we already have the equation $Fe_s = e_i'\varphi_s^i$, it is obvious that

$$Fe_s = f_s.$$ (5.66)

This means that the sth columns of the matrix F which represents the map f, is the value of the canonical basis vectors $e_s \in \mathbb{K}^n$. In addition, it means that the sth column of F gives the coefficients of the value of the sth basis vector b_s in V. These coefficients correspond to the basis B' in V', as expected.

At this stage, we have to explain how F changes by transforming the basis B into the basis C in V and the basis B' into the basis C' in V'. For simplicity's sake we call the new matrix $\bar{F} := f_{C'C}$ and so have to consider the transition from F to $\bar{F}$. This is given by the following proposition.

Proposition 5.2 *Representation change by bases change.*

From the bases B and C in V and B' and C' in V', the matrix $F_{C'C}$ of f is given by $F_{C'C} = T'F_{B'B}T^{-1}$ where the corresponding transition matrices

$$T \equiv T_{CB} \quad and \quad T' \equiv T_{C'B'}$$

are given by

$$B = CT \quad and \quad B' = C'T'.$$

Proof The proof is given by the two commutative diagrams:

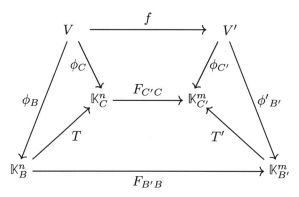

The second diagram leads to $f_{C'C} \circ T = T' \circ f_{B'B}$ and to $f_{C'C} = T' \circ f_{B'B} \circ T^{-1}$.

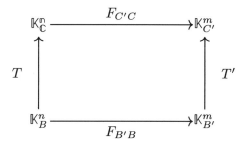

We can also achieve the above result directly using only the tensor formalism: Putting the right indices in the right place, we obtain

$$r, s \in I(n) \quad i, j \in I(m), \quad T = (\tau_s^r), \quad T' = (\tau_i^j),$$
$$f_{B'B} \equiv F_B = (\varphi_s^i) \quad f_{C'C} \equiv F_C = (\eta_i^j).$$

We start with $F_B = (\varphi)_s^i$ and we want to obtain $F_C = (\eta)_r^j$, using $T = (\tau_r^s)$ and $T' = (\tau_i'^j)$. We so obtain

$$\eta_r^j = \tau_i'^j \tau_r^s \varphi_s^i = \tau_i'^j \varphi_s^i \tau_r^s. \tag{5.67}$$

This corresponds to

$$F_C = T' F_B T^{-1}. \tag{5.68}$$

Usually, if the involved bases are evident from the context, they are often not included in the notation. Here, seizing this opportunity, we would like to express all the relevant isomorphisms by the same letter M, using in a way a universal notation:

$$M : \mathrm{Hom}(V, V') \longrightarrow \mathbb{K}^{m \times n},$$
$$f \longmapsto M(f) = F,$$
$$M : V \longrightarrow \mathbb{K}^{n \times 1},$$
$$v \longmapsto M(v) = \vec{v},$$
$$M : V' \longrightarrow \mathbb{K}^{m \times 1},$$
$$w \longmapsto M(w) = \vec{w}.$$

So the equation $f(v) = w$ can be represented by

$$M(f(v)) = M(f)\, M(v) \in \mathbb{K}^{m \times 1}, \tag{5.69}$$

or

$$M(w) = M(f)\, M(v), \tag{5.70}$$

and

$$\vec{w} = F\vec{v}. \tag{5.71}$$

5.6   Linear Equation Systems

We now turn to perhaps the very first applications of linear algebra, which is also one of the most important: solving a system of linear equations. With the results from the previous chapters, tackling this problem and providing the corresponding proofs is quite straightforward.

Definition 5.6 Linear system of equations.
Let A be a coefficient matrix $A \in \mathbb{K}^{m \times n}$ given by $A = (a_s)_s = (\alpha_s^i)$ with $\alpha_s^i \in \mathbb{K}$, $a_s \in \mathbb{K}^m$, $i \in I(m)$ and $s \in I(n)$. A linear system of equation is given by the following three forms:

(i) $\alpha_s^i \xi^s = \beta^i$: tensor form;
(ii) $a_s \xi^s = b$, $b = (\beta^i)_m \in \mathbb{K}^m$: vector form;
(iii) $Ax = b$, $x = \xi \in \mathbb{K}^n$: matrix form.

Every $x_b \equiv \vec{\xi}_b \in \mathbb{K}^n$ with $Ax_b = b$ is called a solution of the linear system. We denote the set of all solutions of the linear system by

$$\mathcal{L}(A, b) := \{x_b : Ax_b = b\}.$$

We call the equation of the form $Ax = 0$ with $b = 0$ homogeneous, and when $b \neq 0$, we call it inhomogeneous. We denote the set of solutions of the homogeneous equation by $\mathcal{L}(A) = \mathcal{L}(A, 0)$.

After formulating this system of equations, the following questions arise:

(i) *Existence.* Under what conditions on A and b is there an x so that $Ax = b$ holds?
(ii) *Universal solvability.* Is the equation $Ax = b$ solvable for all $b \in \mathbb{K}^m$?
(iii) *Unique solvability.* If there is a solution, when is it unique?
(iv) *Representation of solutions of the linear systems of equations.* How do we describe all solutions of the equation system?

With the results provided so far, these questions can already be answered. This leads to the following corresponding propositions. Most proofs are quite direct and are left as exercises to the reader. Denoting by f_A the corresponding map to the matrix or list A, $f_A : \mathbb{K}^n \to \mathbb{K}^m$, we get the following propositions.

Proposition 5.3 *Existence.*
The following assertions are equivalent:

(i) *The equation system $Ax = b$ is solvable.*
(ii) $b \in \operatorname{im} f_A \equiv \operatorname{span} A$.
(iii) $\operatorname{rank} A = \operatorname{rank}(A, b)$.
(iv) *Whenever $y \in \mathbb{K}^m$, $A^\mathsf{T} y = 0 \Leftrightarrow b^\mathsf{T} y = 0$.*

Proposition 5.4 *Universal solvability.*
The following assertions are equivalent:

(i) *The equation system $Ax = b$ is universally solvable.*
(ii) $\operatorname{rank} A = m$.
(iii) $\operatorname{im} f_A = \mathbb{K}^m$.

Proposition 5.5 *Unique solvability.*
The following assertions are equivalent:

(i) *The equation system $Ax = b$ is uniquely solvable.*
(ii) $\ker f_A = \{0\}$.
(iii) $\operatorname{rank} A = n$.

(iv) *The homogeneous equation system has only the trivial solution:*

$$\mathcal{L}(A) = \{0\}.$$

Proposition 5.6 *Representation of solutions of the linear systems of equations.*

(i) *The set $\mathcal{L}(A)$ of the solutions x_0 of the homogeneous system of equations, $Ax_0 = 0$, is a subspace of $\mathbb{K}^n$, $\mathcal{L}(A) \le \mathbb{K}^n$ and its dimension is $\dim \mathcal{L}(A) = n - \operatorname{rank} A$.*

(ii) *If p_b is a solution of $Ax = b$, then all solutions of $Ax = b$ are given by:*

$$x_b = p_b + x_0 \quad \text{with} \quad x_0 \in \mathcal{L}(A) = \ker f_A.$$

So we have

$$\mathcal{L}(A, b) = p_b + \mathcal{L}(A).$$

This means that $\mathcal{L}(A, b)$ is an affine space with the corresponding subspace $\mathcal{L}(A) \le \mathbb{K}^n$, with

$$\dim \mathcal{L}(A) = n - \operatorname{rank} A.$$

Proposition 5.7 *Linear system of equations for $n = m$.*
If A is a coefficient matrix $A \in \mathbb{K}^{n \times n}$, then the following assertions are equivalent:

(i) *$Ax = b$ is solvable for any $b \in \mathbb{K}$.*

(ii) *$Ax = b$ is uniquely solvable for a $b \in \mathbb{K}^n$.*

(iii) *$Ax = 0$ possesses only the trivial solution.*

(iv) *$Ax = b$ is uniquely solvable for any $b \in \mathbb{K}^n$.*

(v) *A is invertible and the solution of $Ax_b = b$ is given by $x_b = A^{-1}b$.*

If we take the opportunity and utilize tailor-made bases for the map

$$f_A : \mathbb{K}^n \to \mathbb{K}^m,$$

with corresponding matrix $A \in \mathbb{K}^{m \times n}$ to give us a decomposition $F'AF = \tilde{A}$ with F' and F invertible matrices of rank m and n respectively, and $\tilde{A}$ of the form

$$\tilde{A} = \begin{bmatrix} \mathbb{1}_r & 0 \\ 0 & 0 \end{bmatrix} \in \mathbb{K}^{m \times n}.$$

This way, one can solve the equation system

$$\tilde{A}\tilde{x}_b = \tilde{b}$$

very comfortably: We first get the result

$$\mathcal{L}(A, b) = \mathcal{L}(F'A, F'b) \equiv \mathcal{L}(F'A, \tilde{b}).$$

The following proposition summarizes the solution for this case.

Proposition 5.8 *Tailor-made solution of the linear equation system.*
Suppose $Ax = b$ is solvable. Let F' and F be as before, and $r = \text{rank } A$, then

$$F'b = \begin{bmatrix} \bar{b} \\ 0 \end{bmatrix} \quad \text{where} \quad \bar{b} \in \mathbb{K}^r.$$

The general solution has the form:

$$x_b = F \begin{bmatrix} \bar{b} \\ \lambda \end{bmatrix} \quad \text{with} \quad \lambda \in \mathbb{K}^{n-r} \quad \text{and} \quad \bar{b} \in \mathbb{K}^r.$$

Proof We have

$$Ax_b = b \Leftrightarrow$$
$$F'Ax_b = F'b \Leftrightarrow$$
$$F'Ax_b = \begin{bmatrix} \bar{b} \\ 0 \end{bmatrix} \quad \text{with} \quad \bar{b} \in \mathbb{K}^r \quad \text{and}$$
$$\tilde{b} := F'b = \begin{bmatrix} \bar{b} \\ 0 \end{bmatrix}. \tag{5.72}$$

This shows the first statement of the proof.
For the general solution of the linear equation, setting

$$\tilde{x}_b := F^{-1}x_b \Leftrightarrow x_b = F\tilde{x}_b, \tag{5.73}$$

leads to

$$Ax_b = b \iff$$
$$F'AF\tilde{x}_b = \tilde{b} \iff$$
$$F'AFF^{-1}x_b = F'b \iff$$
$$\begin{bmatrix} \mathbb{1}_r & 0 \\ 0 & 0 \end{bmatrix} \tilde{x}_b = F'b.$$

The result corresponds to an improved form of the reduced row echelon form. The image of the matrix $\begin{bmatrix} \mathbb{1}_r & 0 \\ 0 & 0 \end{bmatrix}$ is of the required form, so $F'b$ is in the image if and only if $F'b = \begin{bmatrix} b \\ 0 \end{bmatrix}$.

Solving the above equation, we obtain

$$\xi = \bar{b} \quad \text{and} \quad \lambda \in \mathbb{K}^{n-r} \quad \text{with} \quad \lambda \text{ free},$$

This leads to

$$\tilde{x}_b = \begin{bmatrix} \bar{b} \\ \lambda \end{bmatrix}.$$

Using Eqs. (5.72) and (5.73), we obtain

$$x_b = F \begin{bmatrix} \tilde{b} \\ \lambda \end{bmatrix}$$

which proves the proposition.   □

Summary

What is a matrix? This is where we started. By the end of this chapter, hopefully the following became clear: a matrix is what you can do with it! In short, linear algebra is what you can do with matrices.

Initially, we focused on matrix multiplication and its relationship to the composition of linear maps. We explored the four facets of matrix multiplication. A fundamental theorem of linear algebra is the fact that the row rank of a matrix equals its column rank. Many aspects related to this theorem are also considered in the exercises.

The various roles of a matrix, as a linear map, as a basis isomorphism (as a supplier of coordinates), as a representation matrix of abstract vectors and linear maps, along with the associated change of basis formalism, were discussed extensively.

This led to the fundamental theorem of linear maps which relates the kernels and images of a matrix with those if its transpose.

Exercises with Hints

Exercise 5.1 *Preservation of rank under multiplication by invertible matrices (see Comment 5.1).*
For the matrices $A \in \mathbb{K}^{m \times n}$, $F' \in GL(m, \mathbb{K})$, and $F \in GL(n, \mathbb{K})$, show that

$$\text{rank}(F'A) = \text{rank}(A), \text{rank}(AF) = \text{rank}(A) \quad \text{and}$$
$$\text{rank}(F'AF) = \text{rank } A.$$

Exercise 5.2 *As the dot product is positive definite, it leads directly to the following equality for the null space.*
For any matrix $A \in \mathbb{K}^{n \times n}$, show that A and $A^\dagger A$ have the same null space, that is, $\ker A^\dagger A = \ker A$.

> In the next five exercises, you are asked to prove once more the fundamental theorem of linear algebra which yields the equality of row and column ranks of a matrix. All these proofs, for which one needs only very elementary facts, show very interesting aspects of the structure of matrices.

Exercise 5.3 *The row—column—rank theorem.*
Choose a basis B for the column space of a matrix $A \in \mathbb{K}^{m \times n}$. Show that the factorization $A = BC$ is unique. By comparison with $A^\mathsf{T} = C^\mathsf{T} B^\mathsf{T}$, show that the row rank of A equals its column rank.

Exercise 5.4 *The row—column—rank theorem.*
If the column rank of a matrix $A \in \mathbb{K}^{m \times n}$ is r, take the first r columns of A to be linearly independent and write B_r for the corresponding basis of the column space of A. Show that now the factorization $A = B_r C$ gives directly the result row rank C = column rank $A = r$.

Exercise 5.5 *The row—column—rank theorem.*
Choose the row rank of a matrix $A \in \mathbb{K}^{m \times n}$ to be t and rearrange the rows so that the first t rows are linearly independent. Show by inspection that each column of A is a linear combination of the standard basis vector $e'_1, \ldots, e'_t \in \mathbb{K}^m$ and that

$$\text{column rank } A \leq \text{row rank } A$$

and furthermore that

$$\text{row rank } A = \text{column rank } A.$$

Exercise 5.6 *The row—column—rank theorem.*
Use the results of the Exercises 5.1 and 5.2 to show that

$$\operatorname{rank} A = \operatorname{rank} A^{\mathsf{T}} A \le \operatorname{rank} A^{\mathsf{T}}$$

and that

$$\operatorname{rank} A^{\mathsf{T}} \le \operatorname{rank} A ,$$

which leads to

$$\operatorname{rank} A^{\mathsf{T}} = \operatorname{rank} A ,$$

and thus to

$$\operatorname{row} \operatorname{rank} A = \operatorname{column} \operatorname{rank} A .$$

Exercise 5.7 *The change of a basis matrix.*

$$T \equiv T_{CB} = (\tau_s^i)$$

as given in Comments 5.4 and 5.5, can also be directly expressed by the coordinates of the basis vector b_s, $s \in I(n)$ of the basis $B = (b_1, \ldots, b_n)$ in the vector space V. Show that the matrix T_{CB} in the notation of Comments 5.4 and 5.5 is given by

$$T_{CB} = [\phi_C(b_1) \ldots \phi_C(b_n)].$$

Similarly, the matrix T_{BC} is given by

$$T_{BC} = [\phi_B(c_1) \ldots \phi_B(c_n)].$$

Exercise 5.8 *The representation of a linear map $f \in \operatorname{Hom}(V, V')$ by using the canonical basis isomorphism between V and $\mathbb{K}^n$ (V' and $\mathbb{K}^m$) with basis $B = (b_1, \ldots, b_n)$ in a vector space V ($B' = (b'_1, \ldots, b'_m)$ in V') can be expressed symbolically as follows:*

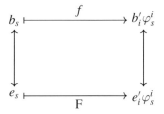

with $s \in I(n)$, $i \in I(m)$, $F = (\varphi_s^i)$, $\varphi_s^i \in \mathbb{K}$, $e_s \in \mathbb{R}^n$, and $e'_i \in \mathbb{R}^m$.
Explain the following relation using the corresponding commutative diagram.

$$f(b)_s = b'_i \varphi_s^i \Leftrightarrow F e_s = e'_i \varphi_s^i.$$

Exercise 5.9 *Rank of a linear map and its representing matrix.*
Suppose the rank of a linear map $f \in \text{Hom}(V, V')$ is r. Show that there exists at least one basis in V and one basis in V' such that, relative to these bases, only the first r columns and the first r rows of the matrix $F = M(f)$ of f are nonzero.

Exercise 5.10 Let A and B be two matrices for which addition and multiplication are defined. Show that

$$(AB)^{\mathsf{T}} = B^{\mathsf{T}} A^{\mathsf{T}} \quad \text{and} \quad ((A)^{\mathsf{T}})^{\mathsf{T}} = A.$$

Exercise 5.11 Let A and B be two invertible matrices in $\mathbb{K}^{n \times n}$.
Show that $(AB)^{-1} = B^{-1} A^{-1}$.

Exercise 5.12 Let $f \in \text{Hom}(V, V')$ and $B = (b_1, \dots, b_n)$ be a basis of V and $B' = (b'_1, \dots, b'_n)$ a basis of V'. Prove that the following are equivalent.

 (i) f is invertible;
 (ii) The columns of the matrix $M(f) = F$ are linearly independent in $\mathbb{K}$;
 (iii) The columns of F span $\mathbb{K}^n$.

> The following five exercises concern systems of linear equations. It is a collection of well-known results. The proof can also be seen as a fairly direct application of Chaps. 2, 3, and this chapter.

Exercise 5.13 *Existence.*
Prove that the following assertions are equivalent:

 (i) The equation system $Ax = b$ is solvable.
 (ii) $b \in \text{im } f_A \equiv \text{span } A$.
 (iii) $\text{rank } A = \text{rank}(A, b)$.
 (iv) If $y \in \mathbb{K}^m$, $A^{\mathsf{T}} y = 0 \Leftrightarrow b^{\mathsf{T}} y = 0$.

Exercise 5.14 *Universal solvability.*
Prove that the following assertions are equivalent:

 (i) The equation system $Ax = b$ is universally solvable.
 (ii) $\text{rank } A = m$.
 (iii) $\text{im } f_A = \mathbb{K}^m$.

Exercise 5.15 *Unique solvability.*
Prove that the following assertions are equivalent:

 (i) The equation system $Ax = b$ is uniquely solvable.
 (ii) $\ker f_A = \{0\}$.
 (iii) $\text{rank } A = n$.
 (iv) The homogeneous equation system has only the trivial solution.

Exercise 5.16 *Representation of solutions of the linear systems of equations.*
Prove the two following assertions:

(i) The set $\mathcal{L}(A)$ of the solutions x_0 of the homogeneous system of equations, $Ax_0 = 0$, is a subspace of $\mathbb{K}^n$, $\mathcal{L}(A) \leq \mathbb{K}^n$ and its dimension is $\dim \mathcal{L}(A) = n - \operatorname{rank} A$.
(ii) If p_b is a solution of $Ax = b$, then all solutions of $Ax = b$ are given by:

$$x_b = p_b + x_0 \quad \text{with} \quad x_0 \in \mathcal{L}(A) = \ker f_A.$$

So we have
$$\mathcal{L}(A, b) = p_b + \mathcal{L}(A).$$

This means that $\mathcal{L}(A, b)$ is an affine space with the corresponding subspace $\mathcal{L}(A) \leq \mathbb{K}^n$. With $r = \operatorname{rank} A$, we have

$$\dim \mathcal{L}(A) = n - r.$$

Exercise 5.17 *Linear system of equations for $n = m$.*
Let A be a coefficient matrix $A \in \mathbb{K}^{n \times n}$. Prove that then the following assertions are equivalent:

(i) $Ax = b$ is solvable for any $b \in \mathbb{K}$.
(ii) $Ax = b$ is uniquely solvable for a $b \in \mathbb{K}^n$.
(iii) $Ax = 0$ possesses only the trivial solution.
(iv) $Ax = b$ is uniquely solvable for any $b \in \mathbb{K}^n$.
(v) A is invertible and the solution of $Ax_b = b$ is given by $x_b = A^{-1}b$.

References and Further Reading

1. S. Axler, *Linear Algebra Done Right* (Springer Nature, 2024)
2. S. Bosch, *Lineare Algebra* (Springer, 2008)
3. G. Fischer, B. Springborn, *Lineare Algebra. Eine Einführung für Stu-dienanfänger*. Grundkurs Mathematik (Springer, 2020)
4. S.H. Friedberg, A.J. Insel, L.E. Spence, *Linear Algebra.* (Pearson, 2013)
5. S. Hassani, *Mathematical Physics: A Modern Introduction to its Foundations* (Springer, 2013)
6. K. Jänich, *Mathematik 1. Geschrieben für Physiker* (Springer, 2006)
7. N. Johnston, *Advanced Linear and Matrix Algebra* (Springer, 2021)
8. N. Johnston, *Introduction to Linear and Matrix Algebra* (Springer, 2021)
9. M. Koecher, *Lineare Algebra und analytische Geometrie* (Springer, 2013)
10. G. Landi, A. Zampini, *Linear Algebra and Analytic Geometry for Physical Sciences* (Springer, 2018)
11. J.M. Lee, *Introduction to Smooth Manifolds*. Graduate Texts in Mathematics (Springer, 2013)
12. J. Liesen, V. Mehrmann, *Linear Algebra* (Springer, 2015)
13. P. Petersen, *Linear Algebra* (Springer, 2012)

14. S. Roman, *Advanced Linear Algebra* (Springer, 2005)
15. B. Said-Houari, *Linear Algebra* (Birkhäuser, 2017)
16. A.J. Schramm, *Mathematical Methods and Physical Insights: An Integrated Approach* (Cambridge University Press, 2022)
17. G. Strang, *Introduction to Linear Algebra* (SIAM, 2022)
18. R.J. Valenza, *Linear Algebra. An Introduction to Abstract Mathematics* (Springer, 2012)

Chapter 6
The Role of Dual Spaces

As mentioned in Sect. 5.3, a linear map f between $\mathbb{K}^n$ and $\mathbb{K}^m$ determines uniquely the four subspaces ker f, coim $f \leq \mathbb{K}^n$, coker f, and im $f \leq \mathbb{K}^m$ which give essential information about this map.

If we consider f between two abstract vector spaces V and V' with no additional structure, coker f and coim f are not uniquely defined. If we want subspaces fixed similarly as for $f \in \mathrm{Hom}(\mathbb{K}^n, \mathbb{K}^m)$, we have to additionally consider the dual spaces of V and V', V^* and V'^*. This is what we are going to show in what follows. Beforehand, we have to note that dual spaces play a unique role in linear algebra and a crucial role in many areas of mathematics, for example, in analysis and functional analysis. They are essential for a good understanding of tensor calculus; they appear in special and general relativity and are ubiquitous in quantum mechanics. The Dirac notation in quantum mechanics is perhaps the best demonstration of the presence of dual spaces in physics.

6.1 Dual Map and Representations

We consider a vector space V and its dual $V^* = \mathrm{Hom}(V, \mathbb{K})$ with $\dim V = \dim V^* = n$. An element of V^* is called a linear form or sometimes a linear function, linear functional, one form or even covector. We denote the elements of V^* by small greek letters $\alpha, \beta, \cdots, \xi, \eta, \cdots$, taking care not to confuse them with the notation of scalars which we also denote by small greek letters. For $\xi \in V^*$ we have

$$\xi : \ V \longrightarrow \mathbb{K},$$
$$v \longmapsto \xi(v),$$

so that ξ is a linear map between the vector spaces V and $V' = \mathbb{K}$. We choose the basis $B = (b_1, \ldots, b_n)$ of V and $B' = (1)$ of $\mathbb{K}$ and the matrix representation of ξ

© The Author(s), under exclusive license to Springer Nature Switzerland AG 2024
N. A. Papadopoulos and F. Scheck, *Linear Algebra for Physics*,
https://doi.org/10.1007/978-3-031-64908-0_6

is given in various notations by

$$\xi_{B'B} \equiv [\xi_1 \dots \xi_n] \equiv (\xi_i)_n \equiv (\vec{\xi})^\mathsf{T} \equiv M(\xi) \in \mathbb{K}^{1 \times n}$$
$$\xi_i \in \mathbb{K}, \vec{\xi} \equiv (\xi^i) \text{ with } \xi^i = \xi_i \text{ and } i \in I(n).$$

Taking $v = b_i v^i$ and using the anonymous or "universal" M notation, we have

$$M[v] = \begin{bmatrix} v^1 \\ \vdots \\ v^n \end{bmatrix} = \vec{v} \in \mathbb{K}^{n \times 1} \ (v^i \in \mathbb{K}) \text{ and}$$

$$M(\xi(v)) = M(\xi)M(v) = [\xi_1 \dots \xi_n] \begin{bmatrix} v^1 \\ \vdots \\ v^n \end{bmatrix} = \xi_i v^i = \langle \vec{\xi} \mid \vec{v} \rangle.$$

The chosen basis B of V uniquely determines a specific basis B^* of V^* which has the simplest possible relation to the basis B. This is expressed in the following proposition:

Proposition 6.1 *Dual basis.*
If $B = (b_1, \dots, b_n)$ is a basis of V, then $B^* = (\beta^1, \dots, \beta^n) \equiv (b_1^*, \dots, b_n^*)$
with $\beta^i(b_j) = \delta_j^i$ or $(b_i^*(b_j) = \delta_{ij})$, $i, j \in I(n)$, is a basis of V^*.
B^* is called the dual basis or cobasis of B. This B^* is uniquely defined.

Proof Since $B^* \subseteq V^*$ is a linearly independent list of cardinality $n = \dim V^*$, it follows that B^* is a basis of V^*: If we set $\lambda_i \beta^i = 0$, it follows that $\forall j \in I(n) \ \lambda_i \beta^i(b_j) = \lambda_j$ and $\lambda_j = 0$. In addition, span $B^* \leq V^*$, and since $\dim V^* = n$, span $B^* = V^*$. So B^* is a basis of V^*. The basis B^* is uniquely defined since every β^i is given by the n numbers $\beta^i(b_j) = \delta_j^i$. $\qquad\square$

Remark 6.1 Canonical dual basis.

For $V = \mathbb{K}^n$ and $V^* = \mathbb{K}^{n*}$, if we take the canonical basis $E = (e_1, \dots, e_n)$, then $E^* = (\varepsilon^1, \dots, \varepsilon^n) \equiv (e_1^*, \dots, e_n^*)$ is given by

$$\varepsilon^i(e_j) = \delta_j^i, \quad \varepsilon^i \equiv e_i^* = e_i^T \in (\mathbb{K}^n)^* = \mathbb{K}^{1 \times n},$$

for example $\varepsilon^1 = [10 \cdots 00], \cdots, \varepsilon^n = [00 \cdots 01]$. It should be clear that any linear map, here the linear function $\xi \in V^* = \text{Hom}(V, \mathbb{K})$, can also have the representation $\xi_{B'B}$. We thereby use the notation previously used to reflect explicitly the basis dependence of the representation.

Comment 6.1 *Basis induced dual isomorphism.*

Proposition 6.1 shows also the following: For every basis B in V, there exists a dual isomorphism Ψ_B from V to V^* which can be regarded as giving V the vector structure of V^*. This isomorphism is given exactly by the dual basis B^*. Using the notation of Proposition 6.1, we have

$$\Psi_B : V \longrightarrow V^*$$
$$b_s \longmapsto \Psi(b_s) = b_s^* \equiv \beta^s.$$

Remark 6.2 Representation of the dual basis.

As expected, with the basis B^* the linear chart Φ_{B^*} is given by:

$$\Phi_{B^*} : V^* \longrightarrow (K^n)^* \equiv \mathbb{K}^{1 \times n},$$
$$\beta^i \longmapsto \Phi_{B^*}(\beta^i) := \varepsilon^i.$$

So we have for $\xi = \xi_i \beta^i$

$$\Phi_{B^*}(\xi) = \Phi_{B^*}(\xi_i \beta^i) = \xi_i \Phi_{B^*}(\beta^i) = \xi_i \varepsilon^i = [\xi_1 \dots \xi_n].$$

This is consistent with the previous result:

$$\Phi_{B^*}(\xi) = \xi_{B'B}.$$

Remark 6.3 Dual basis and coordinates of vectors.

The covector β^i, as an element of the cobasis $B^* = (\beta^1, \ldots, \beta^n)$, determines the ith coordinate of $v \in V$ in the basis $B = (b_1, \ldots, b_n)$.

For $v = b_j v^j$, $v^j \in \mathbb{K}$, we have $\beta^i(v) = \beta^i(b_j v^j) = \beta^i(b_j)v^j = \delta^i_j v^j = v^i$.

As we already saw, the basis B determines similarly the coordinate ξ_i of the covector $\xi \in V^*$ in the cobasis $B^* = (\beta^1, \ldots, \beta^n)$. Taking $\xi = \xi_i \beta^i$, $\xi_i \in \mathbb{K}$, we have

$$\xi(b_i) = \xi_j \, \beta^j(b_i) = \xi_j \, \delta^j_i = \xi_i.$$

We can think that, besides the category of vector spaces, there is also the category of covector spaces, that is, that there exists to each V the associated V^*. Analogously, we can think that to each linear map f there exists the associated dual linear map f^*.

Definition 6.1 Dual map f^*.
If $f \in \mathrm{Hom}(V, W)$, then the dual map $f^* \in \mathrm{Hom}(W^*, V^*)$ is defined by $f^*(\eta) := \eta \circ f \in V^*$.

We can see that f^* is a linear map from:

$$f^*(\eta + \theta) = (\eta + \theta) \circ f = \eta \circ f + \theta \circ f = f^*\eta + f^*\theta \quad \text{and}$$
$$f^*(\lambda\eta) = \lambda\eta \circ f \qquad = \lambda(\eta \circ f) \qquad = \lambda f^*\eta.$$

The map $* : f \mapsto f^*$ is linear as well. It can be verified by

$$f + g \longmapsto (f + g)^* = f^* + g^*$$
$$\lambda f \longmapsto (\lambda f)^* = \lambda f^*.$$

Example: $(f + g)^*(\eta) = \eta \circ (f + g) = \eta \circ f + \eta \circ g = f^*\eta + g^*\eta.$

Comment 6.2 *f and the dual f^*.*

A direct comparison between f and its dual f^* shows that f^* points to the opposite direction:

$$V \xrightarrow{f} V',$$

$$V^* \xleftarrow{f^*} (V')^*.$$

This makes it difficult to handle f^*. The following notation can help. We define a new symbol $\langle\,,\,\rangle$:

$$\langle \xi, v \rangle := \xi(v), \quad \xi \in V^*, \quad v \in V.$$

Thus the above definition $f^*\eta(v) = \eta(fv)$ can also take the form:

$$\langle f^*\eta, v \rangle = \langle \eta, fv \rangle.$$

Proposition 6.2 *Composition of dual maps.*

Let $f \in \mathrm{Hom}(U, V)$ and $g \in \mathrm{Hom}(V, W)$, then $(g \circ f)^* = f^* \circ g^*$.

Proof First proof.
If

$$f^* \in \mathrm{Hom}(V^*, U^*), g^* \in \mathrm{Hom}(W^*, V^*), (g \circ f)^* \in \mathrm{Hom}(W^*, U^*) \text{ and } \eta \in W^*,$$

then

$$(g \circ f)^*(\eta) = \eta \circ (g \circ f) = (\eta \circ g) \circ f$$
$$= \left(g^*(\eta)\right) \circ f = f^*\left(g^*(\eta)\right) = f^* \circ g^*(\eta) \in U^*.$$

Since this holds for all $\eta \in W^*$, we obtain $(g \circ f)^* = f^* \circ g^*$. □

Proof Second proof, using Comment 6.2.

$$\langle (g \circ f)^* \eta, v \rangle =$$
$$= \langle \eta, (g \circ f) v \rangle = \langle \eta, g(f v) \rangle =$$
$$= \langle g^* \eta, f v \rangle = \langle f^*(g^* \eta), v \rangle = \langle (f^* \circ g^*) \eta, v \rangle \Rightarrow (g \circ f)^* \eta = (f^* \circ g^*) \eta.$$

$\square$

Remark 6.4 f^* as pullback.

The map f^* is the pullback associated to f as shown in the commutative diagram:

$$V \xrightarrow{\ f\ } W$$
$$f^*\eta = \eta \circ f \searrow \quad \downarrow \eta$$
$$\mathbb{K}$$

The representation of the dual map f^* is, as expected, deeply connected with the representation of f: $M(f^*) = M(f)^\mathsf{T}$!

Proposition 6.3 *Representation of the dual map.*
Let $f : V \to V'$ be a linear map. Let B, C be bases of V, V', respectively, and B^*, C^* their dual bases, and $F = f_{CB}$. Then the representation matrix of $f^* : (V')^* \to V^*$ is given by $f^*_{B^*C^*} = F^\mathsf{T}$.

Proof For this proof, we use only the corresponding bases and cobases. We have

$$\beta^r(b_s) = \delta^r_s \quad r, s \in I(n) \text{ and}$$
$$\gamma^i(c_j) = \delta^i_j \quad i, j \in I(m)$$

$$\text{with} \quad F = f_{CB} = (\varphi^i_s) \tag{6.1}$$

$$f(b_s) = c_i \varphi^i_s. \tag{6.2}$$

We define

$$f^*(\gamma^i) = \beta^r \chi^i_r. \tag{6.3}$$

For the matrix representation of f^* we write

$$F^* := f_{B^*C^*} = (\chi^i_r). \tag{6.4}$$

We determine F^* by the following sequence of equations:

$$(f^*\gamma^i)(b_s) = \gamma^i \circ f(b_s) = \gamma^i(fb_s) = \gamma^i(c_j\varphi_s^j)$$
$$= \gamma^i(c_j)\varphi_s^j = \delta_j^i\varphi_s^j = \varphi_s^i. \tag{6.5}$$

Equation (6.3) leads to

$$(f^*\gamma^i)(b_s) = (\beta^r\chi_r^i)(b_s) = \chi_r^i\beta^r(b_s) = \chi_r^i\delta_s^r = \chi_s^i \tag{6.6}$$

$$\implies \chi_s^i = \varphi_s^i.$$

So we obtain $f^*(\gamma^i) = \beta^s\varphi_s^i$. If we compare with $f(b_s) = c_i\varphi_s^i$, we see that $F^* = F^\mathsf{T}$ or in a different notation $M(f^*) = M(f)^\mathsf{T}$. $\qquad\qquad\square$

6.2   The Four Fundamental Spaces of a Linear Map

As we saw in Sect. 5.3, the matrix $F \in \mathrm{Hom}(\mathbb{K}^n, \mathbb{K}^m)$ determines uniquely the four subspaces

$$\ker F, \mathrm{im}\ F^\mathsf{T} \leq \mathbb{K}^n \text{ and } \mathrm{im}\ F, \ker F^\mathsf{T} \leq \mathbb{K}^m \tag{6.7}$$

which give important information about the map F. For $f \in \mathrm{Hom}(V, V')$ this is not possible if V and V' have no additional structure. The reason is that only $\ker f$ and $\mathrm{im}\ f$ are uniquely defined by f, but $\mathrm{coim}\ f$ and $\mathrm{coker}\ f$ are not uniquely defined by f. Only if we choose bases B and B' for V and V', the complements of $\ker f$ and $\mathrm{im}\ f$ are also fixed by f and (B, B'). So we may write:

$$V \cong \ker f \oplus \mathrm{coim}_B f \xrightarrow{f} \mathrm{im}\ f \oplus \mathrm{coker}_B f \cong V'. \tag{6.8}$$

We need B and B' since, as mentioned, when V, V' are abstract vector spaces, we do not possess anything analogous to F^T as in the case when we consider $\mathbb{K}^n$ and $\mathbb{K}^m$. As we shall see later in Sect. 6.3, if V and V' are Euclidean or unitary vector spaces, the adjoint f^{ad} will play the role of F^T and we can find a basis-free version of $\mathrm{coim}_B f$ and $\mathrm{coker}_B f$, induced directly from f.

Here, with V and V' abstract vector spaces without further structure, if we want to find from f induced a kind of basis-free decomposition of V and V', we have to make use of the dual point of view and consider $f^* \in \mathrm{Hom}(V'^*, V^*)$.

As we already know, for a given f the dual f^*

$$f^*: V'^* \longrightarrow V^*,$$
$$\eta \longmapsto f^*(\eta) := \eta \circ f. \tag{6.9}$$

is uniquely determined. Now the subspaces im $f^* \leq V^*$ and ker $f^* \leq V'^*$ are also uniquely determined by f^*. These two subspaces, im f^*, ker f^*, which correspond to $\operatorname{coim}_B f$ and $\operatorname{coker}_B f$, are a kind of substitute for im F^T and ker F^T, respectively. So we get the big picture for f as given by the proposition in form of a diagram:

Proposition 6.4 *The four subspaces of a linear map.*

$$
\begin{array}{ccccc}
V^* & \cong \operatorname{coker}_B f^* \oplus \operatorname{im} f^* & \stackrel{f^*}{\longleftarrow} \operatorname{coim}_B f^* \oplus \operatorname{ker} f^* & \cong V'^* \\
\cong\uparrow B & & & B'\uparrow\cong \\
V & \cong \operatorname{ker} f \oplus \operatorname{coim}_B(f) & \stackrel{}{\underset{f}{\longrightarrow}} \operatorname{im} f \oplus \operatorname{coker}_B(f) & \cong V'.
\end{array}
$$

$$\tag{6.10}$$

Proof The proof is obtained straightforwardly almost by inspection, using the dual bases for V and V'. We may also write symbolically for the uniquely defined subspaces:

$$
\begin{array}{ccc}
\cdots & \oplus \operatorname{im} f^* \stackrel{f^*}{\longleftarrow} \cdots & \oplus \operatorname{ker} f^* \\
\operatorname{ker} f \oplus \cdots & \underset{f}{\longrightarrow} \operatorname{im} f \oplus \cdots .
\end{array}
$$

This exhibits the four relevant subspaces that are basis-independent.   □

Comment 6.3 *Isomorphisms of the four fundamental subspaces.*

From the last proportion, we obtain immediately the following isomorphisms.

$$
\operatorname{im} f^* \cong \operatorname{im} f \cong \operatorname{coim}_B(f) \cong \operatorname{coim}_B(f)^* \tag{6.11}
$$
$$
\operatorname{ker} f \cong \operatorname{coker}_B(f)^* \text{ and } \operatorname{ker} f^* \cong \operatorname{coker}_B(f). \tag{6.12}
$$

Proposition 6.4 may also be considered as a synopsis of the results of the second fundamental theorem of linear algebra (see Theorem 5.2) for the general case of an abstract vector space.

If we use the notation of an annihilator, further results are obtained.

Definition 6.2 Annihilator of a subspace.
For $U \leq V$, the annihilator of U, denoted by U^0, is given by $U^0 := \{\xi \in V^* : \xi(u) = 0 \ \forall u \in U\}$.

We can directly verify that U^0 is a subspace of V^*.

Proposition 6.5 *Dimension of the annihilator U^0.*

For $U \subseteq V$ and U^0 as above, we have $\dim U + \dim U^0 = \dim V$. This means that the annihilator of U has the dimension of any complement of U in V. So if we have $U \oplus W \cong V$, then $\dim U^0 = \dim W$.

Proof Let $A := (a_1, \ldots, a_k)$ be a basis of U, and extend it to a basis $C := (a_1, \ldots, a_k, b_1, \ldots, b_l)$ of V. Let $C^* = (\alpha^1, \ldots, \alpha^k, \beta^1, \ldots, \beta^l)$ be its dual basis. In what follows, we set $i, j \in I(k)$ and $r, s \in I(l)$. Note that the choice $B := (b_1, \ldots, b_l)$ is a basis of $W := \operatorname{span}(b_1, \ldots, b_l)$ and $W \leq V$.

We notice that we set $\beta^s(a_i) = 0$ for all $i \in I(k)$ and also $\beta^s(b_r) = \delta_r^s$ for all $s \in I(l)$, as usual. This means that

$$W^* := \operatorname{span}((\beta^1, \ldots, \beta^l) = \{\lambda_s \beta^s : \lambda_s \in \mathbb{K}\}$$

is a subspace of $V^* (W^* \leq V^*)$ and $W^* \cong W$. We also notice that W^* annihilates U:

$$\beta^s(U) = 0 \quad \text{for all} \quad s \in I(l),$$

so that W^* is a subspace of U^0:

$$W^* \leq U^0. \tag{6.13}$$

It is left to show is that $U^0 \leq W^*$ holds: If $w \in U^0$, we have

$$w = \mu_j \alpha^j + \lambda_s \beta^s, \ \mu_j, \ \lambda_s \in \mathbb{K} \tag{6.14}$$

and

$$w(a_i) \overset{!}{=} 0 \quad \text{for all} \quad i \in I(k). \tag{6.15}$$

Equations (6.14) and (6.15) lead to

$$w(a_i) = \mu_j \alpha^j(a_i) + \lambda_s \beta^s(a_i) =$$
$$= \mu_j \delta_i^j + 0 =$$
$$= \mu_i \overset{!}{=} 0. \tag{6.16}$$

This leads to

$$w = \lambda_s \beta^s \in U^0. \tag{6.17}$$

We showed that $U^0 \leq W^*$ and together with Eq. (6.13) we get $W^* = U^0$. Now it is clear that for W^*, the dual of W, we have $\dim W^* = \dim W = \dim U^0$. This proves $\dim U + \dim U^0 = \dim V$. $\qquad \square$

Proposition 6.6 *Annihilators of* im f *and* ker f.

The following equations hold:

$$\ker f^* = (\operatorname{im} f)^0, \tag{6.18}$$

$$\operatorname{im} f^* = (\ker f)^0. \tag{6.19}$$

This follows directly by setting $\ker f = U$ and using tailor-made bases in the proof of Proposition 6.5. Here, we give a basis-independent proof for Eq. (6.18):

Proof In Eq. (6.13), we have

$$\begin{array}{lll} V^* & \xleftarrow{\ f^*\ } & V'^* \geq \ker f^* \\ V & \xrightarrow[\ f\ ]{} & V' \geq \operatorname{im} f. \end{array}$$

We have to show

(a) $\ker f^* \leq (\operatorname{im} f)^0$ and
(b) $\ker f^* \geq (\operatorname{im} f)^0$.

which gives $\ker f^* = (\operatorname{im} f)^0$.
For (a): We consider the sequence of the following implications:

$$\eta_0 \in \ker f^* \Rightarrow f^* \eta_0 = 0^* \in V^*$$

so that for all $v \in V$ this leads to

$$f^* \eta_0(v) = 0^*(v) = 0 \Rightarrow \eta_0(fv) = 0,$$
$$\Rightarrow \eta_0(fV) = 0 \quad \text{or} \quad \eta_0(\operatorname{im} f) = 0$$

which means that

$$\eta_0 \in (\operatorname{im} f)^0 \quad \text{and} \quad \ker f^* \leq (\operatorname{im} f)^0.$$

So (a) is proven.
For (b): We start with $\theta \in (\operatorname{im} f)^0$. Then we have for all $v \in V$

$$\theta(fv) = 0 \quad \text{and} \quad f^* \theta(v) = 0.$$

Hence $f^* \theta = 0^* \in V^*$ and $\theta \in \ker f^*$, so that $(\operatorname{im} f)^0 \leq \ker f^*$. This proves (b).
(a) and (b) both hold so that

$$(\operatorname{im} f)^0 = \ker f^*.$$

$\square$

Proposition 6.7 *Injective and surjective relations between f and f^*. Let $f \in \text{Hom}((V, V')$ and f^* its dual. Then*

(i) *if f is injective, f^* is surjective;*
(ii) *if f is surjective, f^* injective.*

Proof   (i) If f is injective, then we have $\ker f = \{0\}$. Using the propositions 6.5 and 6.6, we obtain

$$(\ker f)^0 = V^* = \dim V^* = n$$

and so

$$\dim(\text{im } f^*) = \dim(\ker f)^0 = n,$$

and so

$$\text{im } f^* = V,$$

thus f^* is surjective.
(ii) Similar to the proof for (i).

$\square$

6.3   Inner Product Vector Spaces and Duality

We consider a linear map $f \in \text{Hom}(V, V')$ between $V = (V, \langle | \rangle)$ and $V' = (V', \langle | \rangle)$, two inner product vector spaces. In our approach, we always mean a finite-dimensional vector space by an inner product vector space, usually a Euclidean or unitary vector space. Here we obtain the same picture of the four relevant subspaces of f as in the case $f \equiv F \in \text{Hom}(\mathbb{K}^n, \mathbb{K}^m)$. The role of the transpose F^T (giving $\ker F^\mathsf{T}$ and $\text{im } F^\mathsf{T}$) is taken over now by the adjoint map f^{ad} of f. We will discuss this new linear algebra notion in this section too. The existence of adjoint and self-adjoint operators is omnipresent in physics. Especially for quantum mechanics, it is interesting to note that self-adjoint operators describe the physical observables on a Hilbert space. It is well-known that for finite dimensions, the notion of a Hilbert space is equivalent to that of a unitary space.

So here we have the opportunity to look at finite dimension spaces first which are much easier than infinite dimensioned spaces, to understand the Hilbert space structure and observe its geometric significance.

The existence of an inner product in V allows a second dual isomorphism which is basis-independent.

Definition 6.3 Dual isomorphism induced by the inner product.

$$\text{Let the map } \quad j : \quad V \longrightarrow V^*,$$
$$v \longmapsto j(v) := \langle v | \cdot \rangle$$

which means that

$$j(v)(u) := \langle v | u \rangle. \tag{6.20}$$

Remark 6.5 Antilinear and semilinear map.

There is here a slight difference between a Euclidean and a unitary vector space. For a Euclidean vector space, j is a linear map, whereas, for a unitary vector space, j is what is called a $\overline{\mathbb{C}}$-linear or antilinear map.

$$j(u + v) = j(u) + j(v) \quad \text{and} \quad j(\lambda v) = \bar{\lambda} j(v). \tag{6.21}$$

To describe both possibilities, we uniformly use the name semilinear map if we mean that a map f is linear or antilinear. For example, f is called semilinear if for

$$f : \quad V \longrightarrow V',$$
$$v \longmapsto f(v),$$
$$f(u + v) \quad = f(u) + f(v) \quad \text{and for } \lambda \in \mathbb{C},$$
$$f(\lambda v) \quad = \bar{\lambda} f(v) \quad \text{or} \quad f(\lambda v) = \lambda f(v). \tag{6.22}$$

It is clear that for a Euclidean vector space, the semilinear map j is a linear map.

Definition 6.4 The adjoint map, f^{ad}.
For $f \in \text{Hom}(V, V')$, the adjoint of f is the map $f^{ad} : V' \rightarrow V$ uniquely defined by the property
$$\langle v | f^{ad} w \rangle := \langle f v | w \rangle \tag{6.23}$$

for all $v \in V$ and all $w \in V'$.

It is equivalent to define the adjoint f^{ad} of f via the commutative diagram:

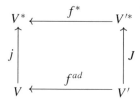

The isomorphism j and J are the corresponding dual isomorphisms as given in Eqs. (6.20) and (6.24) below:

$$J : \quad V' \longrightarrow V'^*$$
$$w \longmapsto J(w) := \langle w | \cdot \rangle \tag{6.24}$$

and we obtain

$$j \circ f^{ad} = f^* \circ J \tag{6.25}$$

or equivalently

$$f^{ad} = j^{-1} \circ f^* \circ J. \tag{6.26}$$

We see that f^{ad} can be considered as the manifestation of f^*.

Equations (6.25) and (6.26) are equivalent to Eq. (6.23). At the level of vector spaces we have:

$$f^{ad} : \quad V' \longrightarrow V,$$
$$w \longmapsto f^{ad}(w).$$

We can obtain the analytic expression (6.23) for f^{ad} from (6.25) as follows:

$$(j \circ f^{ad})(w) = (f^* \circ J)(w)$$
$$\Leftrightarrow j(f^{ad}w) = f^*(Jw) \in V^*$$
$$\Leftrightarrow (j(f^{ad}w))\langle v \rangle = J(w)(fv) \in \mathbb{K} \quad \text{for any} \quad v \in V \quad \text{and} \quad w \in V'$$
$$\Leftrightarrow \langle f^{ad}w | v \rangle = \langle w | fv \rangle$$
$$\Leftrightarrow \langle v | f^{ad}w \rangle = \langle fv | w \rangle \tag{6.27}$$

which is the second definition (see Eq. (6.23)) of f^{ad}.

Remark 6.6 Some properties of f^{ad} and ad-action.

(i) The map f^{ad} is linear.

j and J are semilinear, but in the composition (6.26) both of them appear. Therefore the "semi" is "annihilated" and f^{ad} becomes linear. $\square$

(ii) The map ad is antilinear for $f, g \in \mathrm{Hom}(V, V')$.

$(f + g)^{ad} = f^{ad} + g^{ad}$ and $(\lambda f)^{ad} = \bar{\lambda} f^{ad}, \lambda \in \mathbb{C}$.

The additivity is quite clear because for $v \in V$, $w \in V'$. The semi linearity follows from

$$\langle v \mid (\lambda f)^{ad} w \rangle = \langle \lambda f v \mid w \rangle \quad = \bar{\lambda} \langle f v \mid w \rangle = \bar{\lambda} \langle v \mid f^{ad} w \rangle,$$
$$= \langle v \mid \bar{\lambda} f^{ad} w \rangle \qquad \Rightarrow (\lambda f)^{ad} = \bar{\lambda} f^{ad}. \qquad (6.28)$$

$\square$

(iii) ad is an involution: $(f^{ad})^{ad} = f$. For any $w \in V, v \in V,$

$$\langle w \mid (f^{ad})^{ad} v \rangle = \langle f^{ad} w \mid v \rangle = \langle w \mid f v \rangle. \qquad (6.29)$$

$\square$

(iv) $(g \circ f)^{ad} = f^{ad} \circ g^{ad}.$

Using an obvious notation as above for j, J and K, and (6.26), we obtain

$$(g \circ f)^{ad} = j^{-1} \circ (g \circ f)^* \circ K = j^{-1} \circ f^* \circ g^* \circ K \Rightarrow$$
$$(g \circ f)^{ad} = j^{-1} \circ f^* \circ J \circ (J^{-1} \circ g^* \circ K) = f^{ad} \circ g^{ad}. \qquad (6.30)$$

$\square$

We are now in the position to determine the relation of $\ker f^{ad}$ and $\mathrm{im}\ f^{ad}$ to $\mathrm{im}\ f$ and $\ker f$:

Proposition 6.8 *Kernel image relation between f^{ad} and f.*
Let V and V' be inner product spaces,

$$f : V \longrightarrow V' \text{ be a linear map, and} \qquad (6.31)$$
$$f^{ad} : V' \longrightarrow V \text{ be the adjoint map.} \qquad (6.32)$$

Then:

(i) $\ker f^{ad} = (\mathrm{im}\ f)^{\perp}.$
(ii) $\mathrm{im}\ f^{ad} = (\ker f)^{\perp}.$

Proof   (i) Let $y \in V'$, then $y \in \ker f^{ad}$:

$$\Leftrightarrow \quad f^{ad} y = 0$$
$$\Leftrightarrow \quad \langle v | f^{ad} y \rangle = 0 \quad \forall v \in V$$
$$\Leftrightarrow \quad \langle f v | y \rangle = 0 \quad \forall v \in V$$
$$\Leftrightarrow \quad y \in (\operatorname{im} f)^{\perp}$$

Hence $\ker f^{ad} = (\operatorname{im} f)^{\perp}$. This shows (i).

(ii) Since $(f^{ad})^{ad} = f$, we may replace f with f^{ad} in (i) and we obtain $\ker f = \left(\operatorname{im} f^{ad}\right)^{\perp}$. Its orthogonal complement gives

$$\operatorname{im} f^{ad} = (\ker f)^{\perp}.$$

$\square$

This result also gives a geometric interpretation of f^{ad}:

$$V = \ker f \ominus \operatorname{im} f^{ad} \text{ and } V' = \operatorname{im} \ominus \ker f^{ad}. \tag{6.33}$$

This result shows that f and f^{ad} lead uniquely through "ker" and "im" in an orthogonal decomposition of V and V'.

In this way, we obtained for a general $f \in \operatorname{Hom}(V, V')$, with the inner product vector spaces $(V, \langle | \rangle)$ and $(V', \langle | \rangle')$, the same connection with the four f-relevant subspaces as with $F \in \operatorname{Hom}(\mathbb{R}^n, \mathbb{R}^m)$ in Theorem 5.2. Therefore, it may be considered as another face of the same theorem:

Theorem 6.1 *The fundamental theorem of linear maps for inner product vector spaces.*
Any map $f : V \to w$ decomposes as follows:

$$V = \ker f \ominus \operatorname{im} f^{ad} \xrightarrow{\ f\ } \operatorname{im} f \ominus \ker f^{ad} = V'.$$

$\square$

Furthermore, we know that $\dim(\operatorname{im} f^{ad}) = \dim(\operatorname{im} f) = \operatorname{rank} f = r$. So again if $\dim(\ker f) = k$ and $\dim(\ker f^{ad}) = l$, then

$$\dim V = k + r \text{ and } \dim V' = r + l.$$

Remark 6.7 $\operatorname{im} f = (\ker f^{ad})^{\perp}$.

The orthogonal complement on both sides of (i), in Proposition 6.8, gives $\operatorname{im} f = \left(\ker f^{ad} \right)^{\perp}$.

Proposition 6.9 *Representation of f^{ad}.*

The representation F^{ad} of f^{ad} is given for orthonormal bases in V and V' by the representation F of f:

$$F^{ad} = F^{\dagger} \quad where \quad (F^{\dagger} := \bar{F}^{\mathsf{T}}).$$

Proof For orthonormal bases $B = (v_a)_n$ in V and $C = (w_i)_m$ in V'.

$$\langle v_a \mid v_b \rangle = \delta_{ab} \quad a, b \in I(n) \text{ and } \langle w_i \mid w_j \rangle = \delta_{ij} \quad i, j, \in I(m).$$

We obtain from $f v_a = w_i \varphi_a^i \quad \varphi_a^j \in \mathbb{K}$,

$$\langle w_j \mid f v_a \rangle = \langle w_j \mid w_i \rangle \varphi_a^i = \delta_{ij} \varphi_a^i = \varphi_a^j.$$

Taking $f^{ad} w_i = v_a \chi_i^a \quad \chi_i^a \in \mathbb{K}$, we obtain analogously

$$\langle v_b \mid f^{ad} w_i \rangle = \chi_i^b,$$

$$\langle v_b \mid f^{ad} w_i \rangle = \langle f v_b \mid w_i \rangle = \overline{\langle w_i \mid f v_b \rangle} = \overline{\varphi_b^i}.$$

The comparison with the last two equations leads to the result

$$\chi_i^b = \overline{\varphi_b^i}.$$

So we have $f^{ad} w_i = v_a \bar{\varphi}_a^i$ which means $F^{ad} = F^{\dagger}$. $\square$

Remark 6.8 $F^{ad} = F^{\mathsf{T}}$ for Euclidean vector spaces.

In the case of Euclidean vector spaces,

$$\varphi_a^i \in \mathbb{R} \quad \text{so that} \quad \bar{\varphi}_a^i = \varphi_a^i \quad \text{and} \quad F^{ad} = F^{\mathsf{T}}.$$

This is precisely the result for $f \in \mathrm{Hom}(\mathbb{R}^n, \mathbb{R}^m)$ as discussed in Sect. 5.4 and Theorem 5.2.

6.4   The Dirac Bra Ket in Quantum Mechanics

Quantum mechanics is done in a Hilbert space H, that is, the realm of quantum mechanics. Here, we consider finite-dimensional vector spaces and therefore we also consider finite-dimensional Hilbert spaces. An n-dimensional Hilbert space is a $\mathbb{C}$ vector space with inner product $H = (V, \langle | \rangle)$, $\dim V = n$. If we choose a orthonormal basis $C = (c_1, \ldots, c_n)$, then we have the following isomorphism:

$$H \cong \mathbb{C}^n.$$

So we can identify the Hilbert space H with $\mathbb{C}^n$. The inner product here is also called a Hermitian product, and $\langle | \rangle$ is noting else but the Dirac Bra Ket! But Dirac goes one step further and decomposes the BraKet $\langle | \rangle$ into two maps ($\langle | \rangle \rightsquigarrow \langle || \rangle \rightsquigarrow \langle |$ and $| \rangle$) which is $| \rangle = id_H \in \mathrm{Hom}(H, H)$:

$$| \rangle : H \longrightarrow H,$$
$$v \longmapsto |v\rangle := id_H(v) = v,$$

and $\langle | \in \mathrm{Hom}(H, H^*)$:

$$\langle | : H \longrightarrow H^*,$$
$$v \longmapsto \langle v|$$

with

$$\langle v| : H \longrightarrow \mathbb{C},$$
$$u \longmapsto \langle v|u\rangle = v^\dagger u.$$

So the result is that we have in fact $|v\rangle = v$ and $\langle v| \neq v$, *definitively*. So the new object is only the map $\langle | \in \mathrm{Hom}(H, H^*)$: However, this is nothing else but the well-known canonical isometry between H and H^* (see Chap. 11.2 and also the canonical dual isomorphism as in Definition 6.3). At this point, to facilitate our intuition, we prefer considering a real vector space. So we set now $H = \mathbb{R}^n \equiv \mathbb{R}^{n \times 1}$. This is no restriction for the following considerations.

We notice immediately that because of the equation $\langle v|u\rangle = v^\mathsf{T} u$, the equality $\langle | = (\cdot)^\mathsf{T} : H \to H^*$ holds too. Thus, the transpose T, when restricted to H, is nothing else but the map "Bra" $= \langle |$ taken from Bra Ket. So we have just to call the symbol

$|\rangle = id_H$, Ket, as Dirac did. But what is the difference between the transpose $\top$ and Bra? Bra is only defined on H while the transpose $\top$ is defined on H as well as on H^*. So we have the well-known relations:

$$\top : \mathbb{R}^n \longrightarrow (\mathbb{R}^n)^*$$
$$\text{column} \longmapsto \text{row}$$

$$\text{and} \quad \top : (\mathbb{R}^n)^* \longrightarrow \mathbb{R}^n$$
$$\text{row} \longmapsto \text{column}$$

$$\text{but only} \quad \langle | : \mathbb{R}^n \longrightarrow (\mathbb{R}^n)^*$$
$$\text{column} \longmapsto \text{row}.$$

This means that when we are using $| v \rangle$, we see $v \in \mathbb{R}^n$ but explicitly never $\xi = \xi_v \in (\mathbb{R}^n)^*$. This facilitates the identification of H with H^*. In coordinates (coefficients), using the canonical basis $E = (e_1, \ldots, e_n)$ and the canonical cobases $E^* = (\varepsilon^1, \ldots, \varepsilon^n)$, $\langle e_i | e_s \rangle = \delta_{is}$, $\varepsilon^i(e_s)) = \delta^i_s$, $i, s \in I(n)$, we have

$$| v \rangle = v = e_s v^s \quad v^s \in \mathbb{R} \quad \text{and}$$

$$\langle v | = v^\top = v_i \varepsilon^i \quad \text{with} \quad v_i = v^i.$$

If we consider the standard quantum mechanics case where we take $H = \mathbb{C}^n$, we have, with $\langle v | = v^\dagger = \bar{v}^\top$,

$$\langle v | u \rangle = \bar{v}^\top u = \sum_{i=1}^{n} \bar{v}^i u^i = v_i \, u^i.$$

Then, $\langle |$ corresponds to the conjugate transpose $\dagger$ and we can write for $v \in H$:

$$| v \rangle = e_i v^i, \quad v^i \in \mathbb{C} \quad \text{and}$$
$$\langle v | = v_i \varepsilon^i \quad \text{with} \quad v_i = \bar{v}^i.$$

The conjugate transpose of $v \equiv | v \rangle$ is:

$$\langle v | = \bar{v}^1 \varepsilon^1 + \cdots + \bar{v}^n \varepsilon^n = [v_1 \cdots v_i \cdots v_n] = v^\dagger \in (\mathbb{C}^n)^* = H^*.$$

Remark 6.9 Comparison of $\langle | \rangle$ with $| \rangle \langle |$.

As we know, the symbol $\langle | \rangle$ denotes the Hermitian product for a $\mathbb{C}$ vector space which is a sesquilinear ("one and a half linear") map.

$$\langle | \rangle : H \times H \longrightarrow \mathbb{C}.$$

What is $|\rangle\langle|$? We first consider $|v\rangle\langle u| \in \mathrm{Hom}(H, H) = \mathbb{C}^{n \times n}$. Then $|v\rangle\langle u|$ is a remarkable map as well as a remarkable matrix since $|v\rangle\langle u| = v\bar{u}^{\mathsf{T}} = vu^{\dagger}$ is a matrix with rank $(|v\rangle\langle u|) = 1$. We may think that as a map, $|v\rangle\langle u|$ is acting from the left on $H = \mathbb{C}^n$. On the other hand, $|v\rangle\langle u|$, as a matrix, also acts quite naturally from the right on $H^* = (\mathbb{C}^n)^*$. We can thus also interpret $|v\rangle\langle u| \in \mathrm{Hom}(H^*, H^*)$!

Remark 6.10 Canonical basis in $\mathbb{C}^{n \times n}$ in Dirac's Bra Ket formalism.

The canonical basis in $\mathbb{C}^{n \times n}$ is given by $\mathbb{E} := \{E_{is} : i, s \in I(n)\}$ with basis matrices given by $E_{is} = |e_i\rangle\langle e_s| = (\varepsilon_{is})^j_r = \delta_{ij}\delta_{sr}$. It is interesting that rank$(E_{is}) = 1$ (see also Eq. (7.1)). We thus get for the matrix A the expression:

$$A = \sum_{s,i}^{n} \alpha^i_s | e_i\rangle\langle e_s|.$$

Remark 6.11 The extended identity $\sum_{i=1}^{n} | e_i\rangle\langle e_i|$.

According to remark 6.9, the expression $\sum_{i=1}^{n} |e_i\rangle\langle e_i|$ can be identified with

$$id_H \quad \text{or} \quad id_{H^*}.$$

Remark 6.12 Matrix multiplication.

The present Dirac formalism leads also directly, as in Sect. 5.1, to the expression for matrix multiplication. With $A = (\alpha^i_s)$, $B = (\beta^j_r)$, $C = (\gamma^i_r)$, $i, j, r, s \in I(n)$, using Remark 6.10, we get:

$$\begin{aligned}
C = A\,B &= \alpha^i_s |e_i\rangle\langle e_s|\beta^j_r|e_j\rangle\langle e_r| \\
&= \alpha^i_s\beta^j_r |e_i\rangle\langle e_s|e_j\rangle\langle e_r| = \alpha^i_s\beta^j_r | e_i\rangle\delta^s_j\langle e_r | \\
&= \alpha^i_j\beta^j_r |e_i\rangle\langle e_r|.
\end{aligned}$$

This leads to

$$\gamma^i_r = \alpha^i_j\beta^j_r.$$

$\square$

Summary

The role of the dual vector space and the dual map was thoroughly discussed. This is an area that is often neglected in physics. The last section of this chapter on Dirac formalism in quantum mechanics illustrates that it doesn't have to be the case. Dual maps in situations where only abstract vector spaces are available serve as a certain substitute for adjoint maps, which, as we have seen, are defined on inner product vector spaces.

Here, we also observed the dual version of the four fundamental subspaces of a linear map. The annihilator of a subspace, as a subspace in the dual space, also played an important role. We showed that an inner product space is naturally isomorphic to its dual spaces.

Following this, within duality of inner product vector spaces, the corresponding adjoint to a given linear map, was introduced. The four fundamental subspaces are naturally most visible in the inner product space situation using the adjoint map.

Finally, as mentioned, Dirac formalism was addressed.

Exercises with Hints

Exercise 6.1 Show that transposition

$$T: \quad \mathbb{K}^{m \times n} \longrightarrow \mathbb{K}^{n \times m}$$
$$A \longmapsto A^{\mathsf{T}}$$

is a linear and invertible map.

Exercise 6.2 *Any nontrivial linear function is always surjective.* Show that $\operatorname{im} \xi = \mathbb{K}$ for any $\xi \in \operatorname{Hom}(V, \mathbb{K}) \setminus \{0\}$.

In the following two exercises, we see explicitly the role of the dual space $(\mathbb{K}^m)^*$ in determining the rows of a matrix $A \in \mathbb{K}^{m \times n}$. Similarly, we see that $(\mathbb{K}^n)^*$ determines the rows of A^{T}.

Exercise 6.3 *We consider the matrix $A \in \mathbb{K}^{m \times n}$ as a linear map $A \in \operatorname{Hom}(V, V')$ with $V = \mathbb{K}^n$ and $V' = \mathbb{K}^m$ and we write:*

$$A = (\alpha_s^i) = [a_1' \ldots a_n'] = \begin{bmatrix} \alpha^1 \\ \vdots \\ \alpha^m \end{bmatrix},$$

with $i \in I(m)$, $s \in I(n)$, $\alpha_s^i \in \mathbb{K}$, $a_s \in \mathbb{K}^m$, and $\alpha^i \in (\mathbb{K}^n)^*$.
Show that

$$a'_s = A e_s \quad \text{and} \quad \alpha^i = \varepsilon^i A,$$

where

(e_s, ε^s) is the canonical dual basis pair in $\mathbb{K}^n$;

(e'_i, ε^i) is the canonical dual basis pair in $\mathbb{K}^m$.

Exercise 6.4 *Use a similar notation as in Exercise 5.3.*
Let $A^\mathsf{T} : \mathbb{K}^m \to \mathbb{K}^n$, with

$$A^\mathsf{T} = (\alpha_i^s) = [a_1 \ldots a_m] = \begin{bmatrix} \alpha'^1 \\ \vdots \\ \alpha'^n \end{bmatrix},$$

$$\alpha_i \in \mathbb{K}^n, \alpha'^s \in (\mathbb{K}^m)^*.$$

Show that

$$a_i = A^\mathsf{T} e'_i \quad \text{and} \quad \alpha'^s = \varepsilon^s A^\mathsf{T}.$$

Having more advanced tools in this chapter, it will be even easier to prove the row-column-rank theorem in Exercise 6.6. Beforehand, let us recall the connection between the im f (with $f \in \text{Hom}(V, V')$) and the column rank of $M(f)$ in Exercise 6.5.

Exercise 6.5 Let f be a linear map $f \in \text{Hom}(V, V')$. Show that dim im f equals the column rank of $M(f)$, the matrix of f.

Exercise 6.6 *The row-column-rank theorem.*
Consider the linear map $f = A \in \text{Hom}(\mathbb{K}^n, \mathbb{K}^m)$, then show that the row rank of A equals the column rank of A.

Exercise 6.7 *The form of any rank equal 1 matrix.*
Use the experience made with the proof in Exercise 5.4 to show that for any matrix $A \in \mathbb{R}^{m \times n}$, the rank of A is 1 if and only if the matrix A is of the form $A = w\xi$ with $w \in \mathbb{R}^m$ and $\xi \in (\mathbb{R}^n)^*$. In this case, we can also write:

$$A = wv^\mathsf{T} = |w\rangle\langle v|.$$

Exercise 6.8 *The dual basis covectors select the coordinates of the vectors in V and the basis vectors select the coordinates of covectors in V*.*
If $(b_1, \ldots, b_n)$ is a basis in V and $(\beta^1, \ldots, \beta^n)$ its dual basis, so that $(\beta^s(b_r)) = \delta_r^s$, $s, r \in I(n)$, show that for any vector $v \in V$ and any covector $\xi \in V^*$,

(i) $v = b_s \beta^s(v) \in V$;
(ii) $\xi = \xi(b_s)\beta^s \in V^*$.

Exercise 6.9 *There alway exists an element of the dual space which annihilates any given proper subspace of the corresponding vector space.*
Let U be a subspace of a vector space V. If $\dim U < \dim V$, then show that there exists some $\xi \in V^*$ such that $\xi(U) = 0$.

Exercise 6.10 *The annihilator is a subspace.*
If U is a subspace of a vector space V, then show that the annihilator U^0 of U is a subspace of the dual V^*:

$$U^0 \leq V^*.$$

Exercise 6.11 *Here is another proof of Proposition 6.5. This proof is basis free.*
For U a subspace of a vector space V, the dimension of the annihilator U^0 is given by

$$\dim U + \dim U^0 = \dim V.$$

Prove the above assertion using the inclusion map

$$i : U \longrightarrow V$$
$$u \longmapsto i(u) = u \in V,$$

and the rank-nullity theorem.

> The next four exercises deal with various simple relations between two subspaces and the corresponding annihilators.

Exercise 6.12 If U_1 and U_2 are subspaces of V with $U_1 \leq U_2$, then show that $U_2^0 \leq U_1^0$.

Exercise 6.13 If U_1 and U_2 are subspaces of V with $U_2^0 \leq U_1^0$, then show that $U_1 \leq U_2$.

Exercise 6.14 If U_1 and U_2 are subspaces of V, then show that $(U_1 + U_2)^0 = U_1^0 \cap U_2^0$.

Exercise 6.15 If U_1 and U_2 are subspaces of V, then show that $(U_1 \cap U_2)^0 = U_1^0 + U_2^0$.

The notion of double dual space V^{**} of a vector space V is very important, especially to understand the tensor formalism. This is why it will be present at times in the next chapters.

Definition 6.5 The double dual of V, here denoted by V^{**}, is the dual of the dual space V^*:
$$V^{**} := \mathrm{Hom}(V^*, \mathbb{R}).$$

V^{**} is canonically isomorphic to V:

$$\Psi : V \longrightarrow V^{**}$$
$$v \longmapsto (\Psi(v))(\xi) \equiv v^{\#}(\xi) := \xi(v)$$

for $v \in V$ and $\xi \in V^*$.

Exercise 6.16 Show the following assertion: $\Psi \equiv (\cdot)^*$ is a linear map from V to V^{**}.

Exercise 6.17 Show the following assertion: Ψ is an isomorphism from V to V^{**}.

References and Further Reading

1. S. Axler, *Linear Algebra Done Right*. (Springer Nature, 2024)
2. S. Bosch, *Lineare Algebra*. (Springer, 2008)
3. G. Fischer, B. Springborn, *Lineare Algebra, Eine Einführung für Stu-dienanfänger, Grundkurs Mathematik*. (Springer, 2020)
4. S.H. Friedberg, A.J. Insel, L.E. Spence, *Linear Algebra* (Pearson, 2013)
5. K. Jänich, *Mathematik 1, Geschrieben für Physiker* (Springer, 2006)
6. N. Jeevanjee, *An Introduction to Tensors and Group Theory for Physicists*. (Springer, 2011)
7. J. Liesen, V. Mehrmann, *Linear Algebra*. (Springer, 2015)
8. P. Petersen, *Linear Algebra*. (Springer, 2012)

Chapter 7
The Role of Determinants

The determinant is one of the most exciting and essential functions in mathematics and physics. Its significance stems from the fact that it is a profoundly geometric object. It possesses many manifestations. Its domain is usually the $n \times n$-matrices, and it may also be called the determinant function. Another form is the map from the Cartesian product $V^n = V \times \ldots \times V$ to $\mathbb{K}$ (where dim $V = n$) which is linear in every component with the additional property (alternating) that if two vectors are identical, the result is zero. This is usually called a multilinear alternating form or a determinant form or even a volume form on V. In connection with this, the notion of orientation is illuminated by a determinant.

In what follows, we start with the algebraic point of view for determinants, and in doing so, we derive and discuss most of the properties of determinants. Later we address the geometric point of view. In addition, we define the determinant of a linear operator, which is essentially a third manifestation of determinants.

7.1 Elementary Matrix Operations

From the algebraic point of view, the use of elementary operations and elementary matrices offers some advantages since the expressions and proofs become clearer and shorter. We start with a few remarks on the notations and definitions. We first consider the $m \times n$ canonical basis matrices (see also Comment 3.3, and the Examples 2.5 and 2.6) given by:

$$E_{is} := (\varepsilon_{is})_r^j \tag{7.1}$$

with

$$(\varepsilon_{is})_r^j \equiv (\varepsilon_{is})_r^j := \delta_{ij}\delta_{sr}, \quad \text{for} \quad i, j \in I(m), \quad r, s \in I(n). \tag{7.2}$$

© The Author(s), under exclusive license to Springer Nature Switzerland AG 2024
N. A. Papadopoulos and F. Scheck, *Linear Algebra for Physics*,
https://doi.org/10.1007/978-3-031-64908-0_7

So we can write, as in Comment 3.3,

$$
E_{is} = \begin{bmatrix} & & 0 & & \\ & & \vdots & & \\ & & 0 & & \\ 0 \cdots 0 & 1 & 0 \cdots 0 \\ & & 0 & & \\ & & \vdots & & \\ & & 0 & & \end{bmatrix} \; i\text{th row.} \tag{7.3}
$$

sth column

Now, we consider various matrices $f \in \mathbb{K}^{n \times n}$ with the form:

$$
F_{is} = \mathbb{1}_n + E_{is} \tag{7.4}
$$
$$
F_k(\lambda) = \mathbb{1}_n + (\lambda - 1) E_{kk} \quad \lambda \in \mathbb{K}, i, s, k \in I(n) \tag{7.5}
$$

$$
F_{is} = \mathbb{1}_n + E_{is} = \begin{bmatrix} 1 & 0 & 0 & 0 & 0 \\ 0 & 1 & 0 & 1 & 0 \\ 0 & 0 & 1 & 0 & 0 \\ 0 & 0 & 0 & 1 & 0 \\ 0 & 0 & 0 & 0 & 1 \end{bmatrix} i
$$

$$
F_k(\lambda) = \begin{bmatrix} 1 & & & & \\ & 1 & & 0 & \\ & & \lambda & & \\ & 0 & & 1 & \\ & & & & 1 \end{bmatrix} k
$$

k

Comment 7.1 Inversion and transpose of the "F" matrices.

It is clear that F_{is} and $F_\lambda(k)(\lambda \neq 0)$ belong to $Gl(n, \mathbb{K})$.
Since $F_{is}(\mathbb{1}_n - E_{is}) = \mathbb{1}_n$ and $F_k(\lambda) F_k(\frac{1}{\lambda}) = \mathbb{1}_n$, this is easy to check. The inverses are given by

$$
F_{is}^{-1} = \mathbb{1}_n - E_{is} \text{ and } F_k^{-1}(\lambda) = F_k(\frac{1}{\lambda}). \tag{7.6}
$$

For the transpose, we have

$$F_{ij}^{\mathsf{T}} = F_{ji}.$$
$$F_k(\lambda)^{\mathsf{T}} = F_k(\lambda). \tag{7.7}$$

This follows directly from Eqs. 7.4 and 7.5.

Remark 7.1 The elementary matrices.

The elementary matrices in the literature are usually given by

(i) P_{is} exchange of the two i and s columns or rows,
(ii) $F_k(\lambda)$,
(iii) $F_{is}(\lambda) := \mathbb{1}_n + \lambda E_{is}$.

Elementary operations are obtained using F_{is} and $F_k(\lambda)$. For an $m \times n$-matrix A we obtain:

(a) AF_{is} from A by adding the ith column to the sth column.
(b) $AF_k(\lambda)$ from A by multiplying the kth column with λ.

Analogously, we have $F'_{is} A$ and $F'_k(\lambda)A$, the corresponding row operations for the matrix $A \in \mathbb{K}^{m \times n}$ with $F'_{is}, F'_k(\lambda) \in \mathbb{K}^{m \times m}$.

Remark 7.2 Left and right actions by elementary matrices.

The elementary column (row) operations for a matrix $A \in \mathbb{K}^{m \times n}$ correspond exactly to the right (left) multiplication of A by elementary matrices from $Gl(n, \mathbb{K})$, respectively ($Gl(m, \mathbb{K})$).

Comment 7.2 Elementary row operations and notation.
We use the notation

$$E_s^i \equiv E_{is}, \quad F_s^i \equiv F_{is}, \quad F_s^i(\lambda) \equiv F_{is}$$

and

$$P_{is} \equiv P_s^i.$$

For the elementary row operations with the lower triangular matrix $F_s'^i(\lambda)$, $(i < s)$, we have:

$$\alpha^i \overset{F_s'^i(\lambda)}{\longmapsto} \alpha^i + \lambda\alpha^s \quad \text{and}$$
$$\alpha^j \longmapsto \alpha^j \quad \text{for} \quad j \neq i, \tag{7.8}$$

and with the diagonal matrix $F_j(\lambda)$, we have:

$$\alpha^i \overset{F_i'(\lambda)}{\longmapsto} \lambda\alpha^i \quad \text{and}$$
$$\alpha^j \longmapsto \alpha^j \quad \text{for} \quad j \neq i. \tag{7.9}$$

It is well known that the action on matrices $A \in \mathbb{K}^{m \times n}$ and $B \in \mathbb{K}^{n \times n}$ with rank $B = n$ by a sequence of elementary matrices as given in Remark 7.1, leads to

$$F_1' F_2' \cdots F_{l'}' A = \begin{bmatrix} \mathbb{1}_r & * \\ 0 & * \end{bmatrix} \tag{7.10}$$

and to

$$F_1' F_2' \cdots F_{k'}' B = \mathbb{1}_n. \tag{7.11}$$

Analogously, the column operations lead to the results

$$A\, F_1 \cdots F_l = \begin{bmatrix} \mathbb{1}_r & 0 \\ * & * \end{bmatrix} \tag{7.12}$$

and

$$B\, F_1 \cdots F_k = \mathbb{1}_n. \tag{7.13}$$

Remark 7.3 Normal form of matrices and elementary matrices. Using the tailor-made bases (see Proposition 3.8 and Theorem 3.1), we obtained for an $m \times n$-matrix A with

$$\text{rank}(A) = r, \text{ a representation } \tilde{A} \text{ of } A \text{ in the form}$$
$$\tilde{A} = \begin{bmatrix} \mathbb{1}_r & 0 \\ 0 & 0 \end{bmatrix}. \tag{7.14}$$

The same can be obtained using elementary row and column operations as in Comment 7.1, and in Eqs. (7.10) and (7.12) in Comment 7.2. Hence

$$\tilde{A} = F_1' \ldots F_{l'}' A F_1 \ldots F_l \text{ or } \tilde{A} = F' A F \tag{7.15}$$

with

$$F' := F'_1 F'_2 \ldots F'_{l'} \text{ and } F := F_1 F_2 \ldots F_l. \tag{7.16}$$

$F'_1, F'_2, \ldots, F'_{l'}$ are elementary matrices from $Gl(m)$. $F_1 F_2, \ldots, F_l$ are elementary matrices from $Gl(n)$.

Remark 7.4 Criterion for invertibility.

A is invertible if and only if A is a product of elementary matrices.

Remark 7.5 Equivalent, row equivalent, and column equivalent matrices.

The relation $\tilde{A} \sim A$ defined by $\tilde{A} = F'AF$, as above, is an equivalence relation and we say $\tilde{A}$ and A are equivalent. This was defined in Sect. 3.3 in Definition 3.9. It turns out that in the context of row or column operations, the notion of row equivalent $\left(\underset{r}{\sim}\right)$ or column equivalent $(\underset{c}{\sim})$ are relevant.

So we have the following definition:

Definition 7.1 Row equivalence and column equivalence.
Matrices A and B in $\mathbb{K}^{m \times n}$ are row equivalent, denoted by $A \underset{r}{\sim} B$ if and only if there exists a matrix $F' \in Gl(m)$ such that

$$B = F' A.$$

Similarly, A and B in $\mathbb{K}^{m \times n}$, denoted by $A \underset{c}{\sim} B$, are column equivalent if and only if there exists a matrix $F \in Gl(n)$ such that

$$B = A F.$$

7.2 The Algebraic Aspects of Determinants

We start with an algebraic definition of "determinant".

Definition 7.2 Determinant function.
A map $\Delta : \mathbb{K}^{n \times n} \to \mathbb{K}$ is a determinant function if

$$(\Delta 1) \qquad \Delta(AF_{is}) = \Delta(A) \qquad i \neq s,$$

$$(\Delta 2) \qquad \Delta\left(AF_k(\lambda)\right) = \lambda\Delta(A) \qquad \forall \lambda \in \mathbb{K}.$$

Axiom $\Delta 1$ means that Δ is invariant if we add to a given column another column. This is a specific additive property of Δ. Geometrically means that Δ is shear invariant.

Figure 7.1 shows in 2 dimensions the shear invariance of the area of a parallelogram. As is well-known, this is given by the Euclidean geometry. Here, it corresponds to a certain determinant function which is usually denoted by det (see Definition 7.4).

Property $\Delta 2$ means that if we scale a given column by λ, the same happens to Δ. We may call this homogeneity or the scaling property of Δ.

Using the definition given in Sect. 7.1, we set $B := AF_{is}$ and $B' = AF_k(\lambda)$. It is obvious that A and B' are column equivalent $A \underset{c}{\sim} B$, and similarly A and B' are column equivalent $A \underset{c}{\sim} B'$.

We may also express the properties $\Delta 1$ and $\Delta 2$ using, as usual, the identification between a list and the corresponding matrix:

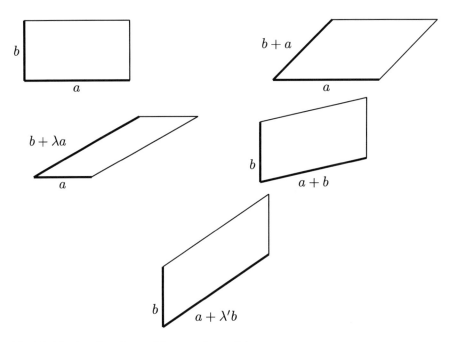

Fig. 7.1 The shear invariance of the area of a parallelogram

$$A = (a_1, a_2, \ldots, a_n) = [a_1 a_2 \ldots a_n] \in \mathbb{K}^{n \times n}$$

shortly, by

$(\Delta 1)$	$\Delta(\ldots a_r + a_s \ldots a_s \ldots)$	$= \Delta(\ldots a_r \ldots a_s \ldots),$
$(\Delta 2)$	$\Delta(\ldots \lambda a_r \ldots)$	$= \lambda \Delta(\ldots a_r \ldots \ldots).$

Remark 7.6 Zero scaling.

The scalar λ is allowed to be zero! This means that if a column in A is zero, then $\Delta(A)$ is zero too.

Remark 7.7 Determinant functions as vector space.

The space of determinant functions $\Lambda^n = \{\Delta \ldots\}$ is a vector space: This is evident since the space of functions $\Delta : K^{n \times n} \to \mathbb{K}$ is a vector space. The additional conditions in axioms $\Delta 1$ and $\Delta 2$ in Definition 7.2 do not affect the vector space structure since they respect linear combinations.

From the axioms $\Delta 1$ and $\Delta 2$, four very important properties follow:

Comment 7.3 Elementary transformations act like scalars.

We can summarize the above axioms by the following characteristic property: Every elementary transformation acts on determinant functions like a scalar. This leads to the next proposition. In the proof of it, we will see explicitly how it works.

Proposition 7.1 *First implications for the determinant functions.*

(i) *Let matrix $A \in \mathbb{K}^{n \times n}$ have rank$(A) < n$, then $\Delta(A) = 0$.*
(ii) *If $\Delta(\mathbb{1}_n) = 0$, then $\Delta = 0$.*
(iii) *If rank$(A) = n$, then there is a scalar $\lambda' \neq 0$ so that $\Delta(A) = \lambda' \Delta(\mathbb{1}_n)$.*
(iv) *dim$(\Lambda^n) = 1$.*

Proof

(i) From Remarks 7.3 and 7.6, for $r = \text{rank } A < n$, that is, $(r \neq n)$ and the properties of elementary matrices, we obtain the sequence of equations with scalars $\lambda_1, \ldots, \lambda_l$ and $\lambda'_1, \ldots, \lambda'_{l'}$:

$$\Delta(A) = \Delta(F'_1 F'_2 \ldots F'_{l'}\begin{bmatrix} \mathbb{1}_r & 0 \\ 0 & 0 \end{bmatrix} F_1 F_2 \ldots F_l),$$

$$\vdots$$

$$\Delta(A) = \Delta(F'_1 \ldots F'_{l'}\begin{bmatrix} \mathbb{1}_r & 0 \\ 0 & 0 \end{bmatrix})\lambda_1, \ldots \lambda_l,$$

$$\text{and} \qquad \Delta(A) = \lambda'_1 \lambda'_2 \ldots \lambda'_{l'} \Delta(\begin{bmatrix} \mathbb{1}_r & 0 \\ 0 & 0 \end{bmatrix})\lambda_1 \ldots \lambda_l.$$

Since for example the last columns are zero, we obtain also $\Delta(A) = 0$. This proves (i).

(ii) In this case, we may assume that $r = n$, for example $\Delta(A) = \Delta(F\mathbb{1}_n)$. Analogously as before, we have $\Delta(A) = \lambda'\Delta(\mathbb{1}_n)$ with $\lambda' \neq 0$. It is now clear that if $\Delta(\mathbb{1}_n) = 0$, it follows $\Delta = 0$ as well. So (ii) is proven.

(iii) In the proof of (ii), we found $\Delta(A) = \lambda'\Delta(\mathbb{1}_n)$ with $\lambda' \neq 0$, so (iii) is already proven.

(iv) The result of (iii) means essentially by itself that the dimension of the space of the determinant functions is 1: $\Delta(A) = \lambda\Delta(\mathbb{1}_n)$ signifies that finally for every matrix A, it is the value $\Delta(\mathbb{1}_n)$ that counts. Since $\Delta(\mathbb{1}_n) \in \mathbb{K}$, we have:

$$\Lambda^n := \{\Delta\} \underset{bij}{\overset{\sim}{=}} \Delta(\mathbb{1}_n)\mathbb{K} \overset{\sim}{=} \mathbb{K}$$

and $\dim(\Lambda^n) = 1$. This proves iv and therefore proposition 7.1.

$\square$

Furthermore, the relation $\Delta(A) = \lambda'\Delta(\mathbb{1}_n)$ leads to the standard determinant, det: We take Δ_0 with

$$\Delta_0(\mathbb{1}_n) = 1. \tag{7.17}$$

Then we have $\Delta(\mathbb{1}_n) = \lambda_\Delta \in \mathbb{K}$ and we may write

$$\Delta(\mathbb{1}_n) = \lambda_\Delta \Delta_0(\mathbb{1}_n). \tag{7.18}$$

This means that we also get

$$\Delta = \lambda_\Delta \Delta_0 \quad \text{and} \quad \Delta(A) = \lambda_\Delta \Delta_0(A). \tag{7.19}$$

This leads to the following definition and proposition.

Definition 7.3 Normalization of the determinant function.
In addition to ($\Delta 1$) and ($\Delta 2$) above, we normalize ($\Delta 3$): For $\Delta_0 \in \Lambda^n$, we take $\Delta_0(\mathbb{1}_n) = 1$.

Proposition 7.2 *Uniqueness of Δ_0.*

The axioms $\Delta 1$, $\Delta 2$, and $\Delta 3$ determine uniquely the function $\Delta_0 \in \Lambda^n$ with $\Delta_0(\mathbb{1}_n) = 1$.

Proof For $\Delta \in \Lambda^n$ with $\Delta(\mathbb{1}_n) = 1$, according to Proposition 7.1 (iii), and Eqs. (7.18) and (7.19,) we have:

$$\Delta = \lambda_\Delta \Delta_0 \quad \text{and}$$
$$\Delta(\mathbb{1}_n) = \lambda_\Delta \Delta_0(\mathbb{1}_n).$$

Since $\Delta(\mathbb{1}_n) \overset{!}{=} 1$ and $\Delta_0(\mathbb{1}_n) = 1$, this gives

$$1 = \lambda_\Delta 1 \quad \text{and} \quad \lambda_\Delta = 1.$$

So $\Delta = \Delta_0$ and Δ_0 is uniquely defined. $\qquad\square$

Definition 7.4 We define the standard det to be Δ_0 and write $\det := \Delta_0$.

Corollary 7.1 Δ *and* det.

Using Eq. 7.19 and Definition 7.4, we can write

$$\Delta = \lambda_\Delta \det. \tag{7.20}$$

$\square$

Having shown the uniqueness of det, we are going now to show also its existence inductively with respect to the dimension n.

> **Proposition 7.3** *The existence of the determinant.*
>
> *Axioms $\Delta 1$, $\Delta 2$, and $\Delta 3$ determine uniquely the determinant function $\Delta = \det : \mathbb{K}^{n \times n} \to \mathbb{K}$.*
> *For low dimensions we have, as is well-known, in an obvious notation the following results:*

Proof For $n = 1$: The existence is clear.

Proof For $n = 2$: If we put $A = (a, b) \equiv [ab] = \begin{bmatrix} \alpha^1 & \beta^1 \\ \alpha^2 & \beta^2 \end{bmatrix}$, we can set $\det_2(A) := \alpha^1 \beta^2 - \alpha^2 \beta^1$.

We verify the axioms $(\Delta 1)$, $(\Delta 2)$, and $(\Delta 3)$:

$\Delta 1$: $\det_2(a + b, b)$ $= \det_2 \begin{bmatrix} \alpha^1 + \beta^1 & \beta^1 \\ \alpha^2 + \beta^2 & \beta^2 \end{bmatrix} = (\alpha^1 + \beta^1)\beta^2 - (\alpha^2 + \beta^2)\beta^1$.

  $= \alpha^1 \beta^2 + \beta^1 \beta^2 - \alpha^2 \beta^1 - \beta^2 \beta^1 = \alpha^1 \beta^2 - \alpha^2 \beta^1 = \det_2(a, b)$.

$\Delta 2$: $\det_2(\lambda a, b)$ $= \det_2 \begin{bmatrix} \lambda\alpha^1 & \beta^1 \\ \lambda\alpha^2 & \beta^2 \end{bmatrix} = \lambda\alpha^1 \beta^2 - \lambda\alpha^2 \beta^1 = \lambda \det_2(a, b)$.

$\Delta 3$: $\det_2(I_2)$ $= \det_2 \begin{bmatrix} 1 & 0 \\ 0 & 1 \end{bmatrix} = 1$.

So the existence of $n = 2$ is proven.

Proof For $n = 3$: It is interesting and useful to proof the existence of $\det_3$ using the well-known iterative expression for $A = [a_1 a_2 a_3] = [abc]$:

$$\det_3 A = \alpha^1 \det_2(\bar{b}, \bar{c}) - \beta^1 \det_2(\bar{a}, \bar{c}) + \gamma^1 \det_2(\bar{a}, \bar{b}), \qquad (7.21)$$

$$\text{with}\quad [a\, b\, c] := \begin{bmatrix} \alpha^1 & \beta^1 & \gamma^1 \\ \alpha^2 & \beta^2 & \gamma^2 \\ \alpha^3 & \beta^3 & \gamma^3 \end{bmatrix} \quad \text{and} \quad [\bar{a}\, \bar{b}\, \bar{c}] := \begin{bmatrix} \alpha^2 & \beta^2 & \gamma^2 \\ \alpha^3 & \beta^3 & \gamma^3 \end{bmatrix}.$$

We have again to show that the axioms $(\Delta 1)$, $(\Delta 2)$, and $(\Delta 3)$ are valid:

$\Delta 3$:   is clear, $\det_3(\mathbb{1}_3) = 1$.

$\Delta 2$:   We have, for example, for the second column

$$\det(a, \lambda b, c) = \begin{bmatrix} \alpha^1 & \lambda\beta^1 & \gamma^1 \\ \alpha^2 & \lambda\beta^2 & \gamma^2 \\ \alpha^3 & \lambda\beta^3 & \gamma^3 \end{bmatrix},$$

$$\det(a, \lambda b, c) = \alpha^1 \det_2(\lambda\bar{b}, \bar{c}) - \lambda\beta^1 \det_2(\bar{a}, \bar{c}) + \gamma^1 \det_2(\bar{a}, \lambda\bar{b}),$$

$$= \alpha^1 \lambda \det_2(\bar{b}, \lambda\bar{c}) - \lambda\beta^1 \det_2(\bar{a}, \bar{c}) + \gamma^1 \lambda \det_2(\bar{a}, \bar{b}).$$

Using Eq. 7.21, we obtain

$$\lambda \det(a, b, c).$$

This proves $\Delta 2$.

What still needs to be shown is that axiom $\Delta 1$ (the shear invariance) is also valid. We proof this in the case of the first column as the other two cases are similar: $\det_3(a + b, b, c) = \det_3(a, b, c)$, using $\det_2(\bar{a} + \bar{b}, \bar{b}) = \det_2(\bar{a}, \bar{b})$:

$$\det(a + b, b, c) =$$
$$= (\alpha^1 + \beta^1) \det_2(\bar{b}, \bar{c}) - \beta^1 \det_2(\bar{b} + \bar{b}, \bar{c}) + \gamma^1 \det_2(\bar{a} + \bar{b}, \bar{b}),$$
$$= \alpha^1 \det_2(\bar{b}, \bar{c}) + \beta^1 \det_2(\bar{b}, \bar{c}) - \beta^1 \det_2(\bar{a}, \bar{c}) - \beta^1 \det_2(\bar{b}, \bar{c}) + \gamma^1 \det_2(\bar{a}, \bar{b}),$$
$$= \alpha^1 \det_2(\bar{b}, \bar{c}) - \beta^1 \det_2(\bar{a}, \bar{c}) + \gamma^1 \det_2(\bar{a}, \bar{b}),$$
$$= \det_3(a, b, c).$$

So the existence of $n = 3$ is also proven.

We now proceed to the proof of the existence of the determinant. For the proof of the existence of det, we use induction with respect to n. We start with an appropriate generalization of Eq. (7.21). Assume $\det_{n-1}$ satisfies axioms $\Delta 1$, $\Delta 2$ and $\Delta 3$.

Proof Proof of Proposition 7.3 for $n \in \mathbb{N}$.

$$\det_n(a_1, \ldots, a_n) := \alpha_1^1 \det_{n-1}(\check{b}_1, b_2, \ldots, b_n) - \alpha_2^1 \det_{n-1}(b_1, \check{b}_2, \ldots, b_n) \ldots +$$
$$(-1)^n \alpha_n^1 \det(b_1, \ldots, b_{n-1}),$$

using

$$(a_1, a_2, \ldots, a_n) = \begin{bmatrix} \alpha_1^1, & \alpha_2^1 & \cdots & \alpha_n^1 \\ b_1, & b_2 & & b_n \end{bmatrix} \text{ with } b_1, \ldots, b_n \in \mathbb{K}^{n-1},$$

where $(b_1, \ldots, \check{b}_i, \ldots, b_n)$ indicates the list $(b_1, \ldots, b_n)$ but omitting b_i. We see that $\det_n$ is linear in every column:

$$\alpha_1^1 \det_{n-1} B_1 - \alpha_2^1 \det_{n-1}, B_2 \ldots + (-1)^n \alpha_n^1 \det_{n-1} B_n$$

with

$$B_1 = (\check{b}_1, b_2, \ldots, b_n), B_2 = (b_1, \check{b}_2, b_3 \ldots b_n), \ldots, B_n = (b_1, \ldots, b_{n-1}),$$

so $\Delta 2$ is valid since linearity contains the homogeneity ($\Delta 2$). In order to show that $\Delta 1$ is valid as well, we proceed in an analogous way as in the case of $n = 3$ and we can show e.g that

$$\det_{n}(a_1 + a_2, a_2, \ldots, a_n) = \det(a_1, \ldots, a_n). \tag{7.22}$$

$\Delta 3$ is evidently also valid: $\Delta(\mathbb{1}_n) = 1$.

So the existence is also proven for $n \in \mathbb{N}$.   □

We therefore consider the existence of the determinant established.

Additionally, we have proven that Δ is a multilinear map with respect to the columns. Written in a symbolic way, we have the property

$$(\Delta 4) \qquad \begin{aligned} \Delta(a_r + a_s) &= \Delta(a_r) + \Delta(a_s), \\ \Delta(\lambda a_r) &= \lambda \Delta(a_r). \end{aligned}$$

A more precise formulation of multi-linearity is given by the following definition: Λ is column-multilinear if for every $i \in I(n)$ and fixed $a_1, \ldots a_{i-1}, a_{i+1}, \ldots a_n$, the map

$$x \rightarrow \Delta(a_1, a_2, \ldots, a_{i-1}, x, a_{i+1} \ldots, a_n)$$

is linear. In the following remark, we discuss the question of multilinearity as example for $n = 3$, using the Leibniz formula.

Remark 7.8 Column linearity.

The expression of Eq. (7.21) has the form

$$\det_{3} A = \varepsilon_{ijk} \alpha^i \beta^j \gamma^k.$$

So every factor is linear in α, β, γ separately, for example $(\alpha + \zeta)\beta\gamma = \alpha\beta\gamma + \zeta\beta\gamma$ and $\beta \mapsto \lambda\beta$ leads to $\lambda\varepsilon_{ijk}\alpha^i\beta^j\gamma^k$. This means that $\det_{3} A$ (as well as $\det_{2} A$) has in addition the following property :

$\Delta 4$: $\det_{3}$ is linear in every column which of course includes the axiom $\Delta 2$.

This leads to a further definition of determinants which is very common in literature.

7.3   Second Definition of the Determinant

The definition is given by the two axioms.

$$(D1) \quad \Delta : \mathbb{K}^{n \times n} \quad \rightarrow \quad \mathbb{K} \text{ is multilinear with respect to the columns,}$$
$$\qquad\qquad A \qquad\quad \mapsto \quad \Delta(A),$$
$$(D2) \quad \text{and} \qquad\qquad : \quad \Delta(A) = 0 \text{ if } \operatorname{rank}(A) < n.$$

The property $D2$ is also called alternating.

Remark 7.9 The definitions $(D1, D2)$ and $(\Delta1, \Delta2)$ are equivalent.

Proof

From $(\Delta1, \Delta2)$ it follows that Δ is multilinear so $D1$ holds.
Axiom $D2$ follows from Proposition 7.1 (i), so we have $(\Delta1, \Delta2) \Rightarrow (D1, D2)$.
Axiom $\Delta2$ follows from $D1$ since homogeneity follows from the linearity of Δ.
From $D2$ and the multi-linearity $(D1)$, since rank $(b, b, \ldots) < n$ gives $\Delta(b, b, \ldots) = 0$, we have for example

$$\Delta(a + b, \ldots) = \Delta(a, b, \ldots) + \Delta(b, b, \ldots) \text{ and } \Delta(a + b, \ldots) = \Delta(a, b).$$

So it follows $\Delta1$ and we have $(D1, D2) \Rightarrow (\Delta1, \Delta2)$. $\qquad\square$

7.4 Properties of the Determinants

We summarize some of the most important properties of det below. Most properties follow directly from the existence of $\Delta1$ and $\Delta2$ (or $D1$ and $D2$) and the normalization

$$\det(\mathbb{1}_n) = 1 \quad (\Delta3).$$

Interestingly, we do not have to explicitly use the permutation group (S_n) at this level.

(i) The determinant of a matrix is linear in every column. This is equivalent to the determinant being a multilinear map on $\mathbb{K}^n$.

(ii) The determinant remains unchanged if we add a linear combination of some columns to a different column. This corresponds geometrically, as we saw in Sect. 7.2 and Fig. 7.1, to the shear transformation invariance of the determinant.

(iii) The determinant is zero if the matrix columns are linearly dependent. This is closely connected with the next property.

(iv) The determinant changes sign if two columns are interchanged. This means that the determinant is an alternating multilinear form.

(v) Multiplication law: If Δ is the normalized determinant (with $\Delta(\mathbb{1}_n) = 1$), then

$$\Delta(AB) = \Delta(A)\Delta(B) \text{ if } \Delta(\mathbb{1}_n) = 1.$$

Proof Let Δ_1 and Δ_2 be determinant functions. From Proposition 7.1 (i), it first follows that

$$\Delta_1(\mathbb{1}_n)\Delta_2(A) = \Delta_2(\mathbb{1}_n)\Delta_1(A) \tag{7.23}$$

since $\Delta_3(A) := \Delta_1(\mathbb{1}_n)\Delta_2(A) - \Delta_2(\mathbb{1}_n)\Delta_1(A)$ is also a det function and with $\Delta_3(\mathbb{1}_n) = \Delta_1(\mathbb{1}_n)\Delta_2(A) - \Delta_2(\mathbb{1}_n)\Delta_1(\mathbb{1}_n) = 0$, we have $\Delta_3 = 0$ so that (7.23) holds.

Setting

$$\Delta_1(B) := \Delta(AB) \quad \text{which is also a det function,}$$
$$\Delta_1(\mathbb{1}_n) = \Delta(A),$$

and using (7.23) for Δ_1 and Δ we obtain

$$\Delta(\mathbb{1}_n)\Delta_1(B) = \Delta_1(\mathbb{1}_n)\Delta(B),$$
$$\Delta_1(B) = \Delta(A)\Delta(B) \text{ and}$$
$$\Delta(AB) = \Delta(A)\Delta(B).$$

$\square$

(vi) Any determinant function Δ is transposition invariant:

$$\Delta(A^\mathsf{T}) = \Delta(A) \quad \text{holds.}$$

Proof This follows from the fact that every invertible matrix A is a product of an elementary matrix (see Comments 7.1 and 7.2, and Remark 7.3):

$$A = F_1, F_2 \dots F_m$$

and for every elementary matrix $F_j, \; j \in I(m)$

$$\det F_j^\mathsf{T} = \det F_j$$

holds (see Comment 7.1). So we have, using the multiplication law,

$$\det A^\mathsf{T} \quad = \det(F_1 \dots F_m)^\mathsf{T} \quad = \det(F_m^\mathsf{T} \dots F_1^\mathsf{T}) = \det F_m^\mathsf{T} \dots \det F_1^\mathsf{T} \text{ and}$$
$$\det A^\mathsf{T} \quad = \det F_m \dots \det F_1 \quad = \det A.$$

$\square$

(vii) The multi-linearity leads to the expression

$$\det(\lambda A) = \lambda^n \det A.$$

(viii) For an upper triangular matrix, the determinant function Δ is given by the product of the diagonal elements:

$$\Delta \begin{bmatrix} a_1 & & * \\ & \ddots & \\ 0 & & a_n \end{bmatrix} = a_1 a_2 \ldots a_n.$$

Proof Using those elementary operations which leave the determinant functions invariant, we obtain

$$\Delta \begin{bmatrix} a_1 & & * \\ & \ddots & \\ 0 & & a_n \end{bmatrix} = \Delta \begin{bmatrix} a_1 & & 0 \\ & \ddots & \\ 0 & & a_1 \end{bmatrix} = a_1 \ldots a_n \Delta(\mathbb{1}_n) = a_1 \ldots a_n.$$

$\square$

(ix) Let $\begin{bmatrix} A & B \\ C & D \end{bmatrix}$ be a block matrix with $A \in \mathbb{K}^{r \times s}$, $B \in \mathbb{K}^{r \times (n-s)}$, $C \in \mathbb{K}^{(m-r) \times s}$, and $D \in \mathbb{K}^{(m-r) \times (n-s)}$, then the following holds: $\det \begin{bmatrix} A & 0 \\ 0 & D \end{bmatrix} = \det A \det D$.

Proof

If we define $\Delta(A) := \det \begin{bmatrix} A & 0 \\ 0 & D \end{bmatrix}$ which is a det function,

using $\quad \Delta(A) = \Delta(\mathbb{1}_n) \det A$ and $\Delta(\mathbb{1}_n) = \det D$,

we obtain $\quad \det \begin{bmatrix} A & 0 \\ 0 & D \end{bmatrix} = \det D \det A = \det A \det D$.

$\square$

(x) Using the same notation as in (ix), the following holds: $\det \begin{bmatrix} A & B \\ 0 & D \end{bmatrix} = \det A \det D$.

Proof If A is invertible, we have:

$$\begin{bmatrix} A & B \\ 0 & D \end{bmatrix} = \begin{bmatrix} A & 0 \\ 0 & D \end{bmatrix} \begin{bmatrix} I & A^{-1}B \\ 0 & I \end{bmatrix} = \det A \det D.$$

$\square$

(xi) A is invertible if and only if $\det A \neq 0$.

Proof If A is invertible, it means that there exists a B so that $AB = \mathbb{1}_n$ and $\det(AB) = \det A \det B = 1 \Rightarrow \det A \neq 0$.

If $\det A \neq 0$ it means that rank $A = n$. Otherwise, rank $A < n$, and we would have $\det A = 0$.

But rank $A = n$ means that ker $A = 0$ and thus A is invertible. $\square$

(xii) All the properties of the determinant that refer to the columns also hold when replacing columns with rows.

(xiii) Cofactor expansion concerning the columns (rows).

From a given matrix $A \equiv (a_s)_n \equiv (\alpha_{is})$, we define various matrices with respect to the fixed position (i, s).

$$A_{is} := (a_1, \dots, a_{s-1}, e_i, a_{s+1}, \dots a_n)$$

$$A(1)_{is} := \begin{bmatrix} A_{11} & 0 & A_{12} \\ 0 & 1 & 0 \\ A_{21} & 0 & A_{22} \end{bmatrix} \text{ where } 1 \text{ is in the } i\text{th row and the } s\text{th column}$$

$$\acute{A}_{is} := \begin{bmatrix} A_{11} & A_{12} \\ A_{21} & A_{22} \end{bmatrix} \text{ is the } (n-1) \text{ by } (n-1) \text{ matrix}$$

where the ith row and the sth column have been deleted.

Let $\gamma_{is} := \det A_{is}$, then $C := (\gamma_{is})$ is called the cofactor of A.

If we use elementary matrix operation, we see that the entry γ_{is} is given by the expressions

$$\gamma_{is} = \det A_{is} = \det(A(1)_{is}) = (-1)^{i+s} \det \acute{A}_{is}. \tag{7.24}$$

Proposition 7.4 *The cofactor expansion (Laplace expansion).*

The adjunct $A_{ij}^\#$ of A is given by $A^\# = (\alpha_{ij}^\#)$ where $\alpha_{ij}^\# = \gamma_{ji}$, or, equivalently, $A^\# = C^\mathsf{T}$. Then the cofactor expansion with respect to the columns is given by

$$A^\# A = A A^\# = (\det A)\mathbb{1}_n$$

or equivalently by

$$\sum_{k=1}^{n} \alpha_{ik}^\# \alpha_{ks} = \delta_{is}(\det A).$$

Proof The calculation of the components of the matrix $A^\# A$ is given by (i fixed)

$$\sum_{k=1}^{n} \alpha_{ik}^\# \alpha_{ks} \qquad = \sum_{k=1}^{n} \det(a_1, \dots, a_{i-1}, e_k, a_{i+1} \dots a_n)\alpha_{ks},$$

$$= \det(a_1, \dots, a_{i-1}, \sum_k e_k \alpha_{ks}, a_{i+1} \dots, a_n),$$

$$= \det(a_1, a_{i-1}, a_s, a_{i+1}, \dots, a_n).$$

If $s \neq i$, we have

$$\det(a_1, \dots, a_{i-1}, a_s, a_{i+1}, \dots a_n) = 0$$

$$\text{since } \det(\dots a_s \dots a_s \dots) = 0.$$

If $s = i$, we have

$$\sum_{k=1}^{n} \alpha_{ik}^\# \alpha_{ks} \qquad = \det(a_1, \dots, a_{i-1}, a_i, a_{i+1}, \dots a_n) = \det A.$$

So we have altogether

$$\sum_{k=1}^{n} \alpha_{ik}^{\#} \alpha_{ks} = \delta_{is}(\det A).$$ □

The corresponding expression for the rows is given by

$$\sum_{k=1}^{n} \alpha_{ik} \alpha_{ks}^{\#} = \delta_{is} \det A.$$

Remark 7.10 Laplace expansion of a determinant.

With $i = 1$ and $s = 1$, we obtain $\sum_{k=1}^{n} \alpha_{1k} \alpha_{k1}^{\#} = \det A$ or equivalently

$$\det A = \sum_{k=1}^{n} (-)^{1+k} \alpha_{1k} \det A_{1k}.$$

This is the recursion formula which was used in the proof of the existence of det.

7.5 Geometric Aspects of the Determinants

As we saw so far, the determinant gives essential information about $(n \times n)$-matrices and subsequently about linear maps between vector spaces of the same dimension, for example, about linear operators (endomorphism).

In addition, determinants also have a deep geometric significance. We restrict ourselves to $\mathbb{R}$- vector spaces to simplify the explanations and support the intuition. The determinant of an operator $f : V \to V$ measures how this f changes the volume of solids in V. In addition, since det f is a scalar with positive and negative values, it also measures how this f changes the orientation in V. The determinant by itself turns out to be essentially a subtle geometric structure that defines the volume and the orientation in V.

It is important to note that this is a new geometric structure on V called a volume form. It may be, or rather has to be defined directly on an abstract vector space (a vector space without a scalar product on it). Despite this, if we already have a Euclidean vector space V, this induces a specific volume form on V. Hence, the volume form is a weaker geometric structure than a scalar product.

We want to demonstrate these ideas in the simplest nontrivial case. We consider the two-dimensional Euclidean space $\mathbb{R}^2$ with its standard basis (e_1, e_2). It is understood that our discussion is also valid for $\mathbb{R}^3, \mathbb{R}^4, \ldots \mathbb{R}^n$.

We start with a parallelogram $P(a_1, a_2)$ given by $a_1 = \begin{bmatrix} \alpha_1^1 \\ \alpha_1^2 \end{bmatrix}$ and $a_2 = \begin{bmatrix} \alpha_2^1 \\ \alpha_2^2 \end{bmatrix}$. The area of $P(a_1, a_2)$ is given by the usual formula. For the square, we have

$$\text{volume}_2(a_1, a_2)^2 = \| a_1 \|^2 \| a_2 \|^2 \sin^2 \alpha, \tag{7.25}$$

with α, the angle $\alpha = \measuredangle(a_1, a_2)$, and $\| a_i \|^2 = \langle a_i \mid a_i \rangle$,

$$\text{volume}_2(a_1, a_2)^2 = \langle a_1 \mid a_1 \rangle \langle a_2 \mid a_2 \rangle - \langle a_1 \mid a_2 \rangle^2. \tag{7.26}$$

Applying the linear map f to a_1, a_2 and volume$_2$, we have from

$$f(a_i) = a_\mu \varphi_i^\mu, \quad \varphi_s^i \in \mathbb{R}, \quad F = (\varphi_s^i), \quad i, \mu, s \in I(2) \tag{7.27}$$

and

$$(fa_1, fa_2) = (a_1, a_2)F \tag{7.28}$$

$$\text{volume}_2(fa_1, fa_2)^2 = \langle fa_1, \mid fa_1 \rangle \langle fa_2 \mid fa_2 \rangle - \langle fa_1 \mid fa_2 \rangle^2. \tag{7.29}$$

A straightforward calculation gives the very interesting result

$$\langle fa_1 \mid fa_1 \rangle \langle fa_2 \mid fa_2 \rangle - \langle fa_1 \mid fa_2 \rangle^2$$
$$= (\varphi_1^1 \varphi_2^2 - \varphi_1^2 \varphi_2^1)^2 (\langle a_1 \mid a_1 \rangle \langle a_2 \mid a_2 \rangle - \langle a_2 \mid a_2 \rangle \langle a_1 \mid a_2 \rangle). \tag{7.30}$$

This is in fact

$$\text{volume}_2(fa_1, fa_2)^2 = (\det F)^2 \text{ volume}_2(a_1, a_2)^2 \tag{7.31}$$

and with the definition $P := P(a_1, a_2)$ for the parallelogram (a_1, a_2), $P' := P(fa_1, fa_2)$, Eq. (7.31) may be written as

$$(\text{volume}_2 P')^2 = (\det F)^2 (\text{volume}_2 P)^2. \tag{7.32}$$

There are a few important remarks to be made:

(i) The ratio volume$_2$ P' / volume$_2$ P is independent of the dot product in $\mathbb{R}^2$.
(ii) A nontrivial result is obtained if both, (a_1, a_2) and (fa_1, fa_2), are linearly independent, that is $B := (a_1, a_2)$ and $B' := (fa_1, fa_2)$ are bases in $\mathbb{R}^2$ and of course f is an isomorphism $(\det F \neq 0)$.
(iii) We remember (see Proposition 3.1) that every arbitrary basis C in $\mathbb{R}^2$ can be obtained from the standard basis $E = (e_1, e_2)$ by applying an isomorphism g in

$\mathbb{R}^2 : C = g(E)$ or equivalently $C = EG$ because $G = g_{CE} = (\gamma_s^i = \gamma_{is})$. This means that if we fix $\text{volume}_2(e_1, e_2) = \text{vol}_2 \in \mathbb{R}$, then $\text{volume}_2(c_1, c_2)$ is given by

$$\text{volume}_2(c_1, c_2)^2 = (\det G)^2 \, \text{volume}_2(e_1, e_2) = (\det G)^2 \, \text{vol}_2^2 . \qquad (7.33)$$

To simplify things, we may put $\text{volume}_2(e_1, e_2) = 1$ and we have

$$\text{volume}_2(c_1, c_2)^2 = (\det G)^2. \qquad (7.34)$$

We can go one step further and define

$$\text{volume}_2(c_1, c_2) := \det G. \qquad (7.35)$$

The result is that we may define on $\mathbb{R}^2$ and in every two-dimensional vector space V, a signed volume which is completely independent of the presence or not of a scalar product. Slightly more generally we may write, as example for $n = 2$, using the notation of Sect. 7.2, the following definition:

Definition 7.5 Volume form on $\mathbb{R}^2$.
A volume form D is a determinant form on $\mathbb{R}^2$.

$$D : \qquad \mathbb{R}^2 \times \mathbb{R}^2 \qquad \longrightarrow \mathbb{R},$$
$$(a_1, a_2) \qquad \longmapsto D(a_1, a_2) := \Delta(A)$$

with $A = [a_1 a_2]$. For $\Delta_o(\mathbb{1}_2) = 1$, we have $\Delta_o = \det$ and

$$D_o(e_1, e_2) = \det(\mathbb{1}_2) = 1.$$

(iv) The interpretation of the signs can be read off from the example

$$D_o(e_1, e_2) \quad = \det[e_1 e_2] \quad = \det \begin{bmatrix} 1 & 0 \\ 0 & 1 \end{bmatrix} \quad = +1$$

and

$$D_o(e_2, e_1) \quad = \det[e_2 e_1] \quad = \det \begin{bmatrix} 0 & 1 \\ 1 & 0 \end{bmatrix} \quad = -1,$$

or more generally

$$D_o(a_1, a_2) \quad = \det[a_1 a_2] \quad = \det A,$$
$$D_o(a_2, a_1) \quad = \det[a_2 a_1] \quad = -\det A.$$

The sign of a given volume form characterizes the "standard" or the "non-standard" orientation of a basis.

(v) From Eq. (7.32), we may write

$$\text{volume}_2\, P' = (\det F)\, \text{volume}_2\, P. \tag{7.36}$$

In this case, if $\det F = -1$, then the basis $B' = (fa_1, fa_2)$ has a different orientation to the basis $B = (a_1, a_2)$. This means that the linear map F changes the orientation.

So far, we actually used the term orientation in a common way. In the next section, we are going to focus our attention on a more profound discussion of this term.

7.6   Orientation on an Abstract Vector Space

The last discussion is a good introduction to the notion of orientation on a real vector space. Orientation is a special structure that can be introduced on an abstract vector space.

Orientation plays a very important role in both physics and mathematics, and has a great impact on daily life. The scientists who are confronted with this concept have a good intuitive understanding of it. Here, we are going to give the precise definition of it. Section 1.2 and our discussion in Sect. 7.5 will be very helpful.

On a vector space V, we first consider all the bases. The reason is that bases, and in particular the relations between them, are the source of additional structures in an abstract vector space. So we choose a basis $A = (a_1, \ldots, a_n)$ and a second basis $B = (b_1, \ldots, b_n)$. There exist certain relations between them, given by the determinant of their transition matrix. It is clear that the transition matrix is invertible and therefore its determinant is nonzero. Here however, we are only interested in whether this determinant is positive or negative. This determines the equivalence relation between the set of bases $B(V)$ of V which we call orientation.

We say two bases are orientation equivalent if and only if the determinant of their transition matrix is positive. In this case, the two bases are consistently oriented. As we learned in Sect. 1.2, this equivalence relation leads to a class decomposition of the set of bases, and so to a quotient space consisting of subsets of bases.

Furthermore, it is evident that this quotient space consists only of two elements, only of two subsets of $B(V)$: The bases consistently oriented to the chosen basis A (with the positive determinant of the corresponding transition matrix), and the bases having opposite orientations to the chosen basis A (with the negative determinant of the corresponding transition matrix).

The next definition summarizes the above considerations.

Definition 7.6 Orientation as an equivalence relation.
Two bases $A = (a_1, \ldots, a_n)$ and $B = (b_1, \ldots, b_n)$ of a real vector space V have the same orientation, denoted by "or", if the transition matrix T, given by $A = BT$ where $T \in Gl(n) \subset \mathbb{R}^{n \times n}$ has positive determinants.

In this case, we write $B \overset{\sim}{\text{or}} A$ and we say also that A and B are consistently oriented. Note that $A = BT$ means equally well

$$[a_1 \cdots a_n] = [b_1 \cdots b_n] \, T$$

or equivalently

$$a_s = b_s \tau_s^i \quad \text{with} \quad T = (\tau_s^i) \quad \tau_s^i \in \mathbb{R} \quad \text{and} \quad i, s \in I(n).$$

Comment 7.4 $\overset{\sim}{\text{or}}$ is an equivalence relation.

Proof We need to show that $\overset{\sim}{\text{or}}$ is (i) reflexive and (ii) symmetric.

(i) $A \overset{\sim}{\text{or}} A$
   since $A = \mathbb{1}_n$, so $\overset{\sim}{\text{or}}$ is reflexive;
(ii) $B \overset{\sim}{\text{or}} A \Leftrightarrow A \overset{\sim}{\text{or}} B$
   since, if $A = BT$, then, with $\det T = $ positive, $B = AT^{-1}$ and $\det T^{-1} = $ positive, so $\overset{\sim}{\text{or}}$ is symmetric;
(iii) $A \overset{\sim}{\text{or}} B$ and $B \overset{\sim}{\text{or}} C \Rightarrow A \overset{\sim}{\text{or}} C$
   since, if $A = BT$ and $B = CT'$, then, with $\det T > 0$ and $\det T' > 0$, we get $A = BT = CT'T$, with $\det TT' = \det T \det T' > 0$, so $\overset{\sim}{\text{or}}$ is transitive.

$\square$

Remark 7.11 $Gl^+(n) < Gl(n)$.

The subset $Gl^+(n) \subseteq Gl(n)$, defined by $Gl^+(n) := \{T \in Gl(n) : \det T > 0\}$, is a subgroup..

$\square$

Using this, we can affirm the following for $A, B \in B(V)$: A has an equal orientation with B if there is some $T \in Gl^+(n)$ with $A = BT$.

Definition 7.7 The quotient space $B(V)/\widetilde{\text{or}}$.
For a given basis $A = (a_1, \ldots, a_n) \in B(V)$, we call the set of bases, given by

$$\text{or}(A) := \{B = (b_1, \ldots, b_n) \in B(V) : B \overset{\sim}{\text{or}} A\}$$

an orientation of V.
    The corresponding quotient space is given by

$$B(V)/\widetilde{\text{or}} := \{\text{or}(A) : A \in B(V)\}.$$

The bases A and B above represent the same equivalence class or coset which we call, as stated, orientation.

Example 7.1 An opposite orientation to a given one.

It is easy to obtain, for example, $\text{or}(\bar{A})$, an opposite orientation of the given $\text{or}(A)$ with $A = (a_1, a_2, \ldots, a_n)$: we take $\text{or}(\bar{A})$ with $\bar{A} = (-a_1, a_2, \ldots, a_n)$ and we observe that $\text{or}(\bar{A}) \neq \text{or}(A)$, since we may write

$$(-a_1, a_2, \ldots, a_n) = (a_1, a_2, \ldots, a_n)\, T'$$

with

$$T' = \begin{bmatrix} -1 & 0 & 0 \\ 0 & 1 & \cdots \\ \vdots & \vdots & 0 \\ 0 & 0 & 1 \end{bmatrix}$$

and $\det T' = -1$.

Remark 7.12 The cardinality of $B(V)/\widetilde{\text{or}}$ is 2.

From Remark 7.11 and Example 7.1, we can see that we have

$$B(V)/\widetilde{\text{or}} = \{[A], [\bar{A}]\}.$$

This means, as expected and widely known, that there are only two orientations on a real vector space.

Now, we can also specify what we specifically mean by an oriented vector space.

Definition 7.8 Oriented vector space.
An oriented vector space is the pair (V, or) with V a real vector space, with
$\dim V = n$ and an orientation denoted by or given by Definitions 7.6 and 7.7.

Comment 7.5 $B(V)/\widetilde{\text{or}}$ as an orbit space.

The last discussion on orientation in linear quotient space was essentially an application of Sect. 1.2.

We now come to a pleasant application of Sect. 1.3 on group actions. It turns out that by using the terminology of that section, the quotient space $B(V)/\widetilde{\text{or}}$ is at the same time a right orbit space of the groups action $Gl^+(n)$ on $Gl(n)$. Taking into account Remark 7.11, we realize that for $g_1 \in Gl(n)$ with for example $\det g_1 = -1$, we can write

$$Gl(n) = Gl^+(n) \cup g_1\, Gl^+(n) \quad \text{with} \quad Gl^+(n) \cap g_1 Gl^+(n) = \emptyset.$$

We so obtain a disjoint composition of $Gl(n)$. As a result, by applying the terminology of Sect. 1.3, we get the following isomorphism:

$$B(V)/\widetilde{\text{or}} \cong Gl(n)/Gl^+(n).$$

7.7 Determinant Forms

There are many ways of defining the determinant of a matrix. In the past (see Sect. 7.2), we first defined determinants as special functions on the set of square matrices

$$\Delta : \mathbb{K}^{n \times n} \to \mathbb{K},$$
$$A \mapsto \Delta(A).$$

In order to make this clear, we used the name determinant function. It was very natural to consider the same object also as a function of the n columns $(a_1, \ldots, a_n)$ of a given matrix $A = [a_1 \ldots a_n]$. This is why we may use now the equivalent definition with the letter D just for distinction:

$$D : \qquad \mathbb{K}^n \times \ldots \times \mathbb{K}^n \qquad\quad \longrightarrow \mathbb{K},$$
$$(a_1, \ldots a_n) \qquad\qquad \longmapsto D(a_1, \ldots, a_1),$$

with D multilinear and alternating. We denote the space of determinant forms by

$$\Lambda^n(\mathbb{K}^m) = \{D \ldots\},$$

and we get

$$\Lambda^n = \Lambda^n(\mathbb{K}^m) \cong \mathbb{K}.$$

This definition can be used to extend the concept of determinant to the case of an abstract vector space V with dim $V = n$. In this case, we are talking about multilinear forms on V or n-linear forms or determinant forms on V. The space of determinant forms on V are denoted similarly by

$$\Lambda^n(V) = \{D \ldots\}.$$

The concept of determinant can be extended further in this direction, as we shall see, also to an endomorphism on V.

It is essential to notice that all the properties of determinant functions are valid for determinant forms if we exchange the words column and matrix with the words vector and endomorphism. In addition, we have to consider the role of permutation in connection with determinants.

7.7.1   The Role of the Group of Permutations

Permutations are appearing here because the determinants are not only multilinear but also alternating. This leads further to the explicit form of the determinant known as the Leibniz formula.

Therefore, it is helpful to summarize some of the relevant properties of the symmetric group (S_n) and the role of the sign of a given permutation.

Definition 7.9 The sign ε_π of a permutation $\pi = \left(\begin{smallmatrix} 1 & 2 & \cdots & n \\ \pi_1 & \pi_2 & \cdots & \pi_n \end{smallmatrix}\right) \in S_n$ is -1 if the number of pairs $\Phi(\pi)$ in the list $(\pi_1, \ldots, \pi_n)$ with $\frac{\pi_j - \pi_i}{j - i} =$ negative is odd, and the sign is $+1$ otherwise.

That is

$$\varepsilon_\pi = (-1)^{\Phi(\pi)}. \tag{7.37}$$

This can also be written as

$$\varepsilon_\pi = \prod_{i<j} \frac{\pi_j - \pi_i}{j - i}. \tag{7.38}$$

Using, without proof, the fact that every permutation is a product of a certain number of transposes, $t(\pi) \in \mathbb{N}_0$,

$$\pi = \tau_1 \tau_2 \ldots \tau_{t(\pi)},$$

and taking into account that every transpose τ has negative sign $\varepsilon_\tau = -1$, we also have

$$\varepsilon_\pi = (-1)^{t(\pi)}. \tag{7.39}$$

We can also show that if we represent π by the matrix P_π, defined as

$$P_\pi := [e_{\pi_1} \cdots e_{\pi_n}] = (\delta_{i,\pi_i}), \tag{7.40}$$

then P is a group homomorphism:

$$
\begin{array}{ccccc}
P & : & S_n & \longrightarrow & Gl(n, \mathbb{N}_0) < Gl(n), \\
& & \pi & \longmapsto & P_\pi
\end{array}
$$

with

$$P_{\pi\sigma} = P_\pi P_\sigma. \tag{7.41}$$

The sign of π is now given by

$$\varepsilon_\pi = \det(P_\pi). \tag{7.42}$$

The above homomorphism shows that the sign ε is also a group homomorphism:

$$\varepsilon : S_n \longrightarrow \mathbb{Z}_2 \cong \{+1, -1\}$$

with

$$\varepsilon_{\pi\sigma} = \varepsilon_\pi \circ \varepsilon_\sigma. \tag{7.43}$$

Comment 7.6 The permutation sign.

$$\varepsilon_\pi = (-1)^{\Phi(\pi)} = (-1)^{t(\pi)} = \det(P_\pi). \tag{7.44}$$

Here, we summarize the various definitions of the permutation sign.

7.7.2   *Determinant Form and Permutations*

Coming back to the determinant terms on V (dim $V = n$), we write

$$D : \qquad V^n \qquad\qquad\qquad \longrightarrow \mathbb{K}$$
$$(v_1, \ldots, v_n) \qquad\qquad \longmapsto D(v_1, \ldots, v_n) \in \mathbb{K},$$

with D n-linear (n-multilinear) and alternating.

We denoted the space of determinant forms by

$$\Lambda^n(V) = \{D \ldots\}.$$

One of the most important result of permutations is the relation

$$D(v_{\pi 1}, \ldots, v_{\pi n}) = \varepsilon_\pi D(v_1, \ldots, v_n) \tag{7.45}$$

which describes the alternating property of D.

By repeating essentially the results of Sect. 7.2 for the determinant function Δ, we obtain altogether the following equivalent conditions for a n-linear form D so that it gets the alternating property.

(i) $D(\ldots v_i \ldots, v_j \ldots) = -D(\ldots v_j \ldots v_i \ldots)$,
(ii) $D(\ldots v_i \ldots, v_i \ldots) = 0$,
(iii) $D(\ldots, v_i = 0, \ldots, \ldots) = 0$,
(iv) if $(v_1, \ldots, v_n)$ are linearly dependent, then $D(v_1, \ldots, v_n) = 0$,
(v)

$$D(v_{\pi 1}, \ldots, v_{\pi n}) = \varepsilon_\pi D(v_1, \ldots, v_n), \ \pi \in S_n. \tag{7.46}$$

As in Sect. 7.2, we obtain here again the result that the set of n-linear alternating forms is a one dimension vector space.

$$\Lambda^n(V) \cong \mathbb{K}. \tag{7.47}$$

Further more, if we use the multi-linearity of D and the above relation (v), we obtain the explicit expression for the determinant. This is a very important Leibniz formula.

7.7.3 The Leibniz Formula for the Determinant

Proposition 7.5 *The Leibniz formula.*
Let $A = (\alpha_s^i)$ be a matrix. Then

$$\det A = \sum_{\pi \in S_n} \varepsilon_\pi \alpha_1^{\pi 1} \alpha_2^{\pi 2} \ldots \alpha_n^{\pi n}. \qquad (7.48)$$

Proof Using the multi-linearity of the determinant, we get

$$\det A = \det(a_1, \ldots, a_n) \quad a_i \in \mathbb{K}^n i,$$
$$\text{for } i_1, \ldots, i_n \in I(n), \text{ we may write}$$
$$\det A = \det(e_{i_1} \alpha_1^{i_1}, e_{i_2} \alpha_2^{i_2}, \ldots, e_{i_n} \alpha^{i_n}),$$
$$\det A = \det(e_{i_1}, e_{i_2}, \ldots, e_{i_n}) \alpha_1^{i_1} \ldots \alpha_n^{i_n}.$$

Using the relation (v) in Eq. (7.46), we obtain immediately

$$\det A = \sum_{\pi \in S_n} \varepsilon_\pi \alpha_1^{\pi 1} \alpha_2^{\pi 2} \ldots \alpha^{\pi n}.$$

$\square$

In Eq. (7.47), we may see immediately that every nontrivial determinant form, also called for good reasons volume form or simply volume, can be used as a basis of $\Lambda^n(V)$.

7.8 The Determinant of Operators in V

Initially, the determinant of an operator in V is a basis dependent extension from the determinant of a given matrix representation of an endomorphism (operator) $f \in \text{Hom}(V, V)$ to the endomorphism f itself. This definition is reasonable since it turns out that it is in fact nonetheless basis independent.

Definition 7.10 The Determinant of an operator.
For a given endomorphism $f \in \text{Hom}(V, V)$ with a matrix representation F_B, with respect to a basis $B \in B(V)$, the determinant of f is given by:

$$\det f := \det F_B.$$

Now, we have to show that this definition is well defined, that it is basis independent which will justify the above notation as $\det f$.

In order to check the basis independence, we choose a second basis $C \in B(V)$ and we expect to show that

$$\det F_C = \det F_B. \tag{7.49}$$

This follows from the commutative diagram given below in an obvious notation, with $T = T_{CB}$:

$$
\begin{array}{ccc}
\mathbb{K}^n_C & \xrightarrow{\ F_C\ } & \mathbb{K}^n_C \\
{\scriptstyle T}\big\uparrow & & \big\uparrow{\scriptstyle T} \\
\mathbb{K}^n_B & \xrightarrow[\ F_B\ ]{} & \mathbb{K}^n_B
\end{array}
$$

This shows instantly that $F_C \circ T = T \circ F_B$ or, equivalently, expressed by matrix multiplication:

$$F_C = T\, F_B\, T^{-1}. \tag{7.50}$$

We therefore obtain from Eq. (7.50), taking the determinant and using the rules mentioned in Sect. 7.4,

$$\det F_C = \det(T\ F_B\ T^{-1}) = \det T \det F_B \det T^{-1} =$$
$$= \det T \det F_B (\det T)^{-1} = \det F_B.$$

This shows that Definition 7.10 is well defined.

An elegant geometric definition of $\det f$ is obtained by the pullback f^* of a given determinant form D:

$$f^* D(v_1, \dots, v_n) := D(f v_1, \dots, f v_n).$$

This may also be written as $f^* D = D \circ (f \times \dots \times f)$.
The determinant is a function on the endomorphisms of V denoting $(\text{End}(V) \equiv \text{Hom}(V, V))$:

$$
\begin{aligned}
\det : \text{End}(V) &\longrightarrow \mathbb{K}, \\
f &\longmapsto \det f
\end{aligned}
\tag{7.51}
$$

given by the relation

$$f^*D = (\det f)D \tag{7.52}$$

which means

$$D(f v_1, \ldots, f v_1) = (\det f)D(v_1, \ldots, v_n), \tag{7.53}$$

or, equivalently

$$\det f = \frac{D(f v_1, \ldots, f v_n)}{D(v_1, \ldots, v_n)}. \tag{7.54}$$

This definition reveals the geometric character of the determinant of a given endo-morphism $f \in \mathrm{End}(V) \equiv \mathrm{Hom}(V, V)$. Equations (7.53) and (7.54) show that $\det f$ characterizes the scaling of the f-transformation on any volume D in V.

Summary

The determinant, alongside the identity, the exponential function, and a few others, is one of the most important maps in mathematics and physics. Therefore, like in most books on linear algebra, we dedicated an entire chapter to determinants.

Initially, we adopted an algebraic approach for defining and understanding the properties of determinants. Here, elementary matrices and elementary row operations were the primary tools. Then, we delved into the geometric aspects of determinants. In doing so, we introduced and extensively discussed the concept of orientation.

The role of permutations and the permutation group was also introduced in a suitable manner.

Finally, the concept of determinants was extended to operators, not just matrices, and we presented the corresponding geometric interpretation.

Exercises with Hints

Elementary matrices induce column and row operations (see Remark 7.2). In connection with the proof of Theorem 5.1, it follows by construction that for instance elementary column operations do not affect the column rank of a matrix. The following Exercise 7.1 shows that they do not affect the row rank either. The same applies when we replace column by row.

Exercise 7.1 Use the result of Exercise 6.2 to show that elementary column operations on a given matrix $A \in \mathbb{K}^{m \times n}$ do not affect the row rank of A.

We now apply the above Exercise 7.1 to prove once more the following theorem.

Exercise 7.2 *The row—column—rank theorem*
Use the fact that elementary column operators affect neither the column nor the rank, and that, similarly, elementary row operations affect neither the row nor the column rank of a matrix to show that the row rank of a matrix $A \in \mathbb{K}^{m \times n}$ equals its column rank.

For the next six exercises one can find proofs in the literature which generally differ from the ones given in this chapter. The reader is asked to chose our proofs, or to try to find different proofs.

Exercise 7.3 Show that the set of determinant functions on a vector space V, or equivalently the set of determinant forms is a vector space of dimension one.

Exercise 7.4 Let Δ_1 and Δ_2 be two determinant functions: $\Delta_1, \Delta_2 : \mathbb{K}^{m \times n} \to \mathbb{K}$ and $A \in \mathbb{K}^{n \times n}$. Show that

$$\Delta_1(\mathbb{1}_n)\Delta_2(A) = \Delta_2(\mathbb{1}_n)\Delta_1(A).$$

Exercise 7.5 *Use the above Exercise 7.4 to prove the following exercise.*
Let Δ be a determinant function with $\Delta(\mathbb{1}_n) = 1$ and $A, B \in \mathbb{K}^{n \times n}$. Show that

$$\Delta(AB) = \Delta(A)\Delta(B).$$

Exercise 7.6 Let Δ be a determinant function and $A \in \mathbb{K}^{n \times n}$. Show that

$$\Delta(A^{\mathsf{T}}) = \Delta(A).$$

Exercise 7.7 *The determinant of block diagonal matrices.*
Let $A \in \mathbb{K}^{s \times s}$, $D \in \mathbb{K}^{r \times r}$, and $s + r = n$. Show that when

$$\begin{bmatrix} A & 0 \\ 0 & D \end{bmatrix} \in \mathbb{K}^{n \times n},$$

then

$$\det \begin{bmatrix} A & 0 \\ 0 & D \end{bmatrix} = \det A \det D.$$

Exercise 7.8 *The determinant of upper block triangular matrices.*
Let $A \in \mathbb{K}^{s \times s}$, $B \in \mathbb{K}^{s \times r}$, $D \in K^{r \times r}$, and $s + r = n$. Show that when

$$\begin{bmatrix} A & B \\ 0 & D \end{bmatrix} \in \mathbb{K}^{n \times n},$$

then

$$\det \begin{bmatrix} A & B \\ 0 & D \end{bmatrix} = \det A \det D.$$

Exercise 7.9 *On the geometric interpretation of determinants.*
Let f be a linear map $f \in \mathrm{Hom}(\mathbb{R}^2, \mathbb{R}^2)$ given by

$$f(a_i) = a_\mu \varphi_i^\mu,$$

with

$$a_i \in \mathbb{R}^2, \varphi_i^\mu \in \mathbb{R}, F = (\varphi_s^i), i, s, \mu \in I(2),$$

so we can write

$$[f(a_1) f(a_2)] = [a_1 a_2] F.$$

Prove

$$vol_2(f a_1, f a_2)^2 = \langle f a_1 | f a_1 \rangle \langle f a_2 | f a_2 \rangle - \langle f a_1 | f a_2 \rangle^2,$$

and hence prove

$$vol_2(f a_1, f a_2)^2 = (\det F)^2 vol_2(a_1, a_2)^2.$$

Exercise 7.10 Let π be a permutation $\pi \in S_n$. Show that π is a product of transpositions. Apply the following induction according to $r(\pi)$, where $r(\pi)$ is the maximal $r(\pi) \in \{0, 1, \ldots, n\}$, with the property $\pi(i) = i$ for $i = 1, \ldots, r(\pi)$. And if $r(\pi) = n$, we get $\pi = id$ and this is a void product of transposition.

Exercise 7.11 Let $P_\pi = [e_{\pi_1} \ldots e_{\pi_n}] = (\delta_{i \pi_i})$ be a permutation matrix $P_\pi \in \mathbb{R}^{n \times n}$. Show that P is a group homomorphism

$$P : \quad S_n \longrightarrow Gl(n, \mathbb{Z}) < Gl(n)$$
$$\pi \longmapsto P_\pi$$

with $P_{\pi \circ \sigma} = P_\pi P_\sigma$.

Note that if we compare the entries of the matrix P_π with the squares of a chessboard, and set on every entry 1 a rook, the zeros correspond exactly to the area of activity of the rook. Therefore we could call P the chess-representation of the group S_n.

Exercise 7.12 Check that for a matrix $A = \begin{bmatrix} \alpha & \beta \\ \gamma & \delta \end{bmatrix} \in \mathbb{R}^{2 \times 2}$ with $\det A \neq 0$,

$$A^{-1} = \frac{1}{\det A} \begin{bmatrix} \delta & -\beta \\ -\gamma & \alpha \end{bmatrix}.$$

References and Further Reading

1. S. Axler, *Linear Algebra Done Right* (Springer Nature, 2024)
2. S. Bosch, *Lineare Algebra* (Springer, 2008)
3. G. Fischer, B. Springborn, *Lineare Algebra. Eine Einführung für Studienanfänger*. Grundkurs Mathematik (Springer, 2020)
4. S.H. Friedberg, A.J. Insel, L.E. Spence, *Linear Algebra* (Pearson, 2013)
5. K. Jänich, *Mathematik 1. Geschrieben für Physiker* (Springer, 2006)
6. N. Johnston, *Introduction to Linear and Matrix Algebra* (Springer, 2021)
7. M. Koecher, *Lineare Algebra und analytische Geometrie* (Springer, 2013)
8. G. Landi, A. Zampini, *Linear Algebra and Analytic Geometry for Physical Sciences* (Springer, 2018)
9. P. Petersen, *Linear Algebra* (Springer, 2012)
10. S. Roman, *Advanced Linear Algebra* (Springer, 2005)
11. G. Strang, *Introduction to Linear Algebra* (SIAM, 2022)

Chapter 8
First Look at Tensors

In Sect. 3.5, we discussed in very elementary terms an origin of tensors. Yet there are many ways to introduce tensors. Some of them are very abstract. We believe that the most direct way, which is also very appropriate for physics, is to use bases of V and V^*. This definition is, of course, basis-dependent, but as we have often seen thus far, this is not a disadvantage. In this chapter, we shall proceed on this path and leave a basis-independent definition for later (see Chap. 14).

Thus far, we have gained considerable experience with bases. We already used the simplest possible tensors (e.g. scalars, vectors, and covectors) with their indices on various occasions. Therefore, it is very instructive to summarize first what we already know from linear algebra about the use of indices and tensors. This leads us to the following section.

8.1 The Role of Indices in Linear Algebra

One important application (one may even say chief application) of bases, is they enable to represent abstract vectors with coordinates, and do concrete calculations with them. One can say, that bases "give indices to vectors" since we label bases vectors with an order. It turns out that indices, as they are used in linear algebra, and in particular in this book, are quite helpful since they give additional information about the properties and structures of the mathematical objects they are related to. We accomplish this using the Einstein Summation Convention and with some straightforward conventions we add. In this way, we obtain valuable indices which we call smart indices.

There are two different kinds of indices corresponding to the vectors in V and the covectors in V^*. These two possibilities appear clearly and efficiently when written as upper or down indices. This meets precisely with the Einstein Convention. Regardless of the Einstein Convention, using indices up or down is much more functional than

© The Author(s), under exclusive license to Springer Nature Switzerland AG 2024   231
N. A. Papadopoulos and F. Scheck, *Linear Algebra for Physics*,
https://doi.org/10.1007/978-3-031-64908-0_8

writing them left or right, as is usually done in the mathematical literature and not seldom in the physics literature.

For the sake of simplicity, we consider a real vector space V with dimension n and its dual V^*, and we are going to summarize our experience with indices in linear algebra till now. Our primary purpose is to revise some subjects and show typical examples of how the indices enter the various expressions. Thus, we also see the positive influence of the chosen conventions on a good understanding of the path to a given expression.

Example 8.1 Change of basis.

We start with two different bases in V.

$$B = (b_r) = (b_1, \ldots, b_n) \text{ and } C = (c_i) = (c_1, \ldots, c_n),$$

together with their dual bases

$$B^* = (\beta^s) = (\beta^1, \ldots, \beta^n) \text{ and } C^* = (\gamma^j) = (\gamma^1, \ldots, \gamma^n).$$

B^* and C^* are the corresponding bases in V^* dual to B and C:

$$\beta^s(b_r) = \delta^s_r \text{ and } \gamma^j(c_i) = \delta^j_i.$$

We use an obvious notation throughout the examples, and we demonstrate how easily and directly the indices themselves lead to the change of relations of bases. We take $r, s, i, j \in I(n)$, then $v \in V$ and $\xi \in V^*$ are given by

$$v = b_r v^r_B = c_i v^i_C \quad \text{and} \tag{8.1}$$

$$\xi = \xi^B_s \beta^s = \xi^C_j \gamma^j. \tag{8.2}$$

with the corresponding coefficients given by $v^r_B, v^i_C, \xi^B_s, \xi^C_j \in \mathbb{R}$. For the change of basis we use a regular matrix

$$T = T_{CB} = (\tau^i_s) \in Gl(n),$$

with its inverses
$$T^{-1} \equiv \bar{T} = (\bar{\tau}^r_i) \qquad \tau^i_r, \bar{\tau}^r_i \in \mathbb{R}.$$

The change of basis is given by

$$b_r = c_j \tau^j_r \quad \text{or} \quad c_i = b_r \bar{\tau}^r_i \quad \text{and} \tag{8.3}$$

$$\gamma^i = \tau^i_s \beta^s \quad \text{or} \quad \beta^r = \bar{\tau}^r_i \gamma^i. \quad \cdot \tag{8.4}$$

The corresponding coefficients are given by

$$v_C^i = \tau_s^i v_B^s \text{ and } \xi_i^C = \xi_s^B \bar{\tau}_i^s. \tag{8.5}$$

Note that the indices r, s correspond to the bases B and B^* and the indices i, j to the bases C and C^*. We use different kinds of indices for different bases even when they correspond to the same vector space. This distinction is not usual in the literature, but it prevents confusion. In connection with this, we point out that coefficients (components) of vectors have upper indices and coefficients of covectors have lower indices. However, vectors themselves have lower indices and covectors themselves upper indices. This is, of course, consistent with the Einstein Convention and in the usual matrix formalism also leads to the following expressions:

$$\vec{v}_B = \begin{bmatrix} v_B^1 \\ \vdots \\ v_B^n \end{bmatrix}, \; \underset{\sim}{\xi}^B = [\xi_1^B \cdots \xi_n^B], \quad \vec{v}_C = \begin{bmatrix} v_C^1 \\ \vdots \\ v_C^n \end{bmatrix}, \; \underset{\sim}{\xi}^C = [\xi_1^C \cdots \xi_n^C], \tag{8.6}$$

$$v = B\vec{v}_B = C\vec{v}_C \text{ and } \xi = \underset{\sim}{\xi}^B B = \underset{\sim}{\xi}^C C, \tag{8.7}$$

$$B := [b_1 \cdots b_n], \quad B^* := \begin{bmatrix} \beta^1 \\ \vdots \\ \beta^n \end{bmatrix}, \quad C = [c_1 \cdots c_n], \quad C^* := \begin{bmatrix} \gamma^1 \\ \vdots \\ \gamma^n \end{bmatrix}. \tag{8.8}$$

$$\vec{v}_B, \vec{v}_C \in \mathbb{R}^n, \; \underset{\sim}{\xi}^B, \underset{\sim}{\xi}^C \in (\mathbb{R}^n)^*, \; v \in V, \; \xi \in V^*. \tag{8.9}$$

Note that by the rotation $\underset{\sim}{\xi}^B$ we express explicitly that ξ^B is a row. Furthermore, this choice is related to the transformation behavior of basis vectors, cobasis elements (covectors), the coefficients of vectors, and the coefficients of covectors as indicated in Eqs. (8.3) (8.4) and (8.5). We may express this very important information in the suggestive expressions in Eqs. (8.9), (8.10) and (8.11) below.

The upper indices are transformed by a matrix $T = (\tau_r^i)$, and the lower indices are transformed by a matrix $T^{-1} \equiv \bar{T} = (\bar{\tau}_i^r)$ which we may express symbolically:

$$(\cdot)^i = \tau_s^i (\cdot)^s, \tag{8.10}$$

$$(\cdot)_i = (\cdot)_s \bar{\tau}_i^s. \tag{8.11}$$

If we want to write ξ^B and ξ^C as columns, $\vec{\xi}_B := \begin{bmatrix} \xi_B^1 \\ \vdots \\ \xi_B^n \end{bmatrix}$ and $\vec{\xi}_C := \begin{bmatrix} \xi_C^1 \\ \vdots \\ \xi_C^n \end{bmatrix}$ with $\xi_B^s = \xi_s^B$

and $\xi_C^i = \xi_i^C$, then we find within the matrix formalism

$$\vec{v}_C = T\vec{v}_B \text{ and } \vec{\xi}_C = (T^{-1})^\mathsf{T}\vec{\xi}_B. \tag{8.12}$$

This also leads to an invariant expression which is very important, not only in physics:

$$(\cdot)_s(-)^s = (\cdot)_i(-)^i = invariant. \tag{8.13}$$

Note that writing the symbol $\underset{\sim}{\xi}$ to indicate the row with entries $\xi_s \in \mathbb{R} : \xi = [\xi_1 \cdots \xi_n]$, we may also write $\underset{\sim}{\xi} \equiv \xi^B \equiv \xi^B \in (\mathbb{R}^n)^*$ and $\vec{v} \equiv \vec{v}_B \equiv v_B \in \mathbb{R}^n$ too.

We would like to iterate that we mostly use the symbol "$\equiv$" to indicate that we use different notations for the same objects. We admit that the use of $\vec{v}_B$ and $\vec{\xi}^B$ instead of v_B and $\underset{\sim}{\xi^B}$ is a pleonasm.

Comment 8.1 *Notation for lists and matrices.*

Note that we often identify the list $B = (b_1, b_2, \ldots, b_n)$ with matrix row $(1 \times n)$ with entries given by vectors $b_1, \ldots, b_n$ which we denote by $[B] := [b_1 b_2 \ldots b_n]$, and we denote both by B. For the cobasis B^*, we analogously write the symbol B^* for the column ($n \times 1$-matrix) with entries the covectors $\beta^1, \ldots, \beta^n$, as in Eq. (8.8). We also apply this to $C, D, \ldots$ and $C^*, D^*, \ldots$.

Sometimes this kind of flexibility in the notation is quite useful to avoid possible confusion.

We are now coming to the next important illustration of using smart indices, by summarizing again our results concerning the representation of linear maps.

Example 8.2 Representation of linear maps.

We consider the map $f \in \text{Hom}(V, V')$ with dim $V' = m$ and the basis $B' = (b'_\varrho) = (b'_1, \cdots, b'_m), \varrho \in I(m)$. Suppose f is given by the equations

$$f(b_r) = b'_\varrho \varphi^\varrho_r. \tag{8.14}$$

We define the matrix F by

$$F = f_{B'B} = (\varphi^\varrho_r) \in \mathbb{R}^{m \times n}. \tag{8.15}$$

Then, using a suggestive notation, we have a commutative diagram.

$$
\begin{array}{ccc}
V & \xrightarrow{\;f\;} & V' \\[2pt]
\psi_B \uparrow & & \uparrow \psi_{B'} \\[2pt]
\mathbb{R}^n_B & \xrightarrow[\;F\;]{} & \mathbb{R}^m_{B'}
\end{array}
$$

The corresponding coefficients $(v^r_B, w^\rho_{B'} \in \mathbb{R})$, taking $v \in V$, $w \in V'$, and $f(v) = w$, are given by

$$
w = b'_\varrho \, w^\varrho_{B'} . \tag{8.16}
$$

We choose our indices, r for v, ρ for w, φ^ρ_r for F, and we write v^r, $w^\varrho \in \mathbb{R}$. This leads directly to the expression:

$$
w^\varrho_{B'} = \varphi^\varrho_r \, v^r_B . \tag{8.17}
$$

This is, of course, also the result of the direct calculation, which in the matrix formalism takes the well-known form:

$$
\vec{w}_{B'} = F \, \vec{v}_B , \tag{8.18}
$$

or even

$$
\vec{w} = \overrightarrow{f(v)} \quad \text{or} \quad \vec{w} = F \, \vec{v},
$$

as mostly used in physics.

Our last example is a change of basis for the representation of the map f.

Example 8.3 Change of basis for the representation of $f \in \mathrm{Hom}(V, V')$.

Here, the change of basis leads us immediately to the correct result with the help of our smart indices. In contrast to the matrix formalism, where this is less straightforward, especially if you are not very familiar with commutative diagrams.

Now, we would like to obtain the representation of f relative to the new bases C and C', as indicated in the following diagram:

$$
\begin{array}{ccc}
V & \xrightarrow{\;f\;} & V' \\[2pt]
\psi_C \uparrow & & \uparrow \psi_{C'} \\[2pt]
\mathbb{R}^n_C & \xrightarrow[\;\tilde{F}\;]{} & \mathbb{R}^m_{C'}
\end{array}
$$

with

$$\tilde{F} = f_{C'C} = (\tilde{\varphi}_i^\alpha) \quad \alpha \in I(m).$$ (8.19)

We take

$$C' = (c'_\alpha) = (c'_1, \ldots, c'_m)$$

and

$$w = b'_\varrho w_{B'}^\varrho = c'_\alpha w_{C'}^\alpha,$$ (8.20)

with

$$T' = T_{C'B'} = (\tau_\varrho^{'\alpha}),$$

$$w_{C'}^\alpha = \tau_\varrho^{'\alpha} w_{B'}^\varrho.$$ (8.21)

In accordance with the convention concerning the choice of indices, we obtain the matrix $\tilde{F}$ in a straightforward manner by following the right indices. Since we want to have the coefficient $\tilde{\varphi}_i^\alpha$ of $\tilde{F}$ in terms of the coefficients φ_r^ϱ of F, we can write:

$$\tilde{\varphi}_i^\alpha = \tau_\varrho^{'\alpha} \varphi_r^\varrho \bar{\tau}_i^r.$$ (8.22)

We have taken into account that the indices i and α correspond to the new basis C and C' and the indices r and ϱ to the bases B and B'. Doing so, the result is uniquely determined. The corresponding expression in the matrix formalism is given from Eq. (8.22) by

$$\tilde{F} = T' \, F \, T^{-1}.$$ (8.23)

This result was also found in Sect. 5.5, "Linear Maps and Matrix Representations".

8.2   From Vectors in V to Tensors in V

To justify the path leading from vectors to tensors, both intuitively and correctly, we need to explain the notion of a free vector space. This will also help to understand better our approach from vectors to tensors via the explicit use of bases. In addition, this will show that both tensors and their indices are something quite natural and, in a way, inevitable.

Starting with an abstract vector space V, we may also choose a basis that is simply a list of $n = \dim V$ linearly independent vectors in V.

We may start now with an arbitrary list of arbitrary objects, that is, elements. Using a list of arbitrary objects in any given set as a basis, we can formally also determine a vector space called a free vector space. This brings us to the following definition:

Definition 8.1 A free vector space.

For a given set $S = \{s(1), \ldots, s(n)\}$ of cardinality n, the free $\mathbb{R}$-vector space on S is

$$\mathbb{R}S := \{s_i v^i : s_i := s(i), v^i \in \mathbb{R}, i \in I(n)\}.$$

It is a vector space with $\dim(\mathbb{R}S) = n$.

We can show immediately that $\mathbb{R}S$ is a vector space and we write $V = \mathbb{R}S$. The set we started with, $S = \{s(1), \ldots, s(n)\}$, is a basis of V so that $\dim V = n$ and we may change the notation and write $B = S$, taking $B = (b_1, \ldots, b_n)$ with $b_1 := s(1), \ldots, b_n := s(n)$.

Repeating the above construction for a new set $S(2) := \{s(i, j) := b_i b_j, i, j \in I(n)\}$, we obtain a new vector space $T^2 V := \mathbb{R}S(2)$ with the basis $B_2 = S(2) = \{b_{ij} := b_i b_j\}$. We call $T^2 V$ a tensor space of rank 2. This is the simplest nontrivial tensor space over V. It is also clear that $T^2 V$ is given by

$$T^2 V = \text{span } S(2) = \{v^{ij} b_i b_j : v^{ij} \in \mathbb{R}\}.$$

For any fixed $k \in \mathbb{N}_0 = \{0, 1, 2, \ldots\}$, we generalize this construction $T^k = \mathbb{R}S(k)$ as follows:

$$T^0 V := \mathbb{R} \quad \text{and} \quad T^1 V := V.$$

So we have, in general, $T^k V := \mathbb{R}S(k)$ with $S(k)$ given by

$$S(k) = \{b_{i_1} b_{i_2}, \ldots, b_{i_k} = i_1, \ldots, i_k \in I(n)\}.$$

It is clear that $S(k)$ is a basis of $T^k V$ and that $\dim T^k V = n^k$. $T^k V$ is given by

$$T^k V = \{v^{i_1 \ldots i_k} b_{i_1} b_{i_2} \ldots b_{i_k} : v^{i_1 \ldots i_k} \in \mathbb{R}\}.$$

An expression like $i_1 i_2 \ldots i_k$ is usually called a multiindex.

It is reasonable to think when we go from $b_i \in B$ to $b_i b_j$ that a kind of multiplication is in action. We may write $b_i b_j = b_i \otimes b_j$ with $\otimes$ a symbol of a product which is called tensor product. This way we may generally write

$$T^k V = \{v^{i_1 \ldots i_k} b_{i_1} \otimes b_{i_2} \ldots \otimes b_{i_k} : v^{i_1 \ldots i_k} \in \mathbb{R}\}.$$

This is also called a contravariant tensor of rank k or k-tensor. Since every contravariant tensor of rank k can be written as a linear combination of tensor products of vectors, we can also justify the following expression for $T^k V$:

$$T^k V = \underbrace{V \otimes \cdots \otimes V}_{\text{k-times}}.$$

It is evident that the same construction can also be made for V^*. So we have $B^* = (\beta^1, \ldots, \beta^n)$ and

$$T^k V^* = \{v_{i_1 \ldots i_k} \beta^{i_1} \otimes \ldots \otimes \beta^{i_k} : v_{i_1 \ldots i_k} \in \mathbb{R}\}.$$

So we have analogously to the above expression for $T^k V^*$:

$$T^k V^* = \underbrace{V^* \otimes \cdots \otimes V^*}_{\text{k-times}}.$$

This is called covariant tensor of rank k or k-tensor. The expressions "contravariant" and "covariant" to $T^k V$ and $T^k V^*$ are purely historical nomenclatures. It is clear that here,

$$S^*(k) := \{\beta^{i_1} \beta^{i_2} \ldots \beta^{i_k} : i_1, i_2, \ldots i_k \in I(n)\},$$

is a basis of $T^k V^*$.

In our construction, there is no restriction to no additional properties on the set S preventing it from being a basis for a vector space. Hence we can also have the set

$$S = S(l, k) = \{b_{i_1} \ldots b_{i_k} \beta^{j_1} \ldots \beta^{j_l} : k, l \in \mathbb{N}_0 \quad \text{and} \quad i_1, \ldots, i_k, j_1 \ldots j_l \in I(n)\}$$

as a basis of a vector space which we denote by $T_l^k V$. We thus obtain another tensor space given by

$$T_l^k V = \{v^{i_1 \ldots i_k}_{ j_1 \ldots j_l} \ b_{i_1} \otimes \cdots \otimes b_{i_k} \otimes \beta^{j_1} \otimes \ldots \otimes \beta^{j_l}\}$$

which we call a mixed tensor space of type (l, k). We can also write

$$T_l^k V = \underbrace{V \otimes \cdots \otimes V}_{\text{k-times}} \otimes \underbrace{V^* \otimes \cdots \otimes V^*}_{\text{l-times}}.$$

A tensor $A \in T_l^k V$ is an element of a vector space (tensor space), as stated above, the representation relative to the basis (B, B^*) of (V, V^*) is given by

$$A = \alpha^{r_1 \ldots r_k}_{ s_1 \ldots s_l} \ b_{r_1} \otimes \cdots \otimes b_{r_k} \otimes \beta^{s_1} \otimes \ldots \otimes \beta^{s_l}.$$

In the basis (C, C^*) of (V, V^*), the corresponding representation is given by

$$A = \alpha^{i_1 \ldots i_k}_{ j_1 \ldots j_k} \ c_{i_1} \otimes \cdots \otimes c_{i_k} \otimes \gamma^{j_1} \otimes \ldots \otimes \gamma^{j_l}.$$

For the corresponding coefficients of A in bases B and C, we may write

$$A_B := (\alpha^{r_1 \cdots r_k}{}_{s_1 \cdots s_2}) \quad \text{and} \quad A_C := (\alpha^{i_1 \cdots i_k}{}_{j_1 \cdots j_2}).$$

Using the notation given in Eqs. (8.1) and (8.2), the change of basis $b_r = c_i \tau_s^i$ leads, for the coefficients of A, to the transformation given as expected by

$$\alpha^{i_1 \cdots i_k}{}_{j_1 \cdots j_l} = \tau_{r_1}^{i_1} \tau_{r_2}^{i_2} \cdots \tau_{r_k}^{i_k} \, \alpha^{r_1 r_2 \cdots r_k}{}_{s_1 s_2 \cdots s_l} \, \bar{\tau}_{j_1}^{s_1} \bar{\tau}_{j_2}^{s_2} \cdots \bar{\tau}_{j_l}^{s_l}.$$

Symmetric and antisymmetric tensors are special kinds of tensors. They are essential for mathematics, especially for differential geometry, and everywhere in physics. We restrict ourselves here to the covariant tensors since similar considerations apply to contravariant tensors too. For mixed tensors, a similar approach is not really relevant.

Symmetric tensors are tensors whose coefficients, in any basis inter-change, stay unchanged. Antisymmetric (totally antisymmetric) or alternating tensors are tensors whose coefficients change signs by interchanging any pair of indices.

Definition 8.2 Symmetric k-tensors.

A k-tensor τ is symmetric if its coefficients $\tau_{i_1 \cdots i_k}$, in any basis, are unchanged by any permutation of the indices $i_1, \ldots, i_k$.

In physics, we have two prominent examples of symmetric tensors, the metric tensor $g_{\mu\nu}$ and the energy impulse tensor $T_{\mu\nu}$.

Definition 8.3 Alternating tensors.

A k-tensor τ is alternating if its coefficients $\tau_{i_n \cdots i_k}$, in any basis, change the sign by interchanging any pair of indices:

$$\tau_{i_1 \cdots i_a \cdots i_b \cdots i_k} = - \tau_{i_1 \cdots i_b \cdots i_a \cdots i_k}.$$

There are two other equivalent definitions:

- T is alternating if any two indices are the same as in $\tau_{i_1 \cdots i_a \cdots i_a \cdots i_k}$, then $\tau_{i_1 \cdots i_a \cdots i_a \cdots i_k} = 0$;
- T is alternating if for any permutation of the indices $\pi \in S_k$, the relation $\tau_{\pi(i_1) \cdots \pi(i_k)} = \text{sgn}(\pi) \tau_{i_1 \cdots i_k}$ holds ($sqn(\pi) \equiv \varepsilon_\pi$).

The most prominent example of an alternating tensor in mathematics and physics, is the volume form in the corresponding dimensions.

Summary

This chapter concludes the section of the book that we consider elementary linear algebra.

Here, we summarized and reiterated our notation for linear algebra and tensor formalism. This notation primarily involves the systematic selection of indices and their positioning, whether upper or lower, in the entries of the representation matrices of linear maps.

Several advantages of this notation were mentioned, and multiple examples facilitated the reader's understanding of using these "smart indices", as we prefer to call them.

Following that, we presented our second elementary introduction to tensors. This introduction, like the one in Chap. 3, is dependent on the basis but represents a certain generalization concerning the dimension and the rank of the tensor space considered and the corresponding tensor notation.

Chapter 9
The Role of Eigenvalues and Eigenvectors

This is one of the most important topics of linear algebra. In physics, the experimental results are usually numbers. In quantum mechanics, these numbers correspond to eigenvalues of observables which we describe with special linear operators in Hilbert spaces or in finite-dimensional subspaces thereof. Eigenvalues are also relevant for symmetries in physics. Eigenvalues and eigenvectors help significantly to clarify the structure of operators (endomorphisms). We recall that an operator in linear algebra is a linear map of a vector space to itself. We denote the set of operators on V by $\text{End}(V) \equiv \text{Hom}(V, V)$.

In this chapter, after some preliminaries and definitions, we will discuss the role of eigenvalues and eigenvectors of diagonalizable and nondiagonalizable operators. For any given operator f on a $\mathbb{C}$ vector space, we get to a corresponding direct sum decomposition of V, using only very elementary notions. This decomposition leads to the more refined Jordan decomposition of V. First, we consider the situation on an abstract vector space without other structures. Later, in Chap. 10, we shall discuss vector spaces with an inner product structure.

9.1 Preliminaries on Eigenvalues and Eigenvectors

The first step towards understanding a part of the structure of a given operator $f \in \text{End}(V)$, was presented in Chap. 6. After choosing a basis, we obtained a decomposition:

$$V = \ker f \oplus \text{coim } f.$$

This also shows the general direction we have to choose to investigate the structure of any arbitrary operator in V. We need to find a finer decomposition of V induced by the operator f. We hope to find a list of subspaces,

$$U_i \quad \text{where} \quad i \in I(\omega) \quad \text{for some} \quad \omega \in \mathbb{N}$$

and a direct sum decomposition of V,

$$V = U_1 \oplus \cdots \oplus U_i \oplus \cdots \oplus U_\omega$$

in such a way that each U_i is as small as possible, that is, their dimensions are as small as possible. The restriction of f to every such subspace must of course be an endomorphism : $f|U_j \in \text{End}(U_j)$. In other words, every U_j must be f invariant. The subspaces ker f and im f are, indeed, both f invariant subspaces. The most we can expect in this situation, is for every f invariant subspace U_i to be one-dimensional. Such one-dimensional f invariant subspaces lead to the specific scalars, the eigenvalues of f, and special vectors, the eigenvectors of f. These are special characteristics and geometric properties of every operator f. Furthermore, if, for example, the operator f is connected with an observable, as in quantum mechanics, then the eigenvalues correspond to the results that the experiments produce. These have to be compared with the theoretical results given from the calculations of such eigenvalues.

However, to clearly understand what is going on, we must consider the most general case without the inner product (metric) structure. This is also justified because eigenvalues and eigenvectors are independent of any isometric structure.

We cannot expect that every operator f will induce such one-dimensional direct sum decompositions of V. The existence of this fundamental property of an operator f is connected with the diagonalization problem we shall discuss below. Diagonalizable operators are a pleasant special case from the mathematical point of view. Fortunately, the most physically relevant operators are diagonalizable too. Almost all diagonalizable operators in physics are so-called normal operators, defined only on inner product vector spaces (see Sect. 10.5).

One more comment has to be made. The theory of eigenvalues and eigenvectors differs when considering a complex or real vector space. The formalism within a complex vector space seems more natural and straightforward than within a real vector space. It should be clear that both real and complex vector spaces, are equally relevant and essential in physics.

In what follows, we start with the theory of eigenvalues and eigenvectors on a $\mathbb{K}$—vector space for $\mathbb{K} \in \{\mathbb{C}, \mathbb{R}\}$ and we may think in most cases of a $\mathbb{C}$—vector space and only when there is a difference to the $\mathbb{R}$—vector space formulation, we shall comment on that appropriately.

9.2   Eigenvalues and Eigenvectors

As already mentioned, the theory of eigenvalues and eigenvectors depends on the field $\mathbb{K} = \{\mathbb{C}, \mathbb{R}\}$. For example, we shall see that the real and complex versions of the spectral theorem are significantly different (see Chaps. 10 and 11). Since the theory with complex vector spaces is easier to deal with, we generally think of complex vector spaces. If there is a difference, we bear this difference in mind when we restrict ourselves to the real vector space framework.

We are led to the notion of eigenvalues and eigenvectors if we think of the smallest possible nontrivial subspace U of V. Then U is of course a one-dimensional subspace and is determined by a nonzero vector u (i.e., $u \neq 0$), so we have $U = \text{span}(u) = \{\alpha u : \alpha \in \mathbb{K}\}$. We would like U to also be consistent with the operator f. That is, we would like U to also be f- invariant,

$$f(U) \leqslant U.$$

It follows that the equation

$$f(u) = \lambda u$$

should be valid. Further, if we take $v \in U$ with $v = \alpha u, \alpha \neq 0$, we have

$$f(v) = f(\alpha u) = \alpha f(u) = \alpha \lambda u = \lambda \alpha u = \lambda v.$$

This leads to the fundamental definition :

Definition 9.1 Eigenvalue and eigenvector.
For a given operator $f \in \text{End}(V)$, a number $\lambda \in \mathbb{K}$ is called eigenvalue of f if there exists a nonzero vector $v \in V \backslash \{0\}$ so that $f(v) = \lambda v$. The vector v is called an eigenvector of f and corresponds to the eigenvalue λ. We may call the pair (λ, v) an eigenelement.

It is important to realize from the beginning that it is the eigenelement, the pair (λ, v), which is uniquely defined. If the eigenvalue λ is given, there are always many eigenvectors belonging to this λ: all the nonzero $u \in U = \text{span}(u) = \{\alpha u : \alpha \in \mathbb{K}\}$. Furthermore, it is also possible that some other vector $w \in V \backslash U$ exists which fulfills the same eigenvalue equations as above.

$$f(w) = \lambda w.$$

This leads us to an f-invariant subspace of V called the λ- eigenspace.

Definition 9.2 Eigenspace $E(\lambda, f)$.
Let $f \in \text{End}(V)$ and $\lambda \in \mathbb{K}$. The eigenspace of f corresponding to the eigenvalue λ denoted by $E(\lambda, f)$ is given by

$$E(\lambda, f) = \{v \in V : f(v) = \lambda v\}$$

or equivalently

$$E(\lambda, f) = \ker(f - \lambda id_V).$$

In other words, $E(\lambda, f)$ is the set of all eigenvectors corresponding to λ with the inclusion of the vector 0 in order $E(\lambda, f)$ to be a vector space.

9.3   Examples

9.3.1   Eigenvalues, Eigenvectors, Eigenspaces

Example 9.1 $f = id_V$

Eigenvectors	Eigenvalue
all $v \in V \setminus \{0\}$	$\lambda = 1$

because $f v = id_V v = 1 v$.

Example 9.2 $f \in \text{Hom}(V, V)$ arbitrary

Eigenvectors	Eigenvalue
$\ker f \setminus \{0\}$	$\lambda = 0$

because $f v = 0 v$.
Hence, the eigenspace $E(\lambda = 0, f) = \ker f$.

Example 9.3 Projections

We consider $f \in \text{Hom}(V, V)$ given by the relation

$$f^2 = f. \tag{9.1}$$

Eigenvectors	Eigenvalue
$\text{im} f \setminus \{0\}$	$\lambda_1 = 1$
$\ker f \setminus \{0\}$	$\lambda_2 = 0$

and V has the decomposition:

$$V = E(1, f) \oplus E(0, f)$$
$$= \text{im } f \quad \oplus \text{ker } f.$$

We prove this as follows:

Proof Let v be an eigenvector with eigenvalue λ so that

$$f v = \lambda v . \tag{9.2}$$

Thus, we have $\quad f^2 v = f(\lambda v) = \lambda^2 v.$ $\tag{9.3}$

Using Eqs. (9.1) and (9.2), we obtain

$$f v = \lambda^2 v \Rightarrow \lambda v = \lambda^2 v. \tag{9.4}$$

Since $v \neq 0$, we get $\lambda = \lambda^2$. This leads to the two eigenvalues $\lambda_1 = 0$ and $\lambda_2 = 1$. The operator f is diagonalizable and thus we are led to a decomposition (spectral decomposition) of V and f: Since $f^2 = f$, we have also $\ker f^2 = \ker f$ and by Proposition 3.12, we get:

$$V = \ker f \oplus \text{im } f. \tag{9.5}$$

The decomposition of $V = \ker f \oplus \text{im } f$ gives a decomposition of the identity into projections as follows. With the decomposition of $id \equiv id_V$,

$$id = f + (id - f). \tag{9.6}$$

We now show that $id - f$ is also a projection operator:

$$\text{Since} \quad (id - f) \circ (id - f) = id \circ id + f \circ f - id \circ f - f \circ id$$
$$= id + f^2 - 2f$$
$$= id + f - 2f$$
$$= id - f.$$

Hence $(id - f) \circ (id - f) = id - f$ and $id - f$ is a projection.

For $\lambda = 0$, we have $\quad f v = 0 v \quad$ and so $\quad E(\lambda = 0) \equiv E(0) = \ker f.$

For $\lambda = 1$, we have $\quad f v = v \quad$ and so $\quad E(\lambda = 1) \equiv E(1) = \text{im } f.$

Using the fact that $\ker f = \text{im}(id - f)$ and $\text{im } f = \ker(id - f)$, we can write alternatively for Eq. (9.5)

$$V = \mathrm{im}(id - f) \oplus \mathrm{im}\, f = E(0) \oplus E(1). \tag{9.7}$$

If we use for both projection operators, the eigenvalue notation $P_0 := id - f$ and $P_1 := f$, we again get from Eqs. (9.5) and (9.7)

$$V = \mathrm{im}\, P_0 \oplus \mathrm{im}\, P_1, \tag{9.8}$$

and for Eq. (9.6)

$$id = P_0 + P_1, \tag{9.9}$$

and for the spectral decomposition of $f \equiv P_1$, the "trivial" relation

$$f = 0\, P_0 + 1\, P_1. \tag{9.10}$$

$\square$

This is what is generally expected of a diagonalizable operator. A further example of an exemplary character as well, especially useful for various symmetries in physics, is the following one.

Example 9.4 Involutions

We consider $f \in \mathrm{Hom}(V, V)$ given by the relation

$$f^2 = id. \tag{9.11}$$

We define the projection $P := \frac{1}{2}(id_V + f)$ and we have

Eigenvectors	Eigenvalues
$\mathrm{im}\, P \setminus \{0\}$	$\lambda_1 = 1$
$\ker P \setminus \{0\}$	$\lambda_2 = -1$

and V has the decomposition

$$V = E(1, f) \oplus E(-1, f).$$

We prove this as follows:

Proof We try, as in Example 9.3, to determine a connection to projections. Using the relation (9.11) with the eigenvalue equation

$$fv = \lambda v \quad \text{and} \quad f^2 v = \lambda^2 v,$$

we obtain $v = \lambda^2 v$ which leads to $1 = \lambda^2$ and to $\lambda_1 = 1$ and $\lambda_2 = -1$. Taking into account the relation (9.11), we define

$$P := \tfrac{1}{2}(id + f)$$

which is a projection operator since:

$$P^2 = \tfrac{1}{2}(id + f) \circ \tfrac{1}{2}(id + f) = \tfrac{1}{4}\{id + 2f + f^2\}$$
$$= \tfrac{1}{4}\{id + 2f + id\}$$
$$= \tfrac{1}{2}(id + f).$$
$$\text{So} \quad P^2 = P.$$

We show that $\operatorname{im} P = E(\lambda = 1, f)$:
Let $v \in E(1)$, then we have $fv = v$ and we obtain:

$$Pv = \tfrac{1}{2}(id + f)v = \tfrac{1}{2}v + \tfrac{1}{2}fv = \tfrac{1}{2}v + \tfrac{1}{2}v = v,$$

hence $v \in \operatorname{im} P$.
Let $v \in \operatorname{im} P$, then we have $Pv = v$ and we obtain:

$$Pv = v \Leftrightarrow \tfrac{1}{2}(id + f)v = v \Leftrightarrow \tfrac{1}{2}v + \tfrac{1}{2}fv - v = 0 \Leftrightarrow \tfrac{1}{2}f - \tfrac{1}{2}v = 0$$
$$\Leftrightarrow fv = v$$

such that $v \in E(1)$. So we established

$$E(1) = \operatorname{im} P. \tag{9.12}$$

□

Similarly, one can show that $E(-1, f) = \ker P$.
Using the procedure of Example 9.3, we see that we can write $\ker P = \operatorname{im}(id - P)$ and we get $E(-1) = \operatorname{im}(id - P)$ such that we have the direct sum decomposition of V:

$$V = E(1) \oplus E(-1). \tag{9.13}$$

Using Eq. (9.12) and the above results, we define $P_1 := P$ and $P_{-1} := id - P$. The spectral decomposition of f is given by

$$f = 1 P_1 + (-1)P_{-1}. \tag{9.14}$$

Example 9.5 Nilpotent operators

Let $f \in \text{Hom}(V, V)$ be nilpotent. We consider a subspace $U \equiv U(f, v_0)$ of V with $v_0 \in V$ and $f^m v_0 = 0$ with $f^{m-1} v_0 \neq 0$. The subspace U is given by

$$U = \text{span}(v_0, f v_0, f^2 v_0, \ldots, f^{m-1} v_0). \tag{9.15}$$

We define $b_s := f^{s-1} v_0$, $s \in I(m)$. By definition, f is a special nilpotent operator and $B = (b_1, \ldots, b_m)$ is a basis of U (see Exercise 9.1). We have $f b_\mu = b_{\mu+1}$ for $\mu \in I(m-1)$ and

$$f b_m = 0.$$

Thus $\lambda = 0$ is the only eigenvalue of f and $E(0) = \ker f$ (see also Definition 9.12, Lemma 9.4, and Proposition 9.6). In this example, there is no basis of eigenvectors and so f is not diagonalizable. As we shall see later, in Theorem (9.4), the fact that f is a nilpotent operator is not an accident, it is the heart of non-diagonalizability.

These examples show that the terms eigenvalue, eigenvector, and eigenspace are very natural ingredients of operators. This applies regardless of how they can be specifically determined.

9.3.2   Eigenvalues, Eigenvectors, Eigenspaces of Matrices

Example 9.6 $A = \begin{bmatrix} 1 & 0 \\ 0 & 0 \end{bmatrix}$

This is a projection, as in Example 9.3. Additionally, we can test directly the relation $A^2 = A$. Further, here A is a symmetric matrix which, as it turns out, leads to an orthogonal projection on the x-axis:

$$A \vec{\xi} = A = \begin{bmatrix} \xi^1 \\ \xi^2 \end{bmatrix} = \begin{bmatrix} \xi^1 \\ 0 \end{bmatrix} = \xi^1 e_1 \in \mathbb{R}_1 \equiv e_1 \mathbb{R}.$$

We further recognize directly that $\ker A = \mathbb{R}_2 \equiv e_2 \mathbb{R}$. So we have $\text{im } A = \mathbb{R}_1$ and $\ker A = \mathbb{R}_2$. Since $\mathbb{1} - A = \begin{bmatrix} 0 & 0 \\ 0 & 1 \end{bmatrix}$, we get $\text{im}(\mathbb{1} - A) = \mathbb{R}_2$. We write, as in Example 9.3,

$$P_1 := A \quad \text{and} \quad P_0 := \begin{bmatrix} 0 & 0 \\ 0 & 1 \end{bmatrix}$$

and we denote the orthogonal decomposition of A given by $\mathbb{R}_1$ and $\mathbb{R}_2$ ($\mathbb{R}_1 \perp \mathbb{R}_2$):

$$A = 1\, P_1 + 0\, P_0.$$

Example 9.7 $A = \frac{1}{2}\begin{bmatrix} 1 & 1 \\ 1 & 1 \end{bmatrix}$

This is again an orthogonal projection since $A^2 = A$ and $A^\mathsf{T} = A$ hold. There are two eigenvalues, $\lambda_1 = 1$ and $\lambda_2 = 0$ with eigenvectors $v_1 = \begin{bmatrix} 1 \\ 1 \end{bmatrix}$ and $v_0 = \begin{bmatrix} 1 \\ -1 \end{bmatrix}$:

$$\frac{1}{2}\begin{bmatrix} 1 & 1 \\ 1 & 1 \end{bmatrix}\begin{bmatrix} 1 \\ 1 \end{bmatrix} = \begin{bmatrix} 1 \\ 1 \end{bmatrix} \quad \text{and} \quad \frac{1}{2}\begin{bmatrix} 1 & 1 \\ 1 & 1 \end{bmatrix}\begin{bmatrix} 1 \\ -1 \end{bmatrix} = \begin{bmatrix} 0 \\ 0 \end{bmatrix}.$$

We see explicitly that $v_1 \perp v_2$. So we get $E(1) = \mathbb{R}v_1 = \operatorname{im} A$ and $E(0) = \mathbb{R}v_0 = \ker A$.

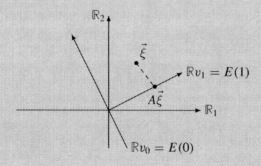

Example 9.8 $A = \begin{bmatrix} 1 & 0 \\ 0 & -1 \end{bmatrix}$

We have

$$\begin{bmatrix} 1 & 0 \\ 0 & -1 \end{bmatrix}\begin{bmatrix} 1 & 0 \\ 0 & -1 \end{bmatrix} = \begin{bmatrix} 1 & 0 \\ 0 & 1 \end{bmatrix}.$$

That is, $A^2 = \mathbb{1}$ and $A^T = A$. This is a direct reflection which is also an involution, see Example 9.9. Here, and in the next example, we see an abstract involution in a concrete situation:

$$A\vec{\xi} = A\begin{bmatrix} \xi^1 \\ \xi^2 \end{bmatrix} = \begin{bmatrix} \xi^1 \\ -\xi^2 \end{bmatrix}.$$

So we have explicitly:

$$\begin{bmatrix} 1 & 0 \\ 0 & -1 \end{bmatrix}\begin{bmatrix} 1 \\ 0 \end{bmatrix} = \begin{bmatrix} 1 \\ 0 \end{bmatrix} \quad \text{and} \quad \begin{bmatrix} 1 & 0 \\ 0 & -1 \end{bmatrix} = \begin{bmatrix} 0 \\ 1 \end{bmatrix} = (-1)\begin{bmatrix} 0 \\ 1 \end{bmatrix}.$$

There are two eigenvalues, $\lambda_1 = 1$ and $\lambda_2 = -1$, with eigenvectors $v_1 = \begin{bmatrix} 1 \\ 0 \end{bmatrix}$ and $v_{-1} = \begin{bmatrix} 0 \\ 1 \end{bmatrix}$ and $E(1) = \mathbb{R}_1$, $E(-1) = \mathbb{R}_2$. For the corresponding projections, we get

$$P_1 : \mathbb{R}^2 \longrightarrow E(1), \ P_{-1} : \mathbb{R}^2 \longrightarrow E(-1).$$

This leads to the spectral decomposition of A:

$$A = 1 \, P_1 + (-1) P_{-1}$$
$$\begin{bmatrix} 1 & 0 \\ 0 & -1 \end{bmatrix} = 1 \begin{bmatrix} 1 & 0 \\ 0 & 0 \end{bmatrix} + (-1) \begin{bmatrix} 0 & 0 \\ 0 & 1 \end{bmatrix}$$

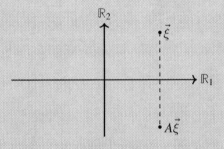

Example 9.9 $A = \begin{bmatrix} \cos\varphi & \sin\varphi \\ \sin\varphi & -\cos\varphi \end{bmatrix}$

A is a reflection (involution). Using almost the same notation as in Example 9.8, the eigenvectors are given here by

$$v_1 = \begin{bmatrix} \cos(\frac{\varphi}{2}) \\ \sin(\frac{\varphi}{2}) \end{bmatrix} \quad \text{and} \quad v_{-1} = \begin{bmatrix} \cos(\frac{\varphi}{2} + \frac{\pi}{2}) \\ \sin(\frac{\varphi}{2} + \frac{\pi}{2}) \end{bmatrix}$$

Formally, we obtain the same result as in Example 9.8:

$$A = 1 \, P_1 + (-1) P_{-1}.$$

Example 9.10 $A = \begin{bmatrix} \cos\varphi & -\sin\varphi \\ \sin\varphi & \cos\varphi \end{bmatrix}$

This is a rotation by the angle φ. Unless $\varphi \in \pi\mathbb{Z}$, there is no subspace of $\mathbb{R}^2$ that stays A-invariant and thus there are neither eigenvalues nor eigenvectors. Observe that this behavior is only because we are working over $\mathbb{R}$, and it is quite different over $\mathbb{C}$!

Example 9.11 $A = \begin{bmatrix} 0 & 1 & 0 \\ 0 & 0 & 1 \\ 0 & 0 & 0 \end{bmatrix}$

A is a nilpotent matrix.

$$A^2 = \begin{bmatrix} 0 & 0 & 1 \\ 0 & 0 & 0 \\ 0 & 0 & 0 \end{bmatrix}, \quad A^3 = \begin{bmatrix} 0 & 0 & 0 \\ 0 & 0 & 0 \\ 0 & 0 & 0 \end{bmatrix}$$

By inspection, we see that there is only one eigenvalue, $\lambda = 0$, with eigenvector $v_0 = e_1 = \begin{bmatrix} 1 \\ 0 \\ 0 \end{bmatrix}$ and eigenspace

$$E(0) = \ker A = \mathbb{R}_1 = \mathbb{R}e_1.$$

Definition 9.3 Geometric multiplicity.
The dimension, n_λ, of $E(\lambda, f)$, is called geometric multiplicity of the eigenvector λ.

In what follows, we consider f fixed and write $E_\lambda \equiv E(\lambda, f)$. As we see, the restriction of f to $E_\lambda \leqslant V$, $f|_{E_\lambda}$ acts only as a multiplication by λ, the simplest nontrivial action an operator can do. It is interesting to notice that for $\lambda_1 \neq \lambda_2$, it follows that $E_{\lambda_1} \cap E_{\lambda_2} = \{0\}$ so that E_{λ_1} and E_{λ_2} are linearly independent (see Definition 3.12). This means that the eigenvectors corresponding to distinct eigenvalues are linearly independent. This is shown in the following proposition.

Proposition 9.1 *Linear independence of eigenvectors.*
Let $f \in \mathrm{End}(V)$. If $\lambda_1, \ldots, \lambda_r$ are distinct eigenvalues of f, then corresponding eigenvectors $v_1, \ldots, v_r$ are linearly independent.

Proof Suppose that the eigenvectors $v_1, \ldots, v_r$ are linearly dependent. This will lead to the contradiction that one of them must be zero. (By definition, every eigenvector is a nonzero vector.) Since the list $(v_1, \ldots, v_r)$ is by assumption linearly dependent, one of the vectors in this list must be a linear combination of the proceeding vectors. Let k be the smallest such index. This means that the vectors $v_1, \ldots, v_{k-1}$ are linearly independent. So v_k is a linear combination of the proceedings:

$$v_k = \alpha^\mu v_\mu, \mu \in I(k-1), \alpha^\mu \in \mathbb{K}. \tag{9.16}$$

Acting by f, we obtain

$$f(v_k) = f(\alpha^\mu v_r) = \alpha^\mu(f v_\mu), \tag{9.17}$$
$$\implies \lambda_k v_k = \alpha^\mu \lambda_\mu v_\mu. \tag{9.18}$$

On the other hand, we multiply Eq. (9.16) by λ_k and we have

$$\lambda_k v_k = \lambda_k \alpha^\mu v_\mu = \alpha^\mu \lambda_k v_\mu. \tag{9.19}$$

Subtracting Eq. (9.18) from (9.19), we obtain

$$0 = \alpha^\mu(\lambda_\mu - \lambda_k)v_\mu. \tag{9.20}$$

Since $(\lambda_\mu - \lambda_k) \neq 0$ and the $(v_1, \ldots, v_{k-1})$ are linearly dependent, it follows that

$$\alpha^\mu = 0 \quad \forall \mu \in I(k-1). \tag{9.21}$$

Equation (9.16) shows that $v_k = 0$, this is in contradiction to the fact that v_k is an eigenvector and therefore nonzero. This completes the proof and the list $(v_1, \ldots, v_r)$ is linearly independent.   □

From the above proposition, we can directly deduce the following corollary.

Corollary 9.1 *Direct sum of eigenspaces.*
If the eigenvalues $(\lambda_1, \ldots, \lambda_r)$ of the operator f are distinct, then

(i) *the sum of the eigenspaces is direct:*

$$E_{\lambda_1} + \cdots + E_{\lambda_r} = E_{\lambda_1} \oplus \cdots \oplus E_{\lambda_r};$$

(ii) *the following inequality is valid:*

$$\sum_{i=1}^{r} \dim E_{\lambda_i} \leqslant \dim V.$$

> **Corollary 9.2** *Number of distinct eigenvalues.*
> *The operator $f \in \text{End}(V)$ can have at most $n = \dim V$ distinct eigenvalues.*

Proof Proof of Corollaries 9.1 and 9.2.
We choose a basis $B_i := (v_1^{(i)}, \ldots v_{n_i}^{(i)})$ of $E_{\lambda_i}, i \in I(r)$. Then the list $(B_1 \cup B_2 \cup \ldots \cup B_r)$ is a basis of

$$W := E_{\lambda_1} \oplus \cdots \oplus E_{\lambda_r}$$

since $E_{\lambda_i} \leqslant V$, $W \leqslant V$ and the above sum is direct which means that the subspaces $(E_{\lambda_1}, \ldots, E_{\lambda_r})$ are linearly dependent. This leads directly to the result:

$$\dim W = \dim E_{\lambda_1} + \cdots + \dim E_{\lambda_r} \leqslant V.$$

$\square$

Everything concerning an operator $f \in \text{End}(V)$ is precisely valid for a matrix F if it is considered as an endomorphism $F \in \text{End}(\mathbb{K}^n)$ and, in particular, if it is regarded as a representation $F \equiv f_B$ of f relative to a basis B. In this sense, the eigenvalue problem of f corresponds precisely to the eigenvalue problem of $F \equiv f_B$. The following statement shows this.

> **Proposition 9.2** *Representation of eigenvectors.*
> *Given an operator $f \in \text{End}(V)$ and a basis $B = (b_1, \ldots, b_n)$ in V. Then v is an eigenvector of F corresponding to the eigenvalue λ if and only if $\phi_B(v) =: v_B \in \mathbb{K}^n$ is an eigenvector of $F \equiv f_B$ corresponding to the same eigenvalue λ. Recall that ϕ_B is the canonical basis isomorphism between V and $\mathbb{K}^n$, as given by the commutative diagram:*
>
> $$\begin{array}{ccc} V & \xrightarrow{\ f\ } & V \\ \phi_B \downarrow & & \downarrow \phi_B \\ \mathbb{K}^n & \xrightarrow{\ F\ } & \mathbb{K}^n \end{array}$$
>
> *So we have :*
>
> $$f v = \lambda v \Leftrightarrow F v_B = \lambda v_B. \tag{9.22}$$

Proof The above diagram explains everything.

$$\phi_B \circ f = F \circ \phi_B. \tag{9.23}$$

Let (λ, v) be an eigenelement of f. Then we have the following equivalences:

$$f v = \lambda v \Leftrightarrow \phi_B \circ f(v) = \phi_B(\lambda v), \qquad (9.24)$$
$$\Leftrightarrow F \circ \phi_B(v) = \lambda \phi_B(v), \qquad (9.25)$$
$$\Leftrightarrow F v_B = \lambda v_B. \qquad (9.26)$$

This proves the equivalence of Eq. (9.22).   □

This means that by using matrices, we can obtain everything we want to know about operators. In particular, the eigenvalues of an operator f are given by the eigenvalues of the corresponding representation $f_B \equiv F$.

How do we find the eigenvalues of a matrix? To answer this, we first notice that λ is an eigenvalue of F if and only if the equation

$$(F - \lambda \mathbb{1}_n) \vec{v} = 0 \qquad (9.27)$$

has a nontrivial solution or, equivalently, if $(F - \lambda \mathbb{1}_n)$ is singular or if $\det(\lambda I_u - F) = 0$. This leads to the following definition:

Definition 9.4 The characteristic polynomial of a matrix.
The polynomial $\chi_F(x) = \det(x \mathbb{1}_n - F)$ is called the characteristic polynomial of the matrix $F \in \mathbb{K}^{n \times n}$ and is given by

$$\chi_F(x) = x^n + \chi_{n-1} x^{n-1} + \cdots + \chi_2 x^2 + \chi_1 x + \chi_0$$

with the coefficients $\chi_0, \chi_1, \ldots, \chi_{n-1} \in \mathbb{K}$.

The roots of χ_F are the eigenvalues of F.

Remark 9.1 Criterion for eigenvalues.
A scalar λ is an eigenvalue of F if and only if

$$\chi_F(\lambda) = 0.$$

The equation $\det(x \mathbb{1} - F) = 0$ holds if and only if the matrix $x \mathbb{1} - F$ is not invertible, which is equivalent to the statement:

$$\ker(x \mathbb{1} - F) \neq \{0\}.$$

This gives us a computational way to calculate the eigenvalues of a matrix F. One simply finds the solutions of the eigenvalue equation:

$$\det[x\mathbb{1} - F] = 0.$$

Definition 9.5 Characteristic polynomial of an operator.
The characteristic polynomial of the operator f is defined by the same polynomial as above if we take $F \equiv f_B$ for some given basis $B \in B(V)$:

$$\chi_f(x) := \chi_{f_B}(x).$$

This definition is well defined since it is independent of the chosen basis B. The following lemma shows this.

Lemma 9.1 *Basis invariance.*
Let $f \in \text{End}(V)$ and $\dim V = n$. For the two bases $B, C \in B(V)$ with the corresponding basis isomorphisms ϕ_B and ϕ_C and the representations of f:

$$f_B = \phi_B \circ f \circ \phi_B^{-1} \quad and \quad f_C = \phi_C \circ f \circ \phi_C^{-1}.$$

Then $\det f_C = \det f_B$.

Proof We define $T := \phi_C \circ \phi_B^{-1}$. We have $f = \phi_B^{-1} \circ f_B \circ \phi_B$. This leads to

$$f_C = \phi_C \circ f\phi_C^{-1} = \phi_C \circ (\phi_B^{-1} \circ f_B \circ \phi_B) \circ \phi_C^{-1},$$
$$f_C = T \circ f_B \circ T^{-1}.$$

Then $\det f_C = \det(Tf_BT^{-1}) = \det T \det f_B \det T^{-1} = \det f_B$. ☐

This proves also that $\det f := \det f_B$ is well defined and that the determinant of the operator $x \, id_V - f$ is the characteristic polynomial of the operator f:

$$\chi_f(x) := \det(x \, id_V - f).$$

Note that the characteristic polynomial is a monic polynomial with $\deg(\chi_f) = n = \dim V$.

Definition 9.6 Algebraic multiplicity of an eigenvalue.
If the characteristic polynomial has the form

$$\chi_f(x) = (x - \lambda)^{m_\lambda} \varrho(x)$$

with $\varrho(\lambda) \neq 0$, the exponent m_λ is called the algebraic multiplicity of the eigenvalue λ.

As expected, there is a relation between the geometric $n_\lambda = \dim E_\lambda$ and the algebraic multiplicity m_λ.

Lemma 9.2 *Geometric and algebraic multiplicity.*
If $f : V \to V$, λ is an eigenvalue and n_λ, m_λ are the geometric and algebraic multiplicity respectively, then $n_\lambda \leqslant m_\lambda$.

Proof We choose a basis of V,

$$B := (c_1, \ldots, c_{n_\lambda}, b_{n_\lambda+1}, \ldots, b_{m'_\lambda}, b_{n_\lambda+m'_\lambda+1}, \ldots, b_n),$$

where $(c_1, \ldots, c_{n_\lambda})$ is an eigenbasis of E_λ. The matrix of f with respect to this basis is given by the block matrix f_B:

$$F \equiv f_B = \left[\begin{array}{c|c} \lambda \mathbb{1}_{n_\lambda} & M_1 \\ \hline 0 & M_2 \end{array} \right]$$

In an obvious notation we have

$$\chi_f(x) = \chi_F(x) = (x - \lambda)^{n_\lambda} \chi_2(x) \quad \text{and} \quad \chi_2(x) = (x - \lambda)^{m'_\lambda} \varrho_2(x) \text{ with } \varrho_2(\lambda) \neq 0$$

so

$$\chi_f(x) = (x - \lambda)^{n_\lambda + m'_\lambda} \varrho_2(x), \quad (m'_\lambda \geqslant 0).$$

So we obtain $m_\lambda = n_\lambda + m'_\lambda$ and $n_\lambda \leqslant m_\lambda$. $\square$

9.3.3 Determining Eigenvalues and Eigenvectors

Example 9.12 $F = \begin{bmatrix} 1 & 2 \\ 2 & 4 \end{bmatrix}$

The eigenvalue equation $\det[x\,\mathbb{1} - F] = 0$ for the above F is given by $\det \begin{bmatrix} x-1 & -2 \\ -2 & x-4 \end{bmatrix} = 0$. This leads to

$$(x - 1)(x - 4) - 4 = 0 \Leftrightarrow x^2 - 5x + 4 - 4 = 0 \Leftrightarrow x^2 - 5x = 0.$$

So the eigenvalues of F are $\lambda_1 = 0$ and $\lambda_2 = 5$. One can calculate an eigenvector corresponding to λ_1 from the matrix equation

$$\begin{bmatrix} 1 & 2 \\ 2 & 4 \end{bmatrix} \begin{bmatrix} \xi^1 \\ \xi^2 \end{bmatrix} = 0 \quad \text{or by the system} \quad \begin{matrix} \xi^1 & +2\,\xi^2 & =0 \\ 2\xi^1 & +4\,\xi^2 & =0 \end{matrix}.$$

The solution gives the eigenvector $v_0 = \begin{bmatrix} 2 \\ -1 \end{bmatrix}$ and the eigenspace $E(0) = \ker F = \mathbb{R}v_0$. Similarly, for the eigenvalue $\lambda_2 = 5$ we have

$$\begin{bmatrix} 1-5 & 2 \\ 2 & 4-5 \end{bmatrix} = 0 \Leftrightarrow \begin{bmatrix} -4 & 2 \\ 2 & -1 \end{bmatrix} = 0.$$

This leads to the system $\begin{matrix} -4\xi^1 & +2\xi^2 & =0 \\ 2\xi^1 & -\xi^2 & =0 \end{matrix}$, with a solution $\xi^2 = 2\,\xi^1$. An eigenvector of $\lambda = 5$ is given by $v_5 = \begin{bmatrix} 1 \\ 2 \end{bmatrix}$. The corresponding eigenspace is given by

$$E(5) = \mathbb{R}v_5.$$

We observe that $v_0 \perp v_5$ since $F^T = F$.

Example 9.13 $F = \begin{bmatrix} 1 & 0 \\ 0 & 0 \end{bmatrix}$

We already know everything from Examples 9.8 and 9.6. Therefore we now check the results $\chi_F(x) = \det \begin{bmatrix} x-1 & 0 \\ 0 & x \end{bmatrix} = (x - 1)x$. This leads as expected to $(x - 1)x = 0$ and to the eigenvalues $\lambda_1 = 0$ and $\lambda_2 = 1$.

Example 9.14 $F = \frac{1}{2}\begin{bmatrix} 1 & 1 \\ 1 & 1 \end{bmatrix}$

Here we proceed as in Example 9.13 and determine the eigenvalues of F.

$$\chi_F(x) = \det \begin{bmatrix} x-\frac{1}{2} & -\frac{1}{2} \\ -\frac{1}{2} & x-\frac{1}{2} \end{bmatrix} = (x - \frac{1}{2})(x - \frac{1}{2}) - \frac{1}{4} =$$

$$= x^2 - \frac{1}{2}x - \frac{1}{2}x + \frac{1}{4} - \frac{1}{4} = x^2 - x = x(x-1).$$

This leads to $x(x-1) = 0$ and to eigenvalues of F, $\lambda_1 = 0$ and $\lambda_2 = 1$.

Example 9.15 $F = \begin{bmatrix} 0 & 1 \\ 1 & 0 \end{bmatrix}$

F is an involution: $F^2 = \mathbb{1}$.
The characteristic polynomial is given by

$$\chi_F(x) = \det \begin{bmatrix} x & -1 \\ -1 & x \end{bmatrix} = x^2 - 1.$$

So we get $\chi_F(x) = 0 \Leftrightarrow x^2 - 1 = 0$ and the solutions are $\lambda_1 = 1$ and $\lambda_2 = -1$.

Example 9.16 $F = \begin{bmatrix} 0 & 1 & 0 \\ 0 & 0 & 1 \\ 0 & 0 & 0 \end{bmatrix}$

This is a nilpotent matrix as in Example 9.11.

$$\chi_F(x) = \det \begin{bmatrix} x & 1 & 0 \\ 0 & x & 1 \\ 0 & 0 & x \end{bmatrix} = x^3.$$

So we have the equation for the eigenvalues

$$x^3 = 0.$$

The only solution is $x = 0$ and we have only one eigenvalue $\lambda = 0$, as expected.

9.4 The Question of Diagonalizability

For a linear map $f \in \text{Hom}(V, V')$, the problem of diagonalization was solved in Sect. 3.3 and in Theorem 3.1. To accomplish this, we simply had to choose two tailor-made bases B_0 and B'_0 in V and V'. For an operator $f \in \text{End}(V)$, the situation is very different and much more difficult. Here, it is natural to look for only one tailor-made basis B_0 (one vector space, one basis) and we expect (or hope) to obtain a diagonal matrix

$$f_{B_0 B_0} \equiv f_{B_0} = \begin{bmatrix} \lambda_1 & & & & 0 \\ & \ddots & & & \\ & & \lambda_s & & \\ & & & \ddots & \\ 0 & & & & \lambda_n \end{bmatrix} \quad \text{with } \lambda_s \in \mathbb{K}.$$

However, we cannot expect to find a diagonal representation for every operator. This leads to the diagonalizability question and to the following equivalent definitions.

Definition 9.7 Diagonalizability 1.
A map $f \in \text{End}(V)$ is diagonalizable if there is a basis B_0 of V so that the representation f_{B_0} is a diagonal matrix.

Definition 9.8 Diagonalizability 2.
The endomorphism $f \in \text{End } V$ is diagonalizable, if the vector space V has a basis C consisting of eigenvectors of f.

Proposition 9.3 *Definitions 9.8 and 9.7 are equivalent.*

Proof If Definition 9.7 holds, f is diagonalizable with a diagonal representative of the equivalence

$$f_{B_0} = \begin{bmatrix} \lambda_1 & & 0 \\ & \ddots & \\ 0 & & \lambda_n \end{bmatrix}. \tag{9.28}$$

Then the entries of this matrix are

$$f_{B_0} = [\varphi_s^i] \quad \text{with} \quad \varphi_s^i = \lambda_s \delta_s^i. \tag{9.29}$$

Suppose the basis B_0 is given by the list $(v_1, \ldots, v_n)$. The values $f v_s$ of the basis vector v_s are given as usual by the expression

$$\begin{aligned} f v_s &= v_i \varphi_s^i = v_i \lambda_s \delta_s^i = v_s \lambda_s, \\ f v_s &= \lambda_s v_s. \end{aligned} \tag{9.30}$$

This shows that the basis vectors of B_0 are eigenvectors of the map f. Hence a tailor-made basis B_0 is an eigenvector basis of f. This is Definition 9.8.

Conversely, if Definition 9.8 holds, f has a basis of eigenvectors $C = (c_1, \ldots, c_n)$ and then we have $f c_s = \lambda_s c_s$. This means $f_C = (\lambda_s \delta_s^i)$, and we see immediately that

$$
f_C = \begin{bmatrix} \lambda_1 & & 0 \\ & \ddots & \\ 0 & & \lambda_n \end{bmatrix}.
$$

This is Definition 9.7. That means $C = B_0$, an eigenvector basis, is a tailor-made basis. This completes the proof of the equivalence of Definition 9.7 and Definition 9.8. $\qquad\square$

The following proposition gives a sufficient condition for the diagonalizability:

> **Proposition 9.4** *Diagonalizability by distinct eigenvectors.*
> *If V (of dimension n) and $f \in \text{End}(V)$ has n distinct eigenvalues, then f is diagonalizable.*

Proof From Proposition 9.1 we know that if the eigenvalues $\lambda_1, \ldots, \lambda_n$ are distinct, so the list of the n corresponding eigenvectors $v_1, \ldots, v_n$ is a linearly independent set. Since dim $V = n$, it follows that $(v_1, \ldots, v_n)$ is a basis of V. This basis consists of eigenvectors. So by Proposition 9.3, f is diagonalizable. $\qquad\square$

Comment 9.1 Abstract decomposition.

The best way to analyze the structure of an operator $f \in \text{Hom}(V, V)$ is to decompose V as finely as possible into f-invariant subspaces:

$$
V = U_1 \oplus \cdots \oplus U_j \oplus \cdots \oplus U_r \tag{9.31}
$$

with corresponding endomorphism decomposition $f_j := f_{|U_j}$

$$
f_j : U_j \longrightarrow U_j \quad \text{so that}
$$

$$
f = f_1 \oplus \cdots \oplus f_j \oplus \cdots \oplus f_r. \tag{9.32}
$$

For this purpose, it is useful to first consider an abstract decomposition of V without the given operator f. Furthermore, if later we want to establish the connection with the operator f, it is helpful to use a corresponding projection system $(P_1, \ldots, P_r)$ when decomposing V with:

$$P_j : V \longrightarrow U_j.$$

This results in a "complete orthogonal system" of idempotents (projections). We describe this with the following definition.

Definition 9.9 Abstract decomposition and projection operators.
We call a list of idempotents (projections) $(P_1, \ldots, P_j, \ldots, P_r)$ a direct decomposition of identity if for $i, j \in I(r)$

(i) $P_i P_j = \delta_{ij} P_i$ and
(ii) $P_1 + \cdots + P_j + \cdots + P_r = id_V$.

The property (i) says that the list $(P_1, \ldots, P_r)$ is a linear independent list of idempotents. Further, it is called "orthogonal" even though there is no mention of an inner product space. The property (ii) is a decomposition of the identity.

The following proposition summarizes the two aspects, the direct sum decomposition of V and the complete orthogonal system of projection.

Proposition 9.5 *Direct sum and direct decomposition of the identity.*

(i) *If $V = U_1 \oplus \cdots \oplus U_j \oplus \cdots \oplus U_r$ is a direct sum of the subspaces $U_1, \ldots, U_r$, then there is a list $(P_1, \ldots, P_r)$ of projections in V with a direct decomposition of the identity such that for each $j \in I(r)$, $U_j = \operatorname{im} P_j$.*
(ii) *Conversely, let $P_1, \ldots, P_r$ be a direct decomposition of identity in V, then there is a direct sum decomposition*

$$V = U_1 \oplus \cdots \oplus U_r$$

with $U_j = \operatorname{im} P_j$.

Proof (i)
The above direct sum allows to write for $v \in V$ and $u_j \in U_j$, $j \in I(r)$:

$$v = u_1 + \cdots + u_j + \cdots + u_r.$$

If we define projectors P_j:

$$P_j : V \longrightarrow U_j$$
$$v \longmapsto P_j v := u_j.$$

We need to show that $(P_1, \ldots, P_r)$ is a direct decomposition of identity, that is, that

- P_j is linear;
- $P_i P_j = \delta_{ij}$;
- $P_1 + \ldots + P_r = id$.

(See Exercise 9.4).

Conditions (i) and (ii) of Definition 9.9 hold and so $P_1, \ldots, P_r$ is the corresponding abstract decomposition. □

Proof (ii)
We have to show that for $v \in V$ there is a unique decomposition:

$$v = u_1 + \cdots + u_j + \cdots + u_r$$

with $u_j \in \operatorname{im} P_j = U_j$, $j \in I(r)$. The property (ii) leads to

$$v = id_V v = (P_1 + \cdots + P_j + \cdots + P_r) \, v = P_1 v + \cdots + P_r v = u_1 + \cdots + u_r.$$

So the decomposition holds. We show the uniqueness of the decomposition. We start with the representation :

$$y_1 + \cdots + y_j + \cdots + y_r = 0.$$

With $y_j \in U_j$, $x_j \in V$ and $y_j = P_j x_j$, using the property (i) of Definition 9.9, we obtain

$$0 = (P_j(y_j + \cdots + y_r)) = P_1 y_j = P_j P_j x_j = P_j^2 x_j = P_j x_j = y_j$$

for all $j \in I(r)$. Therefore, the decomposition is unique so (ii) and with it Proposition 9.5 is proven. □

We are now in the position to give a geometric characterization of diagonalizability.

Theorem 9.1 *Equivalence theorem of diagonalizability.*
If $f \in \operatorname{End}(V)$ and $\dim V = n$, then the following statements are equivalent.

(i) The map f is diagonalizable.
(ii) The characteristic polynomial χ_f decomposes into n linear factors and the geometric multiplicity is equal to the algebraic multiplicity ($n_\lambda = m_\lambda$) for all eigenvalues of f.

> (iii) If $\lambda_1, \ldots, \lambda_r$ are all the eigenvalues of f (assumed distinct), then we
> have the direct sum decomposition of V:
>
> $$V = E_{\lambda_1} \oplus \cdots \oplus E_{\lambda_r}.$$
>
> (iv) There exists a linearly independent list of idempotents $P_1, \ldots, P_r \in$
> End(V) with the following properties:
>
> (a) $P_i P_j = 0$ for $i \neq j$ and $i, j \in I(r)$;
>
> (b) $P_1 + \cdots + P_r = id_V$;
>
> (c) $f = \lambda_1 P_1 + \cdots + \lambda_r P_r$.

According to Definition 9.9, the properties (a) and (b) state that the list $(P_1, \ldots, P_r)$ is a direct decomposition of the identity. Assertion (iv) states that there exists a spectral decomposition of V induced by the operator f.

Proof Part (i) $\Rightarrow$ (ii).
Since f is diagonalizable, the basis B_0 of eigenvectors of f is given by the list

$$B_0 = (B_1, B_2, \ldots, B_r) \tag{9.33}$$

with B_j the basis of E_{λ_j}, $B_j := (b_1^{(j)}, \ldots, b_{n_j}^{(j)})$ $j \in I(r)$ and

$$n_j = \dim E_{\lambda_j}. \tag{9.34}$$

Since $V = \text{span}(B_1, \ldots, B_r)$,

$$n = \dim V = n_1 + \cdots + n_r. \tag{9.35}$$

Since f is diagonalizable, we further get from the characteristic polynomial

$$\chi_f(x) = (x - \lambda_1)^{m_1} \cdots (x - \lambda_r)^{m_r}, \tag{9.36}$$

the equation

$$n = m_1 + \cdots + m_r. \tag{9.37}$$

Since $n_j \leqslant m_j$, we obtain from Eqs. (9.35) and (9.36) $\forall j \in I(r)$, $n_j = m_j$. This proves Part (ii). $\square$

Proof Part (ii) $\Rightarrow$ (iii).
As we saw in Sect. 9.2, when $\lambda_1 \neq \lambda_2$, the sum of E_{λ_1} and E_{λ_2} is direct:

$$E_{\lambda_1} + E_{\lambda_2} = E_{\lambda_1} \oplus E_{\lambda_2}.$$

This can be generalized for all the eigenvalues of f. So we have $E := E_{\lambda_1} \oplus \cdots \oplus E_{\lambda_r}$, a subspace of V. Since dim $E = n_1 + \cdots + n_r = \dim V = n$, it follows that $E = V$ and Part (iii) is proven. □

Proof Part (iii) ⇒ (i).
If $B_j = (b_1^{(j)}, \ldots, b_{n_j}^{(j)})$ is an eigenvector basis of E_{λ_j}, then

$$B = (b_1^{(1)}, \ldots, b_{n_1}^{(1)}, \ldots, b_1^{(r)}, \ldots, b_{n_r}^{(r)})$$

is an eigenbasis of $E = V$. This means that

$$B = B_0 \quad \text{and} \quad f_B = \begin{bmatrix} \lambda_1 & & & & & & \\ & \ddots & & & & & \\ & & \lambda_1 & & & & \\ & & & \ddots & & & \\ & & & & \lambda_r & & \\ & & & & & \ddots & \\ & & & & & & \lambda_r \end{bmatrix}$$

with m_1 the multiplicity of λ_1 and m_r the multiplicity of λ_r.

So f is diagonalizable and Part (i) is proven. □

Proof Part (iii) ⇔ (iv)
Suppose assertion (iii) holds. Then there is an f-invariant decomposition of V with subspaces $U_j = E_{\lambda_j}$. Then Proposition 9.5 implies that there is a direct decomposition of the identity $(P_1, \ldots, P_r)$ with im $P_i = U_i$ for each $i \in I(r)$, which then satisfies condition (iv)(a) and (iv)(b).

For condition (iv)(c), since this decomposition is f-invariant, we have $f(U_j) \subseteq U_j$ for all $j \in I(r)$ and because $U_j = E_{\lambda_j} = \ker(f - \lambda_j id)$, we get

$$f|_{U_j} = \lambda_j id_{U_j} = (\lambda_1 P_1 + \cdots + \lambda_r P_r)|_{U_j}$$

for every $j \in I(r)$. Consequently, we see $f = \lambda_1 P_1 + \cdots + \lambda_r P_r$ and hence (iv) holds.

Conversely, suppose we have (iv). Then by Proposition 9.5, we have a decomposition

$$V = U_1 \oplus \ldots \oplus U_r$$

with $U_i = \text{im } P_i$ for each $i \in I(r)$. Then, since

$$f - \lambda_j id = (\lambda_1 P_1 + \ldots + \lambda_j P_j + \cdots + \lambda_r P_r) - (\lambda_j P_1 + \ldots + \lambda_j P_r)$$
$$= (\lambda_1 - \lambda_j) P_1 + \ldots + (\lambda_{j-1} - \lambda_j) P_{j-1} + (\lambda_{j-1} - \lambda_j) P_j + \ldots + (\lambda_r - \lambda_j) P_r$$

and $\lambda_i - \lambda_j \neq 0$. Whenever $i \neq j$, we see

$$E_{\lambda_j} = \ker(f - \lambda_j id) = \ker((\lambda_1 - \lambda_j) P_1 + \ldots + (\lambda_r - \lambda_j) P_r)$$
$$\subseteq \text{im } P_j$$
$$= U_j$$

because $(\lambda_i - \lambda_j) P_i P_j = 0 \ \forall \ i \neq j$.

Further, if $x \in \ker(f - \lambda_j id)$, there is a unique decomposition of $x = u_1 + \cdots + u_r$ with each $u_i \in U_i$ and so

$$0 = (f - \lambda_j id)(x)$$
$$= (\lambda_1 - \lambda_j) u_1 + \cdots + (\lambda_{j-1} - \lambda_j) u_{j-1} + (\lambda_{j+1} - \lambda_j) u_j + \cdots + (\lambda_r - \lambda_j) u_r.$$

By linear independence, each $(\lambda_i - \lambda_j) u_i = 0$, and thus $u_i = 0$ whenever $i \neq j$. Hence, $x = u_j \in u_j$ and $E_{\lambda_j} \subseteq u_j$. Thus

$$E_{\lambda_j} = U_j,$$
$$V = E_{\lambda_1} \oplus \cdots \oplus E_{\lambda_r}$$

and we have (iii). Thus, (iii) and (iv) are equivalent. So Theorem 9.1 is proven. □

For this theorem, an interesting and clarifying conclusion is given by the following statement.

Corollary 9.3 : *Diagonalizability and direct decomposition.*
If f is diagonalizable, this is equivalent to the direct sum decomposition of V in one-dimensional f-invariant subspaces:

$$V = \overset{n}{\underset{\alpha=1}{\oplus}} V_\alpha$$

with $\dim V_\alpha = 1$ *for all* $\alpha \in I(n)$. *This corresponds to a decomposition of f with appropriate repetitions of the eigenvalues* λ_α :

$$f = \overset{n}{\underset{\alpha=1}{\oplus}} \lambda_\alpha P_{V_\alpha}.$$

Proof We use the notation of Theorem 9.1 and the proof there. The proof is straightforward since we may write $\alpha = (j, \mu)$ and $V_\alpha = V_{j,\mu} := \text{span}(b_\mu^{(j)})$ $\mu \in I(n_j)$. □

Comment 9.2 Diagonalizability of matrices.

All the formalisms mentioned above apply, of course, to matrices. A matrix $F \in \mathbb{K}^{n \times n}$ may also be considered as an operator on the vector space $\mathbb{K}^n$, $F \in \text{End}(\mathbb{K}^n)$. In addition, matrices are needed when using basis isomorphisms to investigate the properties of operators on an abstract vector space V. In this sense, the question of diagonalizability is more direct for matrices.

A matrix F is diagonalizable if there is an invertible matrix P and a diagonal matrix D so that any one of the following equalities is satisfied.

$$F = PDP^{-1} \Leftrightarrow D = P^{-1}FP \Leftrightarrow FP = PD. \qquad (9.38)$$

We immediately see from matrix multiplication on column vectors of P that the diagonalization of F is equivalent to the existence of n linearly independent eigenvectors of F, as proven in the Theorem 9.1. The n columns of P are the n linearly independent eigenvectors of F and the diagonal entries of D are the corresponding eigenvalues of F.

In other words the matrix P considered as a list of n columns (vectors in $\mathbb{K}^n$) is an eigenbasis of F. We can also see it as follows:
For

$$P := (c_1, \ldots, c_n) \quad \text{and} \quad D := \text{diag}(\lambda_1 \ldots, \lambda_n),$$

we may write in the same notation

$$FP = F(c_1, \ldots, c_n) = (Fc_1, \ldots, Fc_n) \qquad (9.39)$$

and

$$PD = [c_1 \ldots c_n] \begin{bmatrix} \lambda_1 & & \\ & \ddots & \\ & & \lambda_n \end{bmatrix} = (\lambda_1 c_1, \ldots, \lambda_n c_n). \qquad (9.40)$$

Equations 9.38, 9.39, and 9.40 lead to the eigenvalue

$$Fc_s = \lambda_s c_s \quad \forall s \in I(n). \qquad (9.41)$$

Again, we see here immediately that the columns of $P = (c_1, \ldots, c_n)$ have to be linearly independent whereas the scalars $(\lambda_1, \ldots, \lambda_n)$ have not to be different from each other. So the diagonal of the diagonal matrix D is given by

$$D = \text{diag}(\lambda_1, \ldots, \lambda_1, \lambda_2, \ldots, \lambda_2, \ldots, \lambda_r, \ldots, \lambda_r), \qquad (9.42)$$

with $(\lambda_1, \ldots, \lambda_r)$ distinct ($r \leqslant n$).

After having discussed the question of diagonalizability, we may now ask what it is suitable for.

There are a lot of reasons. In physics, we sometimes call it the decoupling procedure, which shows an important aspect. In mathematical terms, when we are looking for the most straightforward possible representation of an operator, we have to choose an appropriate basis which we may, in this case, also call a tailor-made basis. It turns out that this basis consists of eigenvectors. So the diagonalization reveals the true face of an operator and its geometric properties. In the case where f corresponds to a physical observable, the eigenvalues of f are exactly the physical values of the experiment. In addition, diagonalization allows a considerable simplification in the calculations.

At this point, it is instructive to compare the situation between $\text{Hom}(V, V')$ and $\text{End}(V)$. In the case of $\text{Hom}(V, V')$, the diagonalization is relatively straightforward to obtain. As we saw in Sect. 3.3, Theorem 3.1, and Remark 3.3, we have at our disposal two bases, one in V and the other one in V'. In the case of $\text{End } V$, it seems natural to work with one basis since the domain and the codomain are identical, and the problem is far more complex. This leads to the question of normal form for endomorphisms. This problem was firstly discussed in Remark 3.4 (On the normal form of endomorphisms) and Proposition 3.12 where we showed some obstacles to using diagonalization.

But there is a chance! The space $\text{End}(V)$ has much more structure than the space $\text{Hom}(V, V')$. $\text{End}(V)$ is an algebra: the operator can be raised to powers of f and to linear combination of powers of f, in contrast to linear maps in $\text{Hom}(V, V')$.

Comment 9.3 The action of polynomials on the algebra $\text{End}(V)$.

We have

$$f^k := f \circ \underbrace{\;\ldots\ldots\ldots\;}_{k - times} \circ f \in \text{End}(V).$$

This also means that we may talk about linear combinations of powers of f, with $f^0 = id$.

$$\alpha_m f^m + \alpha_{m-1} f^{m-1} + \cdots + \alpha_2 f^2 + \alpha_1 f + \alpha_0 id. \tag{9.43}$$

As we immediately see, this is connected to a polynomial $\varphi \in \mathbb{K}[x]$.

$$\varphi(x) = \alpha_m x^m + \cdots + \alpha_2 x^2 + \alpha_1 x + \alpha_0. \tag{9.44}$$

If we consider also the map

$$f \longmapsto \varphi(f) := \alpha_m f^m + \cdots + \alpha_2 f^2 + \alpha_1 f + \alpha_0 f^0, \tag{9.45}$$

we obtain altogether the action of polynomials on the algebra $\text{End}(V)$ which is a further example in our discussion in Sect. 1.3:

$$\mathbb{K}[x] \times \text{End}(V) \longrightarrow \text{End}(V)$$
$$(\varphi, f) \longmapsto \varphi(f). \tag{9.46}$$

This means for the exploration of the subtle properties that we may use any of the operators $\varphi(f)$ in (9.45) with $\varphi \in \mathbb{K}[x]$. This gives us additional information about the one operator f ! In this spirit, we already took advantage of this when we used the linear polynomials $l_{\lambda_j} \in \mathbb{K}[x]$ given by

$$l_{\lambda_j}(x) := x - \lambda_j,$$

to determine the eigenspace E_{λ_j} by

$$E_{\lambda_j} \equiv E(\lambda_j, f) := \ker l_{\lambda_j}(f) = \ker(f - \lambda_j I d_V). \tag{9.47}$$

Now, if we continue along this path, we come to very interesting insights about the operator f. We consider for example the polynomial $p_m(x) = x^m$ and write $f^m := p_m(f)$. So if we have to calculate the power of $F \in \mathbb{K}^{n \times n}$, we may use the diagonalization, as in Eq. (9.38), and we get

$$F = P \, D \, P^{-1}$$

and we have

$$F^2 = FF = P \, D \, P^{-1} \, P \, D \, P^{-1} = P \, D^2 \, P^1$$

and similarly

$$F^m = P \, D^m \, P^{-1}.$$

Since $D^m = diag(\lambda_1^m, \ldots, \lambda_n^m)$, a perfectly simple expression, we obtained every power of F directly.

9.5   The Question of Non-diagonalizability

To understand diagonalizability more thoroughly, we have to consider some aspects of the non-diagonalizability. We start with two examples of matrices, $A = \begin{bmatrix} 0 & -1 \\ 1 & 0 \end{bmatrix}$ and $C = \begin{bmatrix} 0 & 1 \\ 0 & 0 \end{bmatrix}$ and check whether they are diagonizable or not.

Example 9.17 Example of non-diagonalizability: $A = \begin{bmatrix} 0 & -1 \\ 1 & 0 \end{bmatrix}$.
At first, we consider these matrices as real and then as complex matrices. The matrix A leads to the following map:

$$\mathbb{R}^2 \xrightarrow{\ A\ } \mathbb{R}^2$$

$$(e_1, e_2) \longrightarrow (e_2, -e_1).$$

So we have for $u \in \mathbb{R}^2$, $u' := A u$

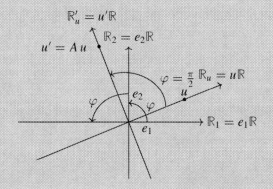

This map is a rotation by $\varphi = \frac{\pi}{2}$. Therefore, there is no one-dimensional linear subspace that is invariant by the map A. There are therefore no eigenvectors or eigenvalues, thus the matrix A is not diagonizable in the real vector space $\mathbb{R}^2$. The fact that there are no real eigenvalues also follows from the solution of the equation:

$$\chi_A(x) = 0, \tag{9.48}$$

given by the characteristic polynomial

$$\chi_A(x) = \det[x\mathbb{1} - A] = \det\begin{bmatrix} x & +1 \\ -1 & x \end{bmatrix} = x^2 + 1. \tag{9.49}$$

Since

$$\chi_A(x) > 0, \tag{9.50}$$

there exist no eigenvalues of this $A \in \mathbb{R}^{2 \times 2}$. Therefore the set of eigenvalues of A is empty. We now consider A as a complex matrix and the map:

$$A: \quad \mathbb{C}^2 \longrightarrow \mathbb{C}^2$$

$$(e_1, e_2) \longmapsto (e_2, -e_1).$$

The eigenvalues of A are now determined by Eqs. (9.48) and (9.49). Hence we obtain the eigenvalues $\lambda_1 = -i$ and $\lambda_2 = i$. The corresponding eigenvectors are given by:

$$b_1 = \begin{bmatrix} 1 \\ i \end{bmatrix} \text{ and } b_2 = \begin{bmatrix} 1 \\ -i \end{bmatrix}.$$

Thus $Ab_1 = -ib_1$ and $Ab_2 = ib_2$. The eigenbasis A is given by:

$$B = [b_1 b_2] = \begin{bmatrix} 1 & 1 \\ i & -i \end{bmatrix}.$$

We notice that in a complex vector space, as expected, the eigenvalue equation $\chi_A(x) = 0$ always has a solution and thus there is always at least one eigenvector.

Example 9.18 Example of non-diagonalizability: $C = \begin{bmatrix} 0 & 1 \\ 0 & 0 \end{bmatrix}$.
For the matrix $C = \begin{bmatrix} 0 & 1 \\ 0 & 0 \end{bmatrix}$ however, we see that even in a complex vector space we have:

$$\mathbb{C}^2 \xrightarrow{\ C\ } \mathbb{C}^2$$
$$(e_1, e_2) \longrightarrow [\vec{0}, \vec{e}_1].$$

There is no way to make the matrix C better, that is, a diagonal one. Here, the eigenvalue equation

$$\chi_C(x) = \det \begin{bmatrix} x & -1 \\ 0 & x \end{bmatrix} = x^2 = 0 \tag{9.51}$$

provides only a single eigenvalue $\lambda_1 = 0$, thus the set of eigenvalues is $EW = \{\lambda_1 = 0\}$.
Here, we obviously do not have enough eigenvectors. We conclude for the moment that if there are not enough eigenvectors to obtain an eigenbasis, so the corresponding matrix is not diagonalizable.
Later, we obtain for C:

$$C^2 = \begin{bmatrix} 0 & 1 \\ 0 & 0 \end{bmatrix} \begin{bmatrix} 0 & 1 \\ 0 & 0 \end{bmatrix} = \begin{bmatrix} 0 & 0 \\ 0 & 0 \end{bmatrix}. \tag{9.52}$$

We call a matrix N nilpotent if for some $m \in \mathbb{N}$ we have $N^m = 0$ (see also Definition 9.12: Nilpotent operator, nilpotent matrix). This result tells us that nilpotent matrices N, as matrix C and Eq. (9.52), are connected to the question of non-diagonalizabilitiy.

The instrument to investigate non-diagonalizability is again the action of polynomials on operators we introduced in Comment 9.3.

Especially, as we shall see, the action of nonlinear polynomials turns out to be very helpful again. To proceed, we have to restrict ourselves to complex vector spaces or, more generally, to operators in $\mathbb{C}$ vector spaces with characteristic polynomials which decompose into linear factors because here, the theory is much easier. In addition, this is also the first and most significant step towards the theory of operators to real vector spaces.

Our aim in this section, is not to develop the whole theory but to give a helpful idea of what we may expect if the operator f is nondiagonalizable.

The theory of diagonalizability leads us to the conclusion that, for a nondiagonalizable operator, the direct sum of the eigenspaces is not sufficient to decompose the whole vector space V. In this case, we obtain strict inequality:

$$\bigoplus_{j=1}^{r} E_{\lambda_j} \not\subseteq V \tag{9.53}$$

where $r < \dim V$ is the number of distinct eigenvalues.

How can we deal with this problem? We first encountered this situation in Sect. 3.4 and Remark 3.4 (On the normal form of endomorphisms). In Proposition 3.12 (ker f − im f decomposition), we saw a possible hint for dealing with this problem. The central point is to first search for the finest possible f-invariant decomposition of V. Here also, f-invariant subspaces play a central role. Therefore, we consider the following proposition:

Lemma 9.3 *Sequence of increasing zero spaces.*
For an operator $g \in \mathrm{Hom}(V, V)$, there is a number $m \in \mathbb{N}$ such that the following sequence holds:

$$\ker g \leq \ker g^2 \leq \ldots \leq \ker g^{m-1} \leq \ker g^m = \ker g^{m+1} = \ker g^{m+2} = \ldots.$$

As we see, at the number m, there is a certain saturation such that thereafter the equality follows endlessly.

Proof In the proof of Proposition 3.12, we saw that the inequality $\ker g \leq \ker g^2$ holds. It is an easy exercise (see Exercise 9.5) that for every power $k \in \mathbb{N}$: $\ker g^k \leq \ker g^{k+1}$. Thus we have the following sequence of inequalities:

$$\ker g \leq \ker g^2 \leq \ldots \leq \ker g^k \leq \ker g^{k+1} \leq \ker g^{k+2} \leq \ldots.$$

So we have to proof that there exists an $m \in \mathbb{N}$ so that the following holds:

$$\ker g \not\leq \ker g^{m-1} \not\leq \ker g^m = \ker g^{m+1} = \ker g^{m+2} = \ldots.$$

First note that

$$\text{if} \quad x \in \ker g^{k+1} \setminus \ker g^k,$$

$$\text{then} \quad g\,x \in \ker g^k \setminus \ker g^{k-1},$$

since

$$x \in \ker g^{k+1} \setminus \ker g^k \Leftrightarrow g^{k+1}x = 0 \quad \text{and} \quad g^k x \neq 0$$

$$\Leftrightarrow g^k(g\,x) = 0 \quad \text{and} \quad g^{k-1}(g\,x) \neq 0$$

$$\Leftrightarrow g\,x \in \ker g^k \setminus \ker g^{k-1}.$$

Conversely,

$$\text{if} \quad \ker g^k = \ker g^{k-1},$$

$$\text{then} \quad \ker g^{k+1} = \ker g^k.$$

Hence, if $\ker g^{k+1} = \ker g^k$, we see through induction that $\ker g^N = \ker g^k \ \forall N \geq k$. Then there are two possibilities. Either there is some $m \in N$ such that

$$\ker g^m = \ker g^{m+1},$$

in which case we are done.

Or we have an infinitely increasing sequence

$$\ker g \leq \ker g^2 \nleq \ldots \nleq \ker g^m \ker g^{m+1} \nleq \ldots$$

of vector spaces of ever increasing dimension. But since the dimensions are bounded above by $\dim V = n$ (as $\ker g^k \leq V \ \forall k \in \mathbb{N}$), this is not possible.   □

Remark 9.2 The $m = \dim V$ case.
One can see from the last paragraph of the proof that in fact $m \leq \dim V$ so $\ker g^n = \ker g^{n+1}$.

Remark 9.3 Criterion for non-diagonalizability.
Theorem 9.1 told us that if there are not enough eigenvectors and correspondingly not enough eigenspaces, then there is no diagonalizability. The eigenspace of a fixed operator f for a fixed eigenvalue λ, is given by $E_\lambda = \ker(f - \lambda \, id_V)$. This shows the relevance of Lemma 9.3. If we set

$g := f - \lambda id_V$, then if we have $\ker g^2 \neq \ker g$ or equivalently if we have $\ker g \lneqq \ker g^2$, then we lose diagonalizability. So if

$$\ker(f - \lambda id_V) \lneqq \ker(f - \lambda id_V)^2$$

holds, then there exists a $w \neq 0$ such that $w \notin E_\lambda$ and $(f - \lambda id_V)w \neq 0$. Therefore

$$(f - \lambda id_V)w = w' \neq 0, \ w \notin E_\lambda$$

and

$$fw = \lambda w + w' \notin E_\lambda.$$

So f is not diagonalizable. However, subspaces of the form $\ker(f - \lambda id_V)^m$ are not only relevant but additionally lead to the answer of the question of an appropriate direct f-invariant decomposition of V. This leads to the Jordan decomposition of V which will be described later, see Theorem 9.2 in Sect. 9.5.

Fortunately, there are the so called "generalized eigenvectors" as elements of the generalized eigenspaces $W(\lambda_j, f)$, given by the following definitions:

Definition 9.10 Generalized eigenvector.
Let V be an n-dimensional vector space, $f \in \mathrm{Hom}(V, V)$ and $\lambda \in \mathbb{K}$ an eigenvalue of f. Then a vector v is a generalized eigenvector corresponding to λ if $v \neq 0$ and $N \in \mathbb{N}$ such that

$$(f - \lambda id_V)^N v = 0. \tag{9.54}$$

Remark 9.4 The $N = \dim V$ case.
From Remark 9.2, one can just as well define

$$W(\lambda, f) \equiv \ker(f - \lambda id_V)^{\dim V}.$$

Definition 9.11 Generalized eigenspace.
For an operator $f \in \text{Hom}(V, V)$ with $\dim V = n$ and $\lambda \in \mathbb{K}$, the generalized eigenspace of f corresponding to the eigenvalue λ is defined by

$$W(\lambda, f) = \ker(f - \lambda\, id_V)^n. \tag{9.55}$$

For a fixed f, we write
$$W_\lambda := W(\lambda, f).$$

Thus we see that W_λ is the set of all generalized eigenvectors of f with respect to λ, the vector 0 included. Additionally, we see that the eigenspace E_λ is contained in the generalized eigenspace W_λ. So we have

$$E_\lambda \le W_\lambda.$$

We now obtain the direct sum decomposition of V:

$$V = \overset{r}{\underset{j=1}{\oplus}} W_{\lambda_j}. \tag{9.56}$$

This decomposition is f-invariant:

$$f : W_{\lambda_j} \longrightarrow W_{\lambda_j} \quad \forall j \in I(r).$$

It is interesting to notice that the nonlinear polynomials $P_{\lambda_j}(x) := (x - \lambda_j)^n$ are relevant. Their mathematical formalism, which led to the above result, is similar to the formalism which led to the diagonalizability theorem. It is worth knowing that in connection with all the generalized eigenspaces W_{λ_j}, a special kind of operators, the nilpotent operators play a crucial role. The problem is reduced to studying the representations of nilpotent operators.

Definition 9.12 Nilpotent operator, nilpotent matrix.
An operator $h \in \text{End}(V)$ is nilpotent if some power of h is zero:

$$h^m = 0 \quad \text{with} \quad m \in \mathbb{N}.$$

Similarly, a matrix $N \in \mathbb{K}^{n \times n}$ is nilpotent if for some $m \in \mathbb{N}$

$$N^m = 0.$$

Lemma 9.4 *Eigenvalues of a nilpotent matrix.*
The only eigenvalue of a nilpotent matrix is zero.

Proof We first show that $\lambda = 0$ is an eigenvalue of a nilpotent matrix $N \in \mathbb{K}^{n \times n}$. Let $N^m = 0$ and $N^{m-1} \neq 0$, then there is some $u \in \mathbb{K}^m$, such that

$$N^{m-1}u = v \neq 0, \quad Nv = N^m u = \vec{0} \quad \text{and} \quad v \in \ker N \neq \{\vec{0}\}.$$

So $\lambda = 0$ is an eigenvalue of N. We further show that $\lambda = 0$ is the only eigenvalue of N: Let λ be an eigenvalue of N, so there is some $v \in \mathbb{K}^n$, $v \neq 0$ such that $Nv = \lambda v$. Then, since $N^m = 0$, we have

$$N^m v = \lambda^m v = 0 \quad \text{and thus} \quad \lambda = 0.$$

$\square$

The sets of eigenvalues of N are given by

$$EW(N) = \{0\}.$$

The result of this lemma is that one nilpotent matrix is similar to a strictly upper triangular matrix:

$$\begin{bmatrix} 0 & & * \\ & \ddots & \\ 0 & & 0 \end{bmatrix}.$$

Definition 9.13 Upper triangular and strictly upper triangular matrix.
A matrix A is called upper triangular if there are only zeros below the main diagonal, that is, $\alpha_s^i = 0$ for $i > s$. Further, A is called strictly upper triangular if there are zeros on and below the main diagonal, that is,

$$\alpha_s^i = 0 \quad \text{for} \quad i \geq s.$$

The next proposition shows that a nilpotent matrix is similar to a strictly upper triangular matrix.

Proposition 9.6 *On the structure of a nilpotent matrix.*
Let $A \in \mathbb{K}^{n \times n}$ be a nilpotent matrix. The following equivalence holds:

   (i)   A is nilpotent,
   (ii) A is similar to a strictly upper triangular matrix.

In this case, we have $A^n = 0$ and the characteristic polynomial, $\chi_A(\xi) = \xi^n$.

Proof

– (i) $\Rightarrow$ (ii)
Let A be nilpotent. Lemma 9.4 shows that $\lambda = 0$ is an eigenvalue and so A is similar to the matrix

$$N = \begin{bmatrix} 0 & * \\ 0 & N_1 \end{bmatrix} \quad \text{with} \quad N_1 \in \mathbb{K}^{(n-1) \times (n-1)}.$$

The matrix N_1 is, after an induction argument n, a strictly upper triangular matrix of the form

$$N_1 = \begin{bmatrix} 0 & & * \\ \vdots & \ddots & \\ 0 & \cdots & 0 \end{bmatrix},$$

and N is also a strictly upper triangular matrix.
– (ii) $\Rightarrow$ (i)
We have just to show that a strictly upper triangular matrix $A \in \mathbb{K}^{n \times n}$ has $A^n = 0$. Again using induction on n, we write

$$A = \begin{bmatrix} 0 & \alpha \\ 0 & B \end{bmatrix} \quad \text{with} \quad \alpha \in (\mathbb{K}^{n-1})^* \quad \text{and} \quad B \in \mathbb{K}^{(n-1) \times (n-1)}$$

with B a strictly upper triangular with $B^{n-1} = 0$. We then take

$$AA = \begin{bmatrix} 0 & \alpha \\ 0 & B \end{bmatrix}\begin{bmatrix} 0 & \alpha \\ 0 & B \end{bmatrix} = \begin{bmatrix} 0 & \alpha B \\ 0 & B^2 \end{bmatrix} \quad \text{such that} \quad A^n = \begin{bmatrix} 0 & \alpha B^{n-1} \\ 0 & B^n \end{bmatrix}.$$

Since, as above, $B^{n-1} = 0$ and $B^n = 0$, we also get $A^n = 0$. This proves (i).

It follows that

$$\chi_A(\xi) = \det(\xi \mathbb{1}_n - A) = \det \begin{bmatrix} \xi & & * \\ & \ddots & \\ 0 & & \xi \end{bmatrix} = \xi^n.$$

So Proposition 9.6 is proven.   □

This proposition shows that a nontrivial nilpotent matrix is not diagonalizable. We will see later that (see Theorem 9.4 (iii)) this is not just an example but one of the two obstructions to diagonalizability. For complex vector spaces, it is even the only obstruction. We therefore supplement this proposition with the following corollary.

Corollary 9.4 *Nilpotent, not diagonalizable.*
Let $N \neq 0$ be a nilpotent matrix. Then it follows that N is not diagonalizable.

Proof This follows directly from Proposition 9.6. □

In our case, the linear polynomial l_{λ_j} with $L_j := l_{\lambda_j}(f) \in \text{End}(V)$, together with the generalized eigenspace, W_{λ_j} leads to the nilpotent operator

$$h_j := (f - \lambda_j id_V)|_{W_{\lambda_j}}. \tag{9.57}$$

This should be compared with the corresponding operator

$$(f - \lambda_j id_V)|_{E_{\lambda_j}}, \tag{9.58}$$

which is by definition the zero operator of the eigenspace E_{λ_j}! Denoting the restrictions of f and id_V on the generalized eigenspace W_{λ_j} by f_j and id_j, we get

$$f_j := f|_{W_{\lambda_j}}, \quad id_j := id|_{W_{\lambda_j}}. \tag{9.59}$$

We may write Eq. (9.57) as

$$f_j = \lambda_j id_j + h_j, \quad j \in I(r). \tag{9.60}$$

This leads us to the expectation that there exists a decomposition of the operator f in partial operators f_j, each one characterized by the eigenvalue λ_j, and that all these f_j should have the same structure.

We may expect a universal structure for the operators f_j, $j \in I(r)$, with r the number of distinct eigenvalues. The following theorem shows that this is the case. We prove it by using very elementary methods, as demonstrated by [7, pg. 238]. Note that most proofs used in the literature use much more advanced techniques for this kind of theorem. For simplicity, the formalism is given at the level of a matrix:

Theorem 9.2 *Structure theorem (pre-Jordan form).*
If the characteristic polynomial of a matrix $F \in \mathbb{K}^{n \times n}$ decomposes into linear factors, the matrix F is similar to a block diagonal matrix:

$$\begin{bmatrix} F_1 & & & 0 \\ & F_2 & & \\ & & \ddots & \\ 0 & & & F_r \end{bmatrix}. \tag{9.61}$$

which we can also represent as follows:

$$F_1 \oplus F_2 \oplus \cdots \oplus F_r.$$

Every block is given by

$$F_j = \lambda_j id_j + H_j \in \mathbb{K}^{m_j \times m_j}, \ j \in I(r) \tag{9.62}$$

with r the number of distinct eigenvalues of F and H_j a strictly upper triangular matrix in the strictly upper triangular form:

$$H_j = \begin{bmatrix} 0 & & * \\ & \ddots & \\ 0 & & 0 \end{bmatrix}. \tag{9.63}$$

We call the upper triangular matrices F_j pre-Jordan matrices.
   This gives the following direct sum decomposition of $\mathbb{K}^n$ into f-invariant subspaces:

$$\mathbb{K}^n \cong \mathbb{K}^{m_1} \oplus \cdots \oplus \mathbb{K}^{m_j} \oplus \cdots \oplus \mathbb{K}^{m_r}.$$

Note that the Jordan approach, using higher level mathematical instruments, continues to decompose each $\mathbb{K}^{m_j}$ into f-invariant, f-irreducible vector spaces.

Proof The proof goes through by induction on $n = \dim V$. For $n = 1$, $F \equiv F_1 = \lambda_1$ and the theorem is trivially valid. We assume that the theorem is proven for all matrices belonging to $\mathbb{K}^{(n-1)\times(n-1)}$. For the matrix $F \in \mathbb{K}^{n \times n}$, the characteristic polynomial is given by

$$\chi_F(x) = (x - \lambda_1)^{m_1} \cdots (x - \lambda_r)^{m_r}. \tag{9.64}$$

F is similar to the matrix

$$\begin{bmatrix} \lambda_1 & * \\ 0 & G \end{bmatrix} \quad \text{and} \quad G \in \mathbb{K}^{(n-1)\times(n-1)} \tag{9.65}$$

and the characteristic polynomial factors as

$$\chi_F(x) = (x - \lambda_1)\chi_G(x),$$

with

$$\chi_G(x) = (x - \lambda_1)^{m_1-1}(x - \lambda_2)^{m_2} \cdots (x - \lambda_r)^{m_r}. \tag{9.66}$$

By induction the matrix G is similar to the matrix

$$\begin{bmatrix} F_1^* & 0 & \cdots & 0 \\ 0 & F_2 & & 0 \\ & & \ddots & \\ 0 & 0 & & F_r \end{bmatrix} \tag{9.67}$$

with

$$F_j = \lambda_j \mathbb{1}_j + H_j \in \mathbb{K}^{(m_j \times m_j)} \quad j \in \{2, \ldots, r\} \tag{9.68}$$

and

$$F_1^* = \lambda_1 \mathbb{1} + H_1 \in \mathbb{K}^{(m_1-1) \times (m_1-1)}. \tag{9.69}$$

From Eqs. (9.65), (9.67) and (9.68), it follows that F is similar to a block matrix given by

$$C = \begin{bmatrix} F_1 & C_2 & \cdots & C_j & \cdots & C_r \\ 0 & F_2 & & & & 0 \\ & & \ddots & & \\ 0 & 0 & & & & F_r \end{bmatrix} \quad \text{and} \quad C_j \in \mathbb{K}^{m_1 \times m_j}. \tag{9.70}$$

Now, we have to show that C is similar to the matrix F. This means, we would like to obtain something of the form

$$F = B^{-1}CB = \begin{bmatrix} F_1 & Y_2 & \cdots & Y_r \\ 0 & F_2 & 0 & 0 \\ & & \ddots & 0 \\ 0 & 0 & & F_r \end{bmatrix} \tag{9.71}$$

with

$$Y_j = 0. \tag{9.72}$$

For this purpose we choose the following invertible matrix B with a similar form as C in Eq. 9.70:

$$B = \begin{bmatrix} \mathbb{1}_{m_1} & B_2 & \cdots & B_r \\ 0 & \mathbb{1}_{m_2} & & 0 \\ & & \ddots & \\ 0 & 0 & & \mathbb{1}_{m_r} \end{bmatrix}. \tag{9.73}$$

Then, when we conjugate C by B, we obtain $F = B^{-1}CB$ as above in Eq. (9.71) with $Y_j = 0$.

$$Y_j = F_1 B_j - B_j F_j + C_j. \tag{9.74}$$

By setting $F_j = \lambda_j \mathbb{1}_j + H_j$, we get

$$Y_j = (\lambda_1 - \lambda_j)B_j + H_1 B_j - B_j H_j + C_j. \tag{9.75}$$

The question is whether we can choose the B_j so that $Y_j = 0$ for every $j \in \{2, \ldots, r\}$. If we divide the expression in Eq. (9.75) by $(\lambda_1 - \lambda_j) \neq 0$, this does not change the form of Eq. (9.75). So we may assume for some fixed j, without loss of generality, that $\lambda_1 - \lambda_j = 1, \forall j \in \{2, \ldots, r\}$, just for our proof. Now the question is whether we can solve the equation

$$0 = B_j + H_1 B_j - B_j H_j + C_j, \tag{9.76}$$

with unknowns the entries of $B_j \equiv X \in \mathbb{C}^{m_1 \times m_j}$. We have to solve the system $m_1 m_j$ of linear equations with $m_1 m_j$ variables:

$$X + H_1 X - X H_j = -C_j. \tag{9.77}$$

Setting

$$L \equiv H_1, \quad R \equiv H_j \quad \text{and} \quad C_0 \equiv -C_j \tag{9.78}$$

we have to solve generically the system

$$X + LX - XR = C_0 \quad \text{with} \quad X \in \mathbb{C}^{m_1 \times m_j}. \tag{9.79}$$

The system of Eq. (9.79) has a unique solution if the homogeneous equation

$$X + LX - XR = 0 \tag{9.80}$$

has only the zero solution $X = 0$. Lemma 9.5, shows that this is true. The solution X obtained for Eq. (9.79) or equivalently B_j for Eq. (9.75), corresponds to $Y_j = 0$ in Eq. (9.71) and F is similar to the block diagonal form in Eq. (9.61). This proves the theorem.   □

The lemma we used ensures that for the nilpotent matrices L and $R(L \equiv H_1, R \equiv H_j)$ the homogeneous Eq. (9.80) has only the trivial solution $(X = 0)$.

Lemma 9.5 *If $L \in \mathbb{K}^{s \times s}, R \in K^{t \times t}, s, t \in \mathbb{N}$ are nilpotent matrices, the system*

$$X - LX + XR = 0 \tag{9.81}$$

has only the solution $X = 0$.

Proof Let $X \in \mathbb{K}^{n \times n}$ and define $X^{(i)}$ iteratively as follows:

$$X^{(1)} = LX - XR$$
$$X^{(2)} = LX^{(1)} - X^{(1)}R$$
$$X^{(2)} = L^2X - 2LXR + XR^2$$
$$X^{(3)} = L^3X - 3L^2XR + 3LXR^3 - XR^3$$

$$X^{(l)} = \sum_{k=0}^{l} (-1)^k \binom{l}{k} L^{l-k} X R^k.$$

Since L and R are nilpotent, we have $L^n = 0$ and $R^n = 0$ where $n = \dim V$. If we take $l = 2n$, we already have $X^{(2n)} = 0$ because

$$X^{(2n)} = L^{2n}X + L^{2n-1}XR + \cdots L^{n+1}XR^{n-1} - L^nXR^n + L^{n-1}XR^{n+1} + \cdots XR^{2n}.$$

We have for L^{l-k} and R^k, $n \leqslant k$ or $n \leqslant l - k$, so we obtain for $L^{l-k}XR^k = 0 \; \forall \, k \in \{0, 1, \ldots, l = 2n\}$. Suppose, now, that X is a solution to $X - LX + XR = 0$.

$$\text{Then} \quad X = LX - XR = X^{(1)}$$
$$= X^{(2)} = \ldots = X^{(2n)} = 0.$$

So X is the trivial solution $X = 0$. $\qquad\qquad\qquad\qquad\qquad\qquad\square$

Comment 9.4 Structure theorem with complex matrices.

The structure theorem is, in particular, valid for all complex matrices since the fundamental theorem of algebra states that every nonconstant polynomial decomposes into linear factors over $\mathbb{C}$. The above formulation of the theorem has the advantage that real and complex matrices are treated uniformly.

Corollary 9.5 *Upper triangular decomposition of V in generalized eigenspaces.*

If the characteristic polynomial χ_f of an operator f decomposes into linear factors, there exists a representation which we may call "pre-Jordan", as in the above structure theorem, and a decomposition of $V \cong \mathbb{K}^n$ in f-invariant generalized eigenspaces $W_{\lambda_j} = \ker(f - \lambda_j id_V)^n$:

$$V = W_{\lambda_1} \oplus W_{\lambda_2} \oplus \cdots \oplus W_{\lambda_r}.$$

A simple proof of the Cayley-Hamilton theorem is a nice application of Theorem 9.2. This proof is also essentially based on properties of nilpotent matrices, together with the action of polynomials on matrices which appear in Proposition 9.2 and which we call pre-Jordan matrices. We first need the following lemma.

Lemma 9.6 *Action of a polynomial on a pre-Jordan matrix.*
For a matrix in the form

$$A = \lambda \mathbb{1} + N \in \mathbb{K}^{m \times m}$$

with N a strictly upper triangular matrix, the action of a polynomial $\varphi \in \mathbb{K}[x]$ is given by

$$\varphi(A) = \varphi(\lambda)\mathbb{1} + N'$$

with N' also a strictly upper triangular matrix.

Example 9.19 Proof by example.
We do not prove the above statement but give an example when $\varphi(x) = x^2$. If we use the fact that

$$(\lambda \mathbb{1} + N)(\lambda \mathbb{1} + N) = \lambda^2 \mathbb{1} + \lambda N + N\lambda + NN$$

with $\bar{N} = NN$ again a strictly upper triangular matrix such that $\varphi(A) = \varphi(\lambda)\mathbb{1} + N'$ is again a pre-Jordan matrix.

Theorem 9.3 *Cayley-Hamilton theorem.*
Let F be a matrix $\in \mathbb{K}^{n \times n}$ with characteristic polynomial $\chi_F(x) = (x - \lambda_1)^{m_1} \cdots (x - \lambda_r)^{m_1}$. Then

$$\chi_F(F) = 0 \quad \in \mathbb{K}^{n \times n}.$$

Proof Since we can use the relations

$$F = T\tilde{F}T^{-1} \qquad\qquad \text{with } T \in Gl(n) \text{ and}$$
$$\varphi(F) = T\varphi(\tilde{F})T^{-1} \qquad\qquad \text{with } \varphi \in \mathbb{K}[x],$$

we can apply Theorem 9.2 to reduce to the case where F is a pre-Jordan matrix. This leads to the following expression:

$$\chi_F(F) = \begin{bmatrix} \chi_F(F_1) & & & 0 \\ & \chi_F(F_2) & & \\ 0 & & \ddots & \\ & & & \chi_F(F_r) \end{bmatrix}.$$

Lemma 9.6 leads for every $\chi_F(F_j) \in \mathbb{K}^{m_j \times m_j}$, $j \in I(r)$ to the result

$$\chi_F(F_j) \quad \text{is proportional to} \quad (F_j - \lambda_j \mathbb{1}_j)^{m_j} \equiv H_j^{m_j}$$

with H_j nilpotent. So we have $H_j^{m_j} = 0$ and $\chi_F(F_j) = 0 \in \mathbb{K}^{(m_j \times m_j)}$ for all $j \in I(r)$. This leads to

$$\chi_F(F) = 0 \in \mathbb{K}^{n \times n}$$

and to the proof of Theorem 9.3. $\qquad\qquad\qquad\qquad\qquad\qquad\qquad\qquad\qquad\qquad\square$

9.6 Algebraic Aspects of Diagonalizability

At this point, two questions arise simultaneously. Firstly, what exactly does the expression

$$\chi_F(F) = 0 \tag{9.82}$$

in the Cayley-Hamilton Theorem 9.2 mean? Secondly, are there other polynomials that fulfill the same relation? In Eq. (9.82), $\chi_F \equiv \chi \in \mathbb{K}(x)$ is the characteristic polynomial of $F \in \mathbb{K}^{n \times n}$,

$$\chi(x) = x^n + \chi_{n-1} x^{n-1} + \cdots + \chi_2 x^2 + \chi_1 x + \chi_0 \in \mathbb{K}, \tag{9.83}$$

and the following expression holds:

$$F^n + \chi_{n-1} F^{n-1} + \cdots + \chi_2 F^2 + \chi_1 F + \chi_0 \mathbb{1}_n = 0 \in \mathbb{K}^{n \times n}. \tag{9.84}$$

This means that the list of matrices

$$(\mathbb{1}_n, F, F^2, \ldots, F^{n-1}, F^n) \tag{9.85}$$

is linearly dependent. Since the space $\mathbb{K}^{n \times n}$ is n^2-dimensional and a list like

$$(\mathbb{1}_n, F, F^2, \ldots, F^{n^2-1}, F^{n^2}) \tag{9.86}$$

is always linearly dependent, we also see that the $n + 1$ elements in (9.85) which are fewer than the $n^2 + 1$ elements in (9.86), are already linearly dependent.

In connection with this, a logical question is whether a list of powers of F with even smaller length than $n + 1$ could also be linearly dependent. It is clear that here we talk about a vector space which we denote by $\mathbb{K}[F]$ and which is generated by the powers of a matrix F:

$$\text{span}(\mathbb{1}_n, F, F^2, F^3, \dots). \tag{9.87}$$

So we define

Definition 9.14 $\mathbb{K}[F]$: The matrix polynomials of F.

$$\mathbb{K}[F] := \{\varphi(F) : \varphi \in \mathbb{K}[x]\}. \tag{9.88}$$

It is further clear that the vector space $\mathbb{K}[F]$ is also a commutative sub-algebra of the matrix algebra $\mathbb{K}^{n \times n}$.

All this leads to the notion of minimal polynomials.

Let

$$I_F := \{\chi \in \mathbb{K}[x] : \chi(F) = 0\} \tag{9.89}$$

be the set of all polynomials which annihilate the matrix F. We call $\chi \in I_F$ an annihilator of the matrix F. Note that $I_F \leqslant \mathbb{K}[x]$ is an ideal in $\mathbb{K}[x]$. This means that I_F is an additive subgroup of $\mathbb{K}[x]$ and that for any $\chi \in I_F$ and $\varphi \in \mathbb{K}[x]$, $\varphi\chi \in I_F$. Thus we can say that any scaling of an annihilator of F by a polynomial, leads back to an annihilator of F.

For a commutative algebra $\mathcal{A}$, we can generally give the following definition for an ideal:

Definition 9.15   Ideal.
An ideal of a commutative algebra $\mathcal{A}$ is a subset I of $\mathcal{A}$ that is closed under addition and subtraction,

$$a, b \in I \Rightarrow a + b, \ a - b \in I,$$

and also under multiplication by elements of $\mathcal{A}$, that is,

$$i \in I, a \in \mathcal{A} \Rightarrow ai \in I.$$

Using the formalism of Sect. 1.3, we can state that there is an action of polynomials on the annihilators of F:

$$\mathbb{K}[x] \times I_F \longrightarrow I_F$$
$$(\chi, \varphi) \longmapsto \varphi\chi \in I_F.$$

It is interesting to see that, using the definition $\chi_F(x) = \det(x\mathbb{1} - F)$, we can write for $\chi_F(F) = 0$:

$$F^n - (\operatorname{tr} F)F^{n-1} + \cdots + (-1)^n \det(F) = 0. \tag{9.90}$$

Hence,

$$\chi_F(x) = x^n + \chi_{n-1}x^{n-1} + \cdots + \chi_2 x^2 + \chi_1 x + \chi_0,$$
$$\text{with} \quad \chi_{n-1} = -\operatorname{Tr}(F) \quad \text{and} \quad \chi_0 = (-1)^n \det(F).$$

The following proposition shows the existence of minimal polynomials.

Proposition 9.7 *Existence of minimal polynomials.*
For a given matrix $A \in \mathbb{K}^{n \times n}$, there exists a unique monic polynomial $\mu \in \mathbb{K}[x]$ with smallest positive degree such that $\mu(A) = 0$.

Proof Since $\mathbb{K}[A]$ is a subspace of $\mathbb{K}^{n \times n}$, the following holds:

$$m := \dim \mathbb{K}[A] \leqslant \dim \mathbb{K}^{n \times n} = n^2. \tag{9.91}$$

Hence the list $(\mathbb{1}_n, A, A^2, \ldots, A^m)$ is linearly dependent and therefore there exist coefficients $\lambda_s \in I(m)$. Not all of them are zero so that

$$\lambda_s A^s = 0 \tag{9.92}$$

holds. This corresponds to the polynomial

$$\chi(x) := \lambda_s x^s, \tag{9.93}$$

with $0 \neq \varphi \in I_A$, such that
$$\chi(A) = 0. \tag{9.94}$$

Since the set of annihilators of A, I_A, is an ideal of $\mathbb{K}[x]$, it follows from the theory of polynomials that
$$I_A = \mathbb{K}[x]\mu,$$

where μ is a uniquely determined monic polynomial with minimal degree and $\deg \mu = m$. $\qquad\square$

There exists another proof without the use of the theory of polynomials.

Definition 9.16 Minimal polynomials.
The minimal polynomial $\mu_F(x)$ of a matrix F is the unique monic polynomial
of minimal positive degree that annihilates F, that is, $\mu_F(F) = 0$.

$$\mu_F(F) = 0, \tag{9.95}$$

and has the smallest degree among such polynomials.

The next proposition shows that the eigenvalues of an operator are also zeros of the
minimal polynomial.

Proposition 9.8 *Eigenvalues and zeros of the minimal polynomial.*
For a matrix $A \in \mathbb{K}^{n \times n}$ and $\lambda \in \mathbb{K}$, the following conditions are equivalent:

(i) λ is an eigenvalue of A.
(ii) λ is a zero of the minimal polynomial of A.

Proof (i) $\Rightarrow$ (ii)
Let $v \neq 0$ be an eigenvector of A with eigenvalue λ. Then $A^k v = \lambda^k v$ and in fact
for a polynomial $\varphi \in \mathbb{K}[x]$, $\varphi(A)v = \varphi(\lambda)v$ holds. Additionally, when $\varphi = \mu$, we
obtain $\mu(A)v = \mu(\lambda)v$, and since $\mu(A) = 0$, we have $\mu(\lambda)v = 0$ and $\mu(\lambda) = 0$. This
is the assertion (ii).

Proof (ii) $\Rightarrow$ (i)
Let $\mu(\lambda) = 0$. We have to show that there is some $v \in V \setminus \{0\}$ so that $Av = \lambda v$ or
$(A - \lambda \mathbb{1})v = 0$. We write

$$\mu(x) = (x - \lambda)\varrho(x). \tag{9.96}$$

Since μ is the minimal polynomial of A, we have $\varrho(A) \neq 0$. Let $W := \operatorname{im} \varrho(A) \leqslant \mathbb{K}^n$
so that $W = \{\varrho(A)u : u \in \mathbb{K}^n\} \neq 0$. We choose $0 \neq v \in W$ so that there is a $u \in \mathbb{K}^n$
with $\varrho(A)u = v$. Hence

$$Av - \lambda v = (A - \lambda \mathbb{1})v = (A - \lambda \mathbb{1})\varrho(A)u. \tag{9.97}$$

Using Eq. (9.96) and $\mu(A) = (A - \lambda \mathbb{1})\varrho(A)$, we obtain from Eq. (9.97):

$$(A - \lambda \mathbb{1})v = \mu(A)u = 0\, u = 0.$$

This shows that (i) holds and Proposition 9.8 is proven. $\square$

The characteristic polynomial and the minimal polynomial have exactly the same
zeros even though they may have different multiplicities. For example, for an operator
A on a $\mathbb{C}$-vector space:

$$\chi_A(x) = (x - \lambda_1)^{m_1}(x - \lambda_2)^{m_2} \cdots (x - \lambda_r)^{m_r} \tag{9.98}$$

and

$$\mu_A(x) = (x - \lambda_1)^{d_1}(x - \lambda_2)^{d_2} \cdots (x - \lambda_r)^{d_r} \tag{9.99}$$

with $d_j \leqslant m_j$, $j \in I(r)$. Further, for all $A \in \mathbb{K}^{n \times n}$,

$$\deg \mu_A \leqslant \deg \chi_A = \dim \mathbb{K}^n = n. \tag{9.100}$$

The characteristic polynomial gives direct information about the eigenvalues of A because $\chi_A(x)$ is equal by definition to $\det(A - x\mathbb{1})$ which is calculable. The minimal polynomial is more difficult to determine. However, it provides indirect information about the eigenvectors and thus about the diagonalizability of A. As we will see (Theorem 9.4), if the minimal polynomial has only simple zeros, then this guarantees the diagonalizability of A.

Theorem 9.2, together with the above preparations such as the commutative algebra $\mathbb{K}[A]$ in Definition 9.14 and the minimal polynomials μ_A, Definition 9.16, provides also a pure algebraic criterion for the diagonalizability of an operator or a matrix. For the sake of simplicity, we treat here the matrix version. We define the term semisimple for a commutative algebra because it is useful for a concise summary of our results.

Definition 9.17 Semisimple commutative algebra.
We call a commutative algebra $\mathcal{A}$ semisimple if $\mathcal{A}$ has no nonzero nilpotent elements. In our case where $\mathcal{A} = \mathbb{K}[A]$, we can say even more concisely that an element $A \in \mathbb{K}[A]$ is semisimple if $\mathbb{K}[A]$ is semisimple.

Theorem 9.4 *Diagonalizability, algebraic perspective.*
If the characteristic polynomials of a matrix $F \in \mathbb{K}^{n \times n}$ decomposes into linear F factors with characteristic polynomials,

$$\chi_F(x) = (x - \lambda_1)^{m_1}(x - \lambda_2)^{m_2} \cdots (x - \lambda_r)^{m_r} \tag{9.101}$$

with $\lambda_j \in \mathbb{K}$, $j \in I(r)$ distinct and with a minimal polynomial μ_F, then the following statements are equivalent:

(i) The matrix F is diagonalizable.
(ii) Every element of $\mathbb{K}[F]$ is diagonalizable.

(iii) *The commutative algebra* $\mathbb{K}[F]$ *contains no nonzero nilpotent elements or, equivalently, the commutative algebra* $\mathbb{K}[F]$ *is semisimple.*

(iv) *The minimal polynomial* μ_F *is*

$$\mu_F(x) = (x - \lambda_1)(x - \lambda_2) \cdots (x - \lambda_r),$$

or, equivalently, all zeros of μ_F *are simple.*

Proof (i) $\Rightarrow$ (ii)

Since F is diagonalizable, there exists a basis B_0 in $\mathbb{K}^n$ such that

$$B_0^{-1} F B_0 = \Lambda \tag{9.102}$$

holds, where Λ is a diagonal matrix with entries $\lambda_1, \ldots, \lambda_r$ (including possible multiplicities). The matrix $\varphi(F)$ for $\varphi \in \mathbb{K}[x]$ is an element of $\mathbb{K}[A]$. The following relation is valid between the polynomial and the adjoint $GL(n)$ action on matrices in $\mathbb{K}^{n \times n}$:

$$B_0^{-1} \varphi(F) B_0 = \varphi(B_0^{-1} F B_0). \tag{9.103}$$

Hence the two actions

$$F \longmapsto \varphi(F) \qquad \text{and}$$
$$F \longmapsto B_0^{-1} F B_0$$

commute.

Using Eqs. (9.102) and (9.103), we obtain

$$B_0^{-1} \varphi(F) B_0 = \varphi(\Lambda). \tag{9.104}$$

Since $\varphi(\Lambda)$ is also diagonal, (ii) holds.

Proof (ii) $\Rightarrow$ (iii)

Assume that $N \in \mathbb{K}[A]$ is nilpotent. By assertion (ii), N is diagonalizable. Corollary 9.4 tells us that N is zero. It follows that no nilpotent nonzero element exists in $\mathbb{K}[F]$. This proves (iii).

Proof (iii) $\Rightarrow$ (iv)

We first show that the polynomial

$$\psi(x) = (x - \lambda_1)(x - \lambda_2) \cdots (x - \lambda_r) \tag{9.105}$$

with $\psi(\lambda_j) = 0$ $j \in I(r)$, leads to a nilpotent matrix $\psi(F)$. According to Theorem 9.2, F is similar to

$$F_1 \oplus \cdots \oplus F_j \oplus \cdots \oplus F_r \tag{9.106}$$

$$= \begin{bmatrix} F_1 & & & & \\ & \ddots & & & \\ & & F_j & & \\ & & & \ddots & \\ & & & & F_r \end{bmatrix} \qquad (9.107)$$

with the pre-Jordan matrix

$$F_j = \lambda_j \, \mathbb{1}_j + N_j \qquad (9.108)$$

with N_j a strictly upper triangular matrix. The action of polynomials on F_j is given by Lemma 9.6:

$$\varphi(F_j) = \varphi(\lambda_j) \, \mathbb{1}_j + N'_j \qquad (9.109)$$

with N'_j again a strictly upper triangular matrix. Taking $\varphi(F_j)$, we obtain

$$\psi(F_j) = \psi(\lambda_j)\mathbb{1}_j + N'' = 0 \, \mathbb{1}_j + N''_j = N''_j \qquad (9.110)$$

with N''_j the strictly upper triangular for all $j \in I(r)$. Thus $\psi(F)$ is nilpotent. Now, since $\psi(F)$ is both nilpotent and diagonalizable, it follows from (iii) that $\psi(F)$ is zero. Therefore, the above polynomial $\psi(x)$ is the minimal polynomial as it is already minimal: $\mu(x) = \psi(x)$. This proves (iv).

Proof (iv) $\Rightarrow$ (i)
By Theorem 9.2, we have as above $F_1 \oplus \cdots \oplus F_j \oplus \cdots \oplus F_r$ with $F_j = \lambda_j \, \mathbb{1}_j + N_j \ j \in I(r)$. We take, without loss of generality, $j = 1$ as an example.

$$\mu_F(F_1) = (F_1 - \lambda_1\mathbb{1}_1)(F_1 - \lambda_2\mathbb{1}_1) \cdots F_1 - \lambda_r\mathbb{1}_1). \qquad (9.111)$$

Consider $F_1 - \lambda_j\mathbb{1}$ with $j \neq 1$. Then $F_1 - \lambda_j\mathbb{1} = \lambda_1\mathbb{1} + N_1 - \lambda_j\mathbb{1} = (\lambda_1 - \lambda_j)\mathbb{1} + N_1$. Hence,

$$(F_1 - \lambda_2\mathbb{1}_1)(F_1 - \lambda_3\mathbb{1}_1) \cdots F_1 - \lambda_r\mathbb{1}_1) =: B, \qquad (9.112)$$

where B is an invertible matrix. From Eqs. (9.111) and (9.112) we obtain

$$\mu_F(F_1) = (F_1 - \lambda_1\mathbb{1}_1)B \qquad \text{and}$$
$$0 = (F_1 - \lambda_1\mathbb{1}_1)B \qquad \text{and}$$
$$(F_1 - \lambda_1\mathbb{1}_1) = 0_1 \equiv 0_{m_1} \in \mathbb{K}^{m_1 \times m_1}.$$

Similarly, we obtain for $j = 2, \ldots, j = r$

$$F_2 - \lambda_2 \, \mathbb{1}_2 = 0_{m_2}, \ldots, F_r - \lambda_r \, \mathbb{K}_r = 0_{m_r}.$$

This proves (i) and thus the diagonalizability Theorem. $\qquad \square$

Theorem 9.4, with the assertion (iii), could also be formulated differently:
A matrix F is diagonalizable if and only if its characteristic polynomial decomposes into linear factors and is semisimple.

This can even be shortened if we write for the decomposability of the characteristic polynomial χ_F of the matrix F:
A matrix F is diagonalizable if and only if F is decomposable and semisimple. This is a pure algebraic point of view for diagonalizability.

9.7   Triangularization and the Role of Bases

The structure theorem in the previous section showed that for every operator in a complex vector space, there exists a basis that leads to an upper triangular matrix representation. As in the case of diagonalization, special bases play a crucial role. But in this section, we will not discuss an f-invariant decomposition of the given vector space V. We are going to use a procedure which allows to consider the vector space V as a whole. This is, in addition, particularly relevant and useful for proving the spectral theorem (see Sect. 10.6) and consequently for a better understanding of that theorem.

Proposition 9.9 *Equivalence definitions for triangularization.*
Let V be a vector space with $\dim V = n$ and $f \in \text{End}(V)$ and $B_0 = (b_1, \ldots, b_n)$ a basis of V with $f b_s = b_i \varphi_s^i$, $s, i, j \in I(n)$ and $F_{B_0 B_0} = (\varphi_s^i) \in \mathbb{K}^{n,n}$. Then the following definitions are equivalent:

(i) f_{B_0, B_0} is upper triangular;
(ii) For each $j \in I(n)$, $f b_j \in \text{span}(b_1, \ldots, b_j)$;
(iii) The vector space $\text{span}(b_1, \ldots, b_j)$ is f-invariant for each $j \in I(n)$;
(iv) There exists an f-invariant "flag":

$$V_0 \lneqq V_1 \lneqq V_2 \cdots \lneqq V_n = V$$

with $\dim V_j = j$ and $f V_j \subseteq V_j$ for all $j \in I(n)$.

It is clear that Proposition 9.9 (iv) is basis-independent. Note that the above f-invariant "flag" is not what we usually mean by an f-invariant decomposition of V. The existence of a basis B_0, as in the above definition, was shown in the structure theorem in Sect. 9.5. It is still instructive and useful to see a second proof.

Proposition 9.10 *Triangularization on a $\mathbb{K}$ vector space.*
There exists a basis B_0 of V so that the matrix $F_{B_0 B_0}$ is upper triangular if and only if the characteristic polynomial χ_f of f decomposes into linear factors over $\mathbb{K}$ (which is always the case in a complex vector space).

Proof If $f_{B_0 B_0}$ is upper triangular, the characteristic polynomial χ_f of f is given by

$$\chi_f(x) = \det \begin{bmatrix} x - \varphi_1^1 & & * \\ & \ddots & \\ 0 & & x - \varphi_n^n \end{bmatrix} = (x - \varphi_1^1) \cdots (x - \varphi_n^n),$$

so it decomposes into linear factors. If, on the other hand, the characteristic polynomial χ_f is given by $\chi_f(x) = (x - \lambda_n) \cdots (\chi - \lambda_n)$, where $\lambda_1 \ldots, \lambda_n$ are the eigenvalues of f (repetition of course is allowed), we show the assertion by induction on n.

For $n = 1$, $f_{BB} \in \mathbb{K}^{1 \times 1}$ is already upper triangular.

We start with an eigenvector $v_1 \equiv b_1$ for $\lambda_1 : f v_1 = \lambda_1 v_1$ and choose a basis $B = (b_1, u_2, \ldots, u_n)$ of V. We set $B_1 = \{v_1\}$, $B_2 = (u_2, u_3, \ldots, u_n)$, $U_1 = \operatorname{span} B_1$, and $U_2 = \operatorname{span} B_2$. So we have

$$V = U_1 \oplus U_2.$$

We define $h : U_2 \to U_1$ and $g : U_2 \to U_2$ with $\mu, \nu \in \{2, \ldots, n\}$ by
$h(u_\mu) = b_1 \varphi_\mu^1$ and $g(u_\mu) = u_\nu \varphi_\mu^\nu$.

$$f : U_2 \to V \quad \text{is given by} \quad f(u) := h(u) + g(u), u \in U_2.$$

We can write f_{BB} as

$$f_{BB} = \begin{pmatrix} \lambda_1 & h_{B_1 B_2} \\ 0 & g_{B_2 B_2} \end{pmatrix}$$

and we get

$$\chi_f(x) = (x - \lambda_1) \chi_g(x) \quad \text{and} \quad \chi_g(x) = (x - \lambda_2) \ldots (x - \lambda_n).$$

The characteristic polynomial of g decomposes into linear factors, so the induction hypothesis goes through, and so a basis $B_2^0 = (b_2, \ldots, b_n)$ exists such that the matrix $g_{B_2^0 B_2^0}$ is upper triangular.

Now we are in the position to define a new tailor-made basis for f, $B_0 := (B_1, B_2^0) = (b_1, \ldots, b_n)$ so that the matrix $f_{B_0 B_0}$ is upper triangular. This proves the proposition. □

Summary

This chapter marks the beginning of the section of the book that we consider advanced linear algebra. From now on, eigenvalues and eigenvectors take center stage.

Initially, we extensively presented the meaning, usefulness, and application of eigenvalues and eigenvectors in physics, facilitating the reader's entry into this sophisticated area of linear algebra with many examples.

The question of diagonalization and the description of this process were the central focus of this chapter. Highlights included two theorems. The first, the equivalence theorem of diagonalizability, addressed the geometric aspects of this question. The second theorem, concerning the algebraic perspective of diagonalizability, required rather advanced preparation.

To understand the question of diagonalization properly, one must also understand what non-diagonalizability means. Here too, we eased access to this question with examples and theory. This theory led to the so-called pre-Jordan form, as we refer to it here, which every diagonalizable and nondiagonalizable operator in a complex vector space possesses. The highlight here was the structure theorem (pre-Jordan form).

At the end of this chapter, we also discussed triangularization.

Exercises with Hints

Exercise 9.1 In Example 9.5, the subspace $U(f, v_o)$, is f-invariant. We call it an f cyclic subspace: Let $f \in \mathrm{Hom}(V, V)$ be nilpotent. We consider a subspace $U \equiv U(f, v_0)$ of V with $v_0 \in V$ and $f^m v_0 = 0$ with $f^{m-1} v_0 \neq 0$. The subspace U is given by

$$U = \mathrm{span}(v_0, f v_0, f^2 v_0, \ldots, f^{m-1} v_0).$$

We define $b_s := f^{s-1} v_0, s \in I(m)$.
Show that $B = (b_1, \ldots, b_n)$ is a basis of U.

Exercise 9.2 Direct sum of eigenspaces. (See Corollary 9.1)
If the eigenvalues $(\lambda_1, \ldots, \lambda_r)$ of the operator f are distinct, then show that

(i) the sum of the eigenspaces is direct:

$$E_{\lambda_1} + \cdots + E_{\lambda_r} = E_{\lambda_1} \oplus \cdots \oplus E_{\lambda_r};$$

(ii) the following inequality is valid:

$$\sum_{i=1}^{r} \dim E_{\lambda_i} \leqslant \dim V.$$

Exercise 9.3 *Linear independence of eigenvectors.*
Prove Proposition 9.1 by induction: Let $f \in \text{End}(V)$. If $\lambda_1, \ldots, \lambda_r$ are distinct eigenvalues of f, then corresponding eigenvectors $v_1, \ldots, v_r$ are linearly independent.

Exercise 9.4 Direct sum and direct decomposition. (See Proposition 9.5)
If $V = U_1 \oplus \cdots \oplus U_j \oplus \cdots \oplus U_r$ is a direct sum of subspaces $U_1, \ldots, U_r$, then show that there is a list $(P_1, \ldots, P_r)$ of projections in V with a direct decomposition of the identity such that for each $j \in I(r)$, $U_j = \text{im } P_j$ (See Definition 9.9). This means that if we define projections P_j:

$$P_j : V \longrightarrow U_j$$
$$v \longmapsto P_j v := u_j.$$

We need to show that $(P_1, \ldots, P_r)$ is a direct decomposition of identity, that is, that

- P_j is linear;
- $P_i P_j = \delta_{ij}$;
- $P_1 + \ldots + P_r = id$.

Exercise 9.5 For an operator $f \in \text{Hom}(V, V)$, show that for every $k \in \mathbb{N}$,

$$\ker f^k \leq \ker f^{k+1}.$$

(See Lemma 9.3)

Exercise 9.6 For an operator $f \in \text{Hom}(V, V)'$, show that for every $k \in \mathbb{N}$,

$$\text{im } f^{k+1} \leq \text{im } f^k.$$

Exercise 9.7 *Sequence of falling ranges.*
Show that for an operator $f \in \text{Hom}(V, V)$, there is a number $m \in \mathbb{N}$ such that the following sequence holds:

$$\text{im } f^{m+2} = \text{im } f^{m+1} = \text{im } f^m \leq \text{im } f^{m-1} \leq \text{im } f^{m-2} \leq \cdots \leq \text{im } f.$$

Exercise 9.8 Let F be the matrix $F = \begin{bmatrix} 0 & 1 \\ 0 & 0 \end{bmatrix}$ (as in Example 9.18), and $U = \text{span}(e_1)$. Show that there is no (complementary) F-invariant subspace $\bar{U}$ of U, such that
$$\mathbb{K}^2 = U \oplus \bar{U},$$

which means that F is not diagonalizable.

> The following five exercises are applications of Comment 9.3 about the action
> of polynomials on the algebra $\mathbb{K}^{n \times n}$ in connection with diagonalization and
> spectral decomposition induced by a diagonalizable matrix, as discussed in
> Theorems 9.1 and 9.4.

Exercise 9.9 Show that the evaluation map ev_A, $A \in \mathbb{K}^{n \times n}$,

$$ev_A : \quad \mathbb{K}[x] \longrightarrow \mathbb{K}[A],$$
$$\varphi \longmapsto \varphi(A),$$
$$\varphi_s x^s \longmapsto \varphi_s A^s, s \in \mathbb{N},$$

is an algebra homomorphism.

Exercise 9.10 For a matrix $A \in \mathbb{K}^{n \times n}$, show that

$$\ker(ev_A) = I_A \leq \mathbb{K}[x]$$

such that we have

$$\mathbb{K}[x]/I_A \cong \mathbb{K}[A]!$$

> For a diagonalizable matrix, we can use the minimal polynomial to obtain
> the corresponding spectral decomposition. This means we determine the
> eigenspaces of a diagonalizable matrix directly from the minimal polynomial
> and the evaluation map.

Exercise 9.11 Let A be a diagonalizable matrix $A \in \mathbb{K}^{n \times n}$ with minimal polynomial
$\mu(x) = (x - \lambda_1) \ldots (x - \lambda_r)$ where $r \leq n$. Use the minimal polynomial to construct
the linearly independent idempotence $P_j \in \mathbb{K}[A]$ $j \in I(r)$ with the following properties:

(i) $P_i P_j = 0$ if $i \in I(r)$ and $i \neq 0$,
(ii) $P_1 + \cdots + P_r = \mathbb{1}_n$ and
(iii) $A = \lambda_1 P_1 + \cdots + \lambda_r P_r$.

For this construction, show that there exist polynomials given by

$$\varphi_j(x) := \frac{\mu(x)}{x - \lambda_j} \text{ and } \psi_j(x) := \frac{\varphi_j(x)}{\varphi_j(\lambda_j)}, x \in \mathbb{K}; j \in I(r)$$

with the following properties:

(i) $\psi_i \psi_j - \delta_{ij} \psi_i \in I_A, i, j \in I(r)$,
(ii) $\psi_1 + \cdots + \psi_r - 1 \in I_A$,
(iii) $\lambda_1 \psi_1 + \cdots + \lambda_r \psi_r - id \in I_A$, with $id(x) = x$.

So the evaluation map ev_A leads directly to the desired result.

Exercise 9.12 Let V be a vector space and $f \in \text{Hom}(V, V)$. Show that the following subspaces are f-invariant:

(i) $\ker f$,
(ii) $\text{im } f$,
(iii) U such that $U \leq \ker f$,
(iv) W such that $W \geq \text{im } f$.

Exercise 9.13 Let V be a vector space and $f, g \in \text{Hom}(V, V)$ with the property $f \circ g = g \circ f$. Show that im f is g-invariant.

Compare the following two exercises with Example 9.17.

Exercise 9.14 Let F be a map $F \in \text{Hom}(\mathbb{R}^2, \mathbb{R}^2)$ given by

$$F = \begin{bmatrix} 0 & +4 \\ 1 & 0 \end{bmatrix}.$$

Find the eigenvectors and eigenvalues of F.

Exercise 9.15 Let V be a map $F \in \text{Hom}(\mathbb{R}^2, \mathbb{R}^2)$ given by

$$F = \begin{bmatrix} 0 & -4 \\ 1 & 0 \end{bmatrix}.$$

Find the eigenvectors and eigenvalues of F.

Compare the following two exercises with Example 9.14.

Exercise 9.16 Determine the eigenvalues and eigenvectors of the matrix

$$F = \begin{bmatrix} 1 & 1 \\ 1 & 1 \end{bmatrix} \in \text{Hom}(\mathbb{K}^2, \mathbb{K}^2).$$

Exercise 9.17 Determine the eigenvalues and eigenvectors of the matrix

$$F = \begin{bmatrix} 1 & 1 & 1 \\ 1 & 1 & 1 \\ 1 & 1 & 1 \end{bmatrix} \in \text{Hom}(\mathbb{K}^3, \mathbb{K}^3).$$

Exercise 9.18 Let $f \in \mathrm{Hom}(V, V)$ be invertible and $\lambda \in \mathbb{K}$. Show that λ is an eigenvalue of f if and only if λ^{-1} is an eigenvalue of f^{-1}!

The following exercise is relevant for Sect. 12.2: SVD.

Exercise 9.19 *Singular Value Decomposition (SVD).*
Let V be a vector space and $f, g \ \mathrm{Hom}(V, V)$. Show that $g \circ f$ and $f \circ g$ have the same eigenvalues.

The next two exercises are relevant for Theorem 9.4.

Exercise 9.20 Show that the evaluation map commutes with the adjoint representation of the linear group: Let $M \in \mathbb{K}^{n \times n}$, $F \in GL(n, \mathbb{K})$ and $\varphi \in \mathbb{K}[x]$. Show that

$$\varphi(FMF^{-1}) = F\varphi(M)F^{-1}.$$

Exercise 9.21 *Example of a diagonalizable nilpotent operator.*
Show that a self-adjoint nilpotent operator $f \in \mathrm{Hom}(V, V)$ is zero: $f = 0$.

Exercise 9.22 *This exercise is significant in connection with the minimal polynomial of an operator.*
Let $f \in \mathrm{Hom}(V, V)$, let $v \in V$ with $v \neq 0$, and let μ be a polynomial of smallest possible degree such that $\mu(f)v = 0$. Show that if $\mu(\lambda) = 0$, this λ is an eigenvalue of f.

Nilpotent operators play a central role in the question of diagonalizability (see Theorem 9.4). Proposition 9.6 showed that a nilpotent operator is similar to a strictly upper triangular matrix. Here, a more direct proof is demanded.

Exercise 9.23 *Representation matrix of a nilpotent operator.*
Let f be a nilpotent operator, $f \in \mathrm{Hom}(V, V)$ and $\dim V = n$. Show that there exists a representation matrix F of f which has the strict upper triangular form:

$$F = \begin{bmatrix} 0 & & * \\ & \ddots & \\ 0 & & 0 \end{bmatrix}.$$

Choose step by step, first a basis of ker f and then extend this to a basis of ker f^2 and so on. The result is a basis of V, and show that with respect to this, the basis matrix F has the desired form.

The following two exercises are first a wrong and then a more direct proof of the Cayley-Hamilton theorem (see Theorem 9.3).

Exercise 9.24 Explain why the following proof is incorrect:

$$\chi_F(F) = \det(F\mathbb{1} - F) = \det(F - F) = \det(0) = 0 \,!$$

Exercise 9.25 *Cayley-Hamilton theorem: second proof.*
Let F be a matrix $\mathbb{K}^{n \times n}$ with characteristic polynomial

$$\chi_F(x) = \det(x\mathbb{1}_n - F).$$

Show, using the expression $(x\mathbb{1}_n - F)(x\mathbb{1}_n - F)^{\#} = \det(x\mathbb{1}_n - F)\mathbb{1}_n$ (see Proposition 7.4), by a direct calculation:

$$\chi_F(F) = 0 \in \mathbb{K}^{n \times n}.$$

Exercise 9.26 Check by explicit calculation that the characteristic polynomial $\chi_A(x) = \det(x\mathbb{1}_n - A)$ of a matrix $A \in \mathbb{K}^{n \times n}$, the expression $\chi_A(B) \in \mathbb{K}^{n \times n}$, with $B \in \mathbb{K}^{n \times n}$ does not mean that

$$\chi_A(B) \quad \text{equals} \quad \det(A - B).$$

Exercise 9.27 This exercise answers the question whether for every given polynomial $\phi(x) = x^n + \varphi_{n-1}x^{n-1} + \cdots + \varphi_1 x + \varphi_0$, $\varphi_0, \cdots, \varphi_{n-1} \in \mathbb{K}$ a corresponding matrix $F \in \mathbb{K}^{n \times n}$ exists, such that the characteristic polynomial of F is exactly the given polynomial $\phi(x)$.
Check that the matrix

$$F = \begin{bmatrix} 0 & & & -\varphi_0 \\ 1 & 0 & & -\varphi_1 \\ & \ddots & & \vdots \\ & & 0 & -\varphi_{n-2} \\ & & 1 & -\varphi_{n-1} \end{bmatrix}$$

has characteristic polynomial of F:

$$\chi_F(x) := \det(x\mathbb{1}_n - F) = \phi(x).$$

References and Further Reading

1. S. Axler, *Linear Algebra Done Right* (Springer Nature, 2024)
2. S. Bosch, *Lineare Algebra* (Springer, 2008)
3. G. Fischer, B. Springborn, *Lineare Algebra. Eine Einführung für Stutdienanfünger*. Grundkurs Mathematik (Springer, 2020)
4. S.H. Friedberg, A.J. Insel, L.E. Spence, *Linear Algebra* (Pearson, 2013)
5. K. Jänich, *Mathematik 1. Geschrieben für Physiker* (Springer, 2006)
6. N. Johnston, *Advanced Linear and Matrix Algebra* (Springer, 2021)
7. M. Koecher, *Lineare Algebra und analytische Geometrie* (Springer, 2013)
8. G. Landi, A. Zampini, *Linear Algebra and Analytic Geometry for Physical Sciences* (Springer, 2018)
9. J. Liesen, V. Mehrmann, *Linear Algebra* (Springer, 2015)
10. P. Petersen, *Linear Algebra* (Springer, 2012)
11. S. Roman, *Advanced Linear Algebra* (Springer, 2005)
12. B. Said-Houari, *Linear Algebra* (Birkhäuser, 2017)
13. A.J. Schramm, *Mathematical Methods and Physical Insights: An Integrated Approach* (Cambridge University Press, 2022)
14. G. Strang, *Introduction to Linear Algebra* (SIAM, 2022)
15. R.J. Valenza, *Linear Algebra. An Introduction to Abstract Mathematics* (Springer, 2012)
16. R. Walter, *Lineare Algebra und analytische Geometrie* (Springer, 2013)

Chapter 10
Operators on Inner Product Spaces

In this chapter, we summarize and complete some important facts about inner product spaces, real and complex ones, mostly on a more advanced level than in Chaps. 2 and 6. In this context, the notions of orthogonality, orthogonal compliment, orthogonal projection, and orthogonal expansion are discussed once again.

In Sect. 10.4, we motivate and discuss normal operators in some detail. We give both an algebraic and a geometric definition of normal operators. We explain a surprising and gratifying analogy of normal operators to complex and real numbers. The highlight of this chapter and in fact of linear algebra altogether, are the spectral theorems that are treated at the end of this chapter.

10.1 Preliminary Remarks

Vector spaces with additional structures are especially welcome in physics and mathematics. One such structure which we particularly like, is an inner product space. This has to do with our surrounding space being a Euclidean space. It seems natural, especially in applications, to prefer spaces with more structure than a pure abstract vector space in mathematics. In connection with inner product spaces, there is the marvelous Pythagorean theorem. In physics, whenever a mathematical space is needed as a model for physical reality, there is the tendency to burden it immediately with as many structures as possible and often with even more structures than required. Therefore, in physics, when we are talking about vector spaces, we usually mean an inner product vector space or, as it is also called, a metric vector space or a vector space with a metric.

As we saw, we distinguish between real and complex vector spaces for abstract vector spaces. The same is true for inner products; in the real case, we talk about Euclidean vector spaces, and in the complex case, about unitary vector spaces. The standard Euclidean vector space in n-dimensions is, as we already know, $\mathbb{R}^n$

© The Author(s), under exclusive license to Springer Nature Switzerland AG 2024
N. A. Papadopoulos and F. Scheck, *Linear Algebra for Physics*,
https://doi.org/10.1007/978-3-031-64908-0_10

with the usual real dot product. The standard unitary vector space in n-dimensions (n-complex dimensions) is $\mathbb{C}^n$ with the usual complex dot product (Hermitian inner product). Inner product vector spaces are characterized by the property of having positive definite scalar products. In physics, especially in the real case, there are also nondegenerate scalar products, and so we also talk about semi-Euclidean or pseudo-Euclidean vector spaces. This is the case in special and general relativity.

In this chapter, before coming to operators, we review a few facts about inner product vector spaces and discuss some critical applications of the metric structure connected with orthogonality.

10.2   Inner Product Spaces Revisited

For a vector space V with dimension n over a field $\mathbb{K} = \mathbb{R}$ or $\mathbb{C}$, the inner product $\langle -|-\rangle$ was defined in Sect. 10.3. For the standard vector space $\mathbb{K}^n$, the standard (canonical) inner product $\langle -|-\rangle_0$ is given in the form of the complex dot product:

$$\langle u|v\rangle_0 := \bar{u}^1 v^1 + \cdots + \bar{u}^n v^n = u_i v^i,$$

with $u, v \in \mathbb{K}^n$, $u^i, v^i \in \mathbb{K}$, $u_i := \bar{u}^i$ and $i \in I(n)$. We can always construct an isomorphism of inner product spaces:

$$(V, \langle | \rangle) \cong (\mathbb{K}^n, \langle | \rangle_0).$$

The quadratic form, $|| \cdot ||^2$, is given by

$$\| \cdot \|^2 : \qquad V \longrightarrow \qquad \mathbb{R}^+ \cup \{0\},$$
$$v \longmapsto \qquad \|v\|^2 = \langle v|v\rangle.$$

At this point, it is interesting to notice the so called "polarization identities" which connects the inner product to the corresponding quadratic form: If V is a $\mathbb{R}$ vector space, we have
$$4\langle u|v\rangle = \|u+v\|^2 - \|u-v\|^2.$$

If V is a $\mathbb{C}$ vector space, we have

$$4\langle u|v\rangle = \|u+v\|^2 - \|u-v\|^2 - i\{\|u+iv\|^2 - \|u-iv\|^2\}.$$

As we see, the inner product can be expressed in terms of its quadratic form.

10.3 Orthonormal Bases

We first recall the definition of the orthogonality for vectors $u, v \in V$. We say u and v are orthogonal if $\langle u|v \rangle = 0$ and we write $u \perp v$.

Definition 10.1 Orthogonal complement.
If M is a subset of V, the set $M^\perp$ of all vectors in V which are orthogonal to M is called the orthogonal complement of M in V:

$$M^\perp = \{v \in V : \langle u|v \rangle = 0 \ \ \forall u \in M\}.$$

Definition 10.2 Orthogonal and orthonormal basis.
A basis $B = (b_1, \ldots, b_n)$ of V of dimension n is called orthogonal if $\langle b_i|b_j \rangle = \beta_j \delta_{ij}$ with $\beta_j > 0 \ \ \forall i, j \in I(n)$.
A basis $C = (c_1, \ldots, c_n)$ of V is called orthonormal if $\langle c_i|c_j \rangle = \delta_{ij} \ \ \forall i, j \in I(n)$.

Proposition 10.1 *Orthogonality and linear independence.*
A list of nonzero orthogonal vectors is linearly independent.

Proof If $(v_1, \ldots, v_k), k \in I(n)$ is a list of nonzero orthogonal vectors and if $v_i \lambda^i = 0$ with $\lambda^i \in \mathbb{K}$, then

$$\|v_i \lambda^i\|^2 = \langle v_i \lambda^i|v_j \lambda^j \rangle = |\lambda^i|^2 \|v_i\|^2 = 0.$$

Since each $v_i \neq 0$, positive definiteness tells us that $\lambda^i = 0 \ \forall \, i \in I(h)$. Thus, $(v, \ldots, v_n)$ is linearly independent. $\qquad \square$

Proposition 10.2 *Orthonormal expansion.*
For $v \in V$ and an orthonormal basis (c_i), the expansion $v = c_1\langle c_1|v \rangle + \cdots + c_n\langle c_n|v \rangle$ holds. This is the widely used notation in quantum mechanics (see Sect. 6.4):
$$|v\rangle = |c_1\rangle\langle c_1|v\rangle + \cdots + |c_n\rangle\langle c_n|v\rangle.$$

Proof Since $(c_1, \ldots, c_n)$ is a basis, we have $v = c_i v^i$. Thus,

$$\langle c_j | v \rangle = \langle c_j | c_i v^i \rangle = \langle c_j | c_i \rangle v^i = \delta_{ji} v^i = v^i.$$

So we get $v = c_i \langle c^i | v \rangle$, $(c^i \equiv c_i)$.   □

Proposition 10.3 (Gram-Schmidt orthogonalization)
If $(a_1, \ldots, a_n)$ is a basis of V, then there exists an orthonormal basis $(c_1, \ldots, c_n)$ of V with

$$\mathrm{span}(a_1, \ldots, a_j) = \mathrm{span}(c_1, \ldots, c_j) \quad \text{for each} \quad j \in I(n).$$

Proof Consider the following inductive definitions:

$$b_1 := a_1 \quad \text{with}$$

$$c_i := \frac{b_i}{\|b_i\|} \quad i \in I(n) \quad \text{and}$$

$$b_k := a_k - \sum_{\mu=1}^{k-1} c_\mu \langle c_\mu | a_k \rangle, \ k \in \{2, \ldots, n\}.$$

Set

$$A_i := (a_1, \ldots, a_i),$$
$$B_i := (b_1, \ldots, b_i),$$
$$C_i := (c_1, \ldots, c_i),$$
$$\text{and} \quad V_i := \mathrm{span}\, A_i.$$

We are going to show the result by induction on $n = \dim V$.

For $n = 1$ we have $V_1 = \mathrm{span}\, C_1 = \mathrm{span}\, B_1 \equiv \mathrm{span}(a_1)$. By induction hypothesis, $C_{n-1} = (c_1, \ldots, c_{n-1})$ is an orthonormal basis of $V_{n-1} = \mathrm{span}(a_1, \ldots, a_{n-1})$, with $C_j = (c_1, \ldots, c_j)$ orthonormal basis of V_j, $j \in I(n-1)$.

We have still to prove that $C_n \equiv C = (c_1, \ldots, c_n)$ is also an orthonormal basis to $V_n \equiv V$. Since by assumption $a_n \notin V_{n-1}$, we have $b_n \neq 0$ and so $c_n = \frac{b_n}{\|b_n\|}$ is well defined with $\|c_n\| = 1$.

To show that c_n is orthogonal to V_{n-1}, we have for $k = n$, setting for simplicity $\langle c^\mu | \equiv \langle c_\mu |$ in order to use the Einstein convention, , $b_n = a_n - c_\mu \langle c^\mu | a_n \rangle$ and we consider $\mu, j \in I(n-1)$:

$$\langle c_j | b_n \rangle = \langle c_j | c_n \|b_n\| \rangle$$
$$= \langle c_j | c_n \rangle \|b_n\|.$$

On the other hand,

$$\langle c_j | b_n \rangle = \langle c_j | (a_n - c_\mu \langle c^\mu | a_n \rangle) \rangle$$
$$= \langle c_j | a_n \rangle - \langle c_j | c_\mu \rangle \langle c^\mu | a_n \rangle.$$

So

$$\langle c_j | b_n \rangle = \langle c_j | a_n \rangle - \delta_{j\mu} \langle c_\mu | a_n \rangle$$
$$= \langle c_j | a_n \rangle - \langle c_j | a_n \rangle = 0.$$

Since c_n and b_n are colinear, we have also $\langle c_j | c_n \rangle = 0$ and C is indeed an orthonormal basis of V.   □

Corollary 10.1 *Schur's Theorem.*
Suppose V is an inner product $\mathbb{K}$-vector space and $f \in \mathrm{End}(V)$. If the characteristic polynomial χ_f of f decomposes into linear factors over $\mathbb{K}$, then there exists an orthonormal basis C of f so that f_{CC} is an upper triangular matrix. This is, of course, especially valid for a $\mathbb{C}$-vector space.

Proof Section 9.7 and the proposition made there showed that there exists a basis $B_0 = (b_1, \ldots, b_n)$ of V so that $f_{B_0 B_0}$ is triangular, and that span $B_j = \mathrm{span}(b_1, \ldots, b_j)$ is f-invariant for all $j \in I(n)$. We apply the above proposition concerning the Gram-Schmidt orthogonalization to the basis B_0, with $\mathrm{span}(C_j) = \mathrm{span}(c_1, \ldots, c_j) = \mathrm{span}(b_1, \ldots, b_j)$ for all $j \in I(n)$. So we conclude that span C_j is also f-invariant for all $j \in I(n)$ and that f_{CC} is triangular.   □

10.4 Orthogonal Sums and Orthogonal Projections

As we saw, metric structures (inner products) on vector spaces lead to the notion of orthogonality and the orthogonal complement. This allows a refinement of the direct product and the parallel projection to the orthogonal sum and orthogonal projection. These are pure geometric properties well-known from Euclidean geometry and show once more the entanglement between geometry and algebra in linear algebra. We are first going to study some elementary properties of orthogonal complements in the following propositions:

Proposition 10.4 *Properties of orthogonal complements.*
For the subspaces U and W of V, the following is valid:

(i) $U^\perp \leqslant V$;

(ii) $U \cap U^\perp = \{0\}$;

(iii) $U \leqslant W \Rightarrow W^\perp \leqslant U^\perp$;

(iv) $\{0\}^\perp = V$ and $V^\perp = \{0\}$.

The symbol $\leqslant$ stands for "subspace of" as throughout this book.

Proof (i)
Since $\langle 0|u \rangle = 0$ for $u \in U$, it then follows that $0 \in U^\perp$. By the linearity of the scalar product $\langle v|\cdot \rangle$, we have, when $w, z \in U^\perp$, $\langle w|u \rangle = 0$, and $\langle z|u \rangle = 0$ for every $u \in U$. It then follows that $\langle w + z|u \rangle = \langle w|u \rangle + \langle z|u \rangle = 0 + 0 = 0$ so that $w + z \in U^\perp$. Similarly $\lambda w \in U^\perp$. $\qquad\square$

Proof (ii)
Let $z \in U \cap U^\perp$. Then $z \in U$, $z \in U^\perp$ so that $\langle z|z \rangle = 0$. Hence $z = 0$ and thus $U \cap U^\perp = \{0\}$. $\qquad\square$

Proof (iii)
Let $\bar{w} \in W^\perp$. Then, as $U \subseteq W$, we have $\langle \bar{w}, u \rangle = 0 \,\forall\, u \in U$. Thus $\bar{w} \in U^\perp$ and so $W^\perp \subseteq U^\perp$. $\qquad\square$

Proof (iv)
For all $v \in V$, we have $\langle v|0 \rangle = 0$ and so $v \in \{0\}^\perp$ which means that $\{0\}^\perp = V$. For $v \in V^\perp$, we have $\langle v|v \rangle = 0$ and so $v = 0$ which means that $V^\perp = \{0\}$. $\qquad\square$

Definition 10.3 Orthogonal sum.
Let $U_1, \ldots, U_k \leq U$ be subspaces of U. We say $U_1 + \cdots + U_k$ is an orthogonal sum if:

(i) $U_i \cap U_j = \{0\}$ whenever $i \neq j$ (so the sum is a direct sum),
(ii) Whenever $i \neq j$, $U_i \perp U_j$.

We denote orthogonal sums with $\ominus$ as follows:

$$U_1 \ominus U_2 \ominus \cdots \ominus U_k.$$

Proposition 10.5 *Orthogonal complement.*
If U is a subspace of V, then:

$$U \oplus U^{\perp} = V.$$

This means that every subspace U uniquely induces a direct sum decomposition of V, an orthogonal decomposition. We may also write as above $U \ominus U^{\perp} = V$.

Proof Let $v \in V$. We write $v = u + v - u$ and we set $\tilde{u} := v - u$ or $v = u + \tilde{u}$. Let $(u_1, \ldots, u_k)$ be an orthonormal basis of U, we set $\langle u^i| = \langle u_i|$, and

$$u = u_i \langle u^i | v \rangle.$$

Let

$$\tilde{u} = v - u_i \langle u^i | v \rangle.$$

We now show that $\tilde{u} \in U^{\perp}$ or, equivalently, that $\langle u_j | \tilde{u} \rangle = 0$ for all $j \in I(k)$:

$$
\begin{aligned}
\langle u_j | \tilde{u} \rangle &= \langle u_j | v - u \rangle \\
&= \langle u_j | v \rangle - \langle u_j | u^i \rangle \langle u^i | v \rangle \\
&= \langle u_j | v \rangle - \delta_{ji} \langle u_i | u \rangle \\
&= \langle u_j | v \rangle - \langle u_j | v \rangle = 0.
\end{aligned}
$$

$\square$

So we may use the notation $\tilde{u} \equiv u^{\perp}$ and $v = u + u^{\perp}$. Of course u and $u^{\perp}$ depend on v.

Corollary 10.2 $\dim U + \dim U^{\perp} = \dim V$.

As expected, the property of "$\perp$" is an involution:

Proposition 10.6 *Orthogonality as involution. If U is a subspace of V, then*

$$(U^{\perp})^{\perp} = U.$$

Proof

(i) We show that $U \leqslant (U^{\perp})^{\perp}$.
    Let $u \in U$. Then whenever $w \in U^{\perp}$, $\langle u | w \rangle = 0$. Hence $u \in (U^{\perp})^{\perp}$ and so $U \subseteq (U^{\perp})^{\perp}$.

(ii) We show that $(U^{\perp})^{\perp} \subset U$:
    Let $\bar{w} \in (U^{\perp})^{\perp}$. The orthogonal decomposition of $\bar{w} \in V$ relative to U is given (see Proposition 10.5) by $\bar{w} = u + z$ (so $\bar{w} - u = z$) with $u \in U$ and $z \in U^{\perp}$. Since $u \in U$ from (i), we have $u \in (U^{\perp})^{\perp}$ and so $z = \bar{w} - u \in (U^{\perp})^{\perp}$. As we see, $z \in U^{\perp} \cap (U^{\perp})^{\perp} = \{0\}$ (Proposition 10.4). So we have $z = 0$ and $\bar{w} - u = 0 \Rightarrow \bar{w} = u \in U$ which means that $(U^{\perp})^{\perp} \subset U$. So with (i) and (ii) we obtain $(U^{\perp})^{\perp} = U$.

□

Analogous to orthogonal sums, orthogonal projections also lead to more refined and "perfect" projections than parallel projections.

Definition 10.4 Orthogonal projections.
Suppose U is a subspace of V. As $U \ominus U^{\perp} = V$, every element $v \in V$ can be decomposed as $v = u + u^{\perp}$ with $u \in U$, $u^{\perp} \in U^{\perp}$. The orthogonal projection P_U is the projection given by:

$$P_U : V \longrightarrow U$$

$$v \longmapsto P_U(v) := u.$$

If we take into account an orthogonal or an orthonormal basis in U, we can express P_U more directly. In Sect. 10.3 we already saw that for $U := \mathrm{span}(u)$, P_U is given by

$$P_U(v) = u \frac{\langle u | v \rangle}{\langle u | u \rangle}.$$

More generally, we notice that if we choose an orthonormal basis $B_U = (u_1, \ldots, u_k)$ with $\langle u_i | u_j \rangle = \delta_{ij}$ $i, j \in I(k)$ in U, then

$$P_U(v) = \sum_{j=1}^{k} u_j \langle u_j | v \rangle$$

or equivalently

$$P_U = \sum_{j=1}^{k} |u_j\rangle\langle u_j|.$$

Here, we use a formalism which is routine, especially in quantum mechanics (see Sect. 6.4), $(|u_j\rangle \equiv u_j$ and $\langle u_j| \in U^*$, the dual of U). For the orthogonal projection P_U, the same relations as those for the parallel projections also hold. They are given in the next proposition.

Proposition 10.7 *Properties of orthogonal projections.*

(i) $P_U|_U = id_U;$
(ii) $P_U|_{U^\perp} = \hat{0}$: *zero operator;*
(iii) $\ker P_U = U^\perp$, $\operatorname{im} P_U = U$;
(iv) $P_U^2 = P_U$;
(v) $P_U + P_{U^\perp} = id_V.$ □

Comment 10.1 Comparison with parallel projections.

Perhaps the most important property of an orthogonal projection which makes the difference to parallel projections, is not clearly visible in this last proposition. To define P_U, we just have to fix one subspace U since the complement $U^\perp$ is uniquely defined. On the other hand, we need two subspaces to determine a parallel projection. In a similar notation as above, we have to write for a parallel projection

$$P_{U,W} : V \to U \quad \text{with } V = U \oplus W.$$

Comment 10.2 Cauchy-Schwarz inequality and projections.

The expression (v) in Proposition 10.7 leads us directly to a kind of generalization of the Cauchy-Schwartz inequality of Sect. 10.3.

$$\|P_U v\| \leqslant \|v\|.$$

A very interesting application of orthogonal projections in various fields in mathematics, physics, and many other sciences, refers to the minimization problem as given in the following proposition.

Proposition 10.8 *Minimization.*
Let U be a subspace of V and $u \in U$. Then

(i) $\|v - P_U v\| \leqslant \|v - u\|$ *for* $v \in V$;
(ii) $\|v - P_U v\| = \|v - u\|$ *if and only if* $u \equiv u_0 = P_U v.$

Proof As $\langle v - P_U v | P_U v - u \rangle = 0$, we have by Pythagoras

$$\|v - u\|^2 = \|v - P_U v + P_U v - u\|^2 = \|v - P_U v\|^2 + \|P_U v - u\|^2.$$

Then

(i) $\|v - u\|^2 \geqslant \|P_U v - u\|^2$ since $\|v - P_U v\|^2 \geqslant 0$.
(ii) $\|v - u\|^2 = \|v - P_U v\|^2$

  – iff $\|P_U v - u\|^2 = 0$,
  – iff $\|P_U v = u$.

$\square$

10.5   The Importance of Being a Normal Operator

One could say that normal operators, particularly those on a complex vector space, are mathematically the nicest operators one can hope to have: They are diagonal-izable (even if not the only diagonalizable operators) and they have an orthogonal eigenbasis. As we shall see later, normal operators are the only diagonalizable oper-ators with an orthonormal basis. They lead to the so-called complex and real spectral theorems, which underlines their beauty and usefulness. One could also say that they are the most important operators to a physicist. Self-adjoint operators and isometries are both normal. We call self-adjoint operators on complex and real vector spaces also Hermitian and symmetric, respectively. Isometries for complex and real vector spaces (always finite-dimensional in our approach), are also known as unitary and orthogonal, respectively.

There is no doubt that an inner product vector space has much more structure than an abstract vector space. This is why it is, at least in physics, much more pleasant to have at our disposal an inner product vector space than having only an abstract vector space with the same dimension. In addition, when dealing with operators in an inner product vector space, it is natural to be interested in operators that interact well with the inner product structures. Stated differently, if an operator has nothing to do with the inner product, then the inner product is not relevant to this operator, so there is no necessity to introduce a metric in addition to the linear structure. This justifies dealing with inner product spaces and special operators with a characteristic intrinsic connection to the inner product. It turns out that these operators are the "normal" operators and it may appear rather surprising that the normal operators are not exclusively those operators which preserve the inner product. This is what we would expect from our experience in similar situations. It may also partially explain why, in physics, the notion of a normal operator is often absent.

The operator "adjoint",

$$ad : \mathrm{Hom}(V, V) \longrightarrow \mathrm{Hom}(V, V)$$
$$f \longmapsto f^{ad}$$

is, first, at the technical level, what is needed for the formulation of the specific connection of the normal operator with the help of the inner product. At the same time it is quite clear that the structure "inner product" is a necessary prerequisite for the definition of what is "adjoint". In the case of $V = \mathbb{K}^n$ and $V' = \mathbb{K}^m$, the inner product is the canonically given dot product. So we have $f^{ad} = f^{\dagger}$ as was shown in Sect. 6.3 and Proposition 6.9. From the discussion in Sect. 6.3, it follows in addition that f^{ad} is a kind of a substitution for the inverse of f. Now if we take this assumption concerning "inverse" seriously and look for a weaker condition for invertibility, it turns out that this idea can lead us to the definition of normal operators: If f^{ad} was really an inverse operator ($f^{ad} = f^{-1}$), we would have $f \circ f^{ad} = id_V$ and $f^{ad} \circ f = id_V$. A weaker condition would then be: $f^{ad} \circ f = f \circ f^{ad}$. This is exactly the definition of a normal operator! There is in addition a surprisingly pleasant analogy to complex and real numbers which leads also to the notion of normal, unitary and self-adjoint operators. For complex and real numbers we recall the following well-known relation: if $z \in \mathbb{C} \setminus \{0\}$ and $x, y, r, \varphi \in \mathbb{R}$,

$$z = x + iy, \quad \bar{z}z = z\bar{z}, \quad |z| := \sqrt{\bar{z}z}, \quad z = \frac{z}{|z|}|z|, \quad |z| = r = \sqrt{x^2 + y^2}$$

positive or zero (nonnegative) and $\frac{z}{|z|} = e^{i\varphi}$ with $\left|\frac{z}{\bar{z}}\right| = |e^{i\varphi}| = 1$.

If we ask for an operator f in $(V, \langle|\rangle)$ in a $\mathbb{C}$ vector space with metric structure which corresponds to z and its properties shown above, using the analogy $z \mapsto \bar{z}$ with $f \mapsto f^{ad}$, we are led to the normal operators which in addition contain, again in analogy to the complex numbers, the nonnegative and self-adjoint operators. So for a normal operator f, we expect the relation $f^{ad} \circ f = f \circ f^{ad}$. This produces immediately two obvious special cases for a normal operator: $f^{ad} \circ f = id_V (f^{ad} = f^{-1})$ and $f^{ad} = f$ which are also the most important ones, the isometries and the self-adjoint operators. Further, if we define $f := h_1 + ih_2$ with the two self-adjoint operators $h_1^{ad} = h_1$ and $h_2^{ad} = h_2$, it follows that f normal is equivalent to the commutative relation $h_1 h_2 = h_2 h_1$. All displayed above may show that the introduction of the notions "adjoint" and "normal" is not made by accident but are connected with deep structures and are of tremendous relevance for mathematics and physics.

In order to proceed, we shortly recall the definitions of adjoint and self-adjoint operators given in Sect. 6.3 and Definition 6.4. The adjoint operator of $f \in \mathrm{End}(V)$ is given by:

$$\langle f^{ad} w | v \rangle := \langle w | f v \rangle \quad \text{for all } v, w \in V.$$

We say in addition, f is self-adjoint if $f^{ad} = f$.

We are now in a position to define a normal operator.

Definition 10.5 Normal operator (algebraic definition).
The operator $f \in \text{End}(V)$ is normal if it commutes with its adjoint: $f^{ad} \circ f = f \circ f^{ad}$.

It is clear that for the notion adjoint, self-adjoint and normal operator, the existence of a metric structure on V is required which in our case is expressed by the inner product $s = \langle -|\cdot\rangle$:

$$s : V \times V \longrightarrow \mathbb{K}$$
$$(u, v) \longmapsto s(u, v) \equiv \langle u|v\rangle$$

as defined in Scct. 10.3.

For normal operators, we can also give the following equivalent definition:

Proposition 10.9 *Normal operator (geometric definition).*
$f \in \text{End}(V)$ *is normal if and only if*

$$\|f^{ad}v\| = \|fv\| \quad \text{for all} \quad v \in V.$$

Proof The following sequence of equivalences leads to a pleasant result.

$$f \text{ is normal} \Leftrightarrow f^{ad}f - ff^{ad} = 0$$
$$\Leftrightarrow \langle w|(f^{ad}f - ff^{ad})v\rangle \; \forall \; w, v \in V$$
$$\Leftrightarrow \langle w|(f^{ad}fV - \langle w|(ff^{ad}v\rangle$$
$$\Leftrightarrow \langle fw|fv\rangle = \langle f^{ad}w|f^{ad}v\rangle.$$

The expression $\langle fw|fv\rangle = \langle f^{ad}w|f^{ad}v\rangle \; \forall \; w, v \in V$ is equivalent to $\|fv\|^2 = \|f^{ad}v\|^2 \; \forall v \in V$ because $\langle w|v\rangle$ can be written in terms of its norm $\| \cdot \|$ through the polarization identity (see Sect. 10.2). $\qquad\square$

In what follows, we need to specify more of what we know about f and f^{ad}. For this reason, we have to consider separately for $f \in \text{End}(V)$ whether V is a complex or real vector space. This means that we have to take into account whether f is a $\mathbb{C}$-linear or a $\mathbb{R}$-linear operator. Since $\mathbb{C}$-linearity is a stronger condition than $\mathbb{R}$ linearity, it leads as expected to more substantial results. This fact may be helpful in understanding the results that follow. These are also extremely useful for understanding crucial aspects of quantum mechanics theory.

> **Proposition 10.10** *Zero expectation value of an operator.*
> *Suppose V is a complex vector space and $f \in \text{End}(V)$. If $\langle v|fv \rangle = 0$ for all v in V, then $f = \hat{0}$.*

Proof Suppose $u, v \in V$. We use a modified version of the polarization identity (see Exercise 2.31):

$$4\langle v|fu \rangle = \langle v + u|f(v+u) \rangle + \langle v - u|f(v-u) \rangle -$$
$$- i(\langle v + iu|f(v+iu) \rangle - \langle v - iu|f(v-iu) \rangle).$$

We observe that the terms on the right-hand side are of the form required by the condition. This implies that the right-hand side is zero and so the left-hand side is zero too. Now set $v = fw$. Then $\langle fw|fw \rangle = 0$ and so $fw = 0$. Since w is arbitrary, we obtain $f = \hat{0}$. $\qquad\square$

We are here in the situation where we have to distinguish between a complex and a real vector space. For a real vector space (i.e., for a linear operator f), Proposition 10.10 is not valid: In $\mathbb{R}^2$ we have a rotation of $90°$ by $\begin{bmatrix} 0 & -1 \\ 1 & 0 \end{bmatrix} \neq \hat{0}$ and moreover we have $\langle v|fv \rangle = 0 \quad \forall v \in \mathbb{R}^2$.

It is interesting to notice that the above proposition is also valid for a real vector space in the special case of a self-adjoint operator.

Comment 10.3 $\mathbb{C}$-linearity versus $\mathbb{R}$-linearity.

The assertion here that V is a complex inner product space is important. The condition $\langle v|fv \rangle = 0$ means that for every matrix representation f_B, the diagonal is zero. Stated geometrically, the vector fv is always orthogonal to v or, as in quantum mechanics, the expectation of the observable f is zero. It then follows that the operator f itself is zero.

> **Proposition 10.11** *Zero expectation values for a self-adjoint operator. If f is a self-adjoint operator and if*
>
> $$\langle v|fv \rangle = 0 \text{ for all } v \in V, \text{ then } f = \hat{0}.$$

Proof For a self-adjoint operator in a complex vector space, it was already proven in Proposition 10.10. Therefore, we can assume that V is a real inner product vector space (f is only $\mathbb{R}$-linear), then we have for the appropriate modified version of the polarization identity in Sect. 10.2,

$$4\langle v|fu\rangle = \langle v+u|f(v+u)\rangle + \langle v-u|f(v-u)\rangle$$

which holds if f is symmetric and $\langle v|fv\rangle$ real (see Exercise 2.28). So we obtain the desired result as in Proposition 10.10.   □

Before discussing the spectral theorems, it is helpful to study some properties of eigenvalues and eigenvectors of normal and self-adjoint operators. This leads first to Propositions 10.12, 10.13, and 10.14.

Proposition 10.12 *Eigenvalues of self-adjoint operators.*
Every eigenvalue of a self-adjoint operator is real.

Proof Let $(V, \langle -|-\rangle)$ be an inner product space, $f = f^{ad}$ a self-adjoint operator on V, and $v \in V \setminus \{0\}$ an eigenvector v with eigenvalue λ. Then

$$\lambda\langle v|v\rangle = \langle v|\lambda v\rangle = \langle v|fv\rangle = \langle fv|v\rangle = \bar{\lambda}\langle v|v\rangle.$$

Since $\langle v|v\rangle \neq 0$, we have $\bar{\lambda} = \lambda$.   □

Comment 10.4 Characteristic polynomials of self-adjoint operators.

The characteristic polynomial of self-adjoint operators decomposes into linear factors.

This follows essentially from above Proposition 10.12: When for $\mathbb{K} = \mathbb{C}$, the fundamental theorem of algebra states that the characteristic polynomial χ_f of a self-adjoint operator in a $\mathbb{C}$ vector space factorizes into linear factors. Proposition 10.12 states that all the zeros of this χ_f are real.

For $\mathbb{K} = \mathbb{R}$ we get $\chi_f \in \mathbb{R}[x]$. The matrix F of f is real and symmetric: $\bar{F} = F$ and $F^{\mathsf{T}} = F$. This indicates in addition that $F^\dagger = F$ holds. So the result of $\mathbb{K} = \mathbb{C}$ above also applies to the $\mathbb{K} = \mathbb{R}$ case.

A comparison between the eigenelements of f^{ad} and f is given in the following relation.

Proposition 10.13 *Eigenvalues and eigenvectors of a normal operator, and its adjoint.*
Suppose f is normal and (λ, v) is an eigenelement of f, then $(\bar{\lambda}, v)$ is an eigenelement of f^{ad}.

Proof If f is normal, $f - \lambda Id_V$ is also normal since $f - \lambda Id_V$ and $f^{ad} - \bar{\lambda} Id_V$ commute as we see after a direct calculation. From Proposition 10.9, we obtain

$$\|(f - \lambda Id_V)v\| = \|(f^{ad} - \bar{\lambda} Id_V)v\| \quad \text{so that}$$
$$(f - \lambda Id_V)v = 0 \quad \Leftrightarrow \quad (f^{ad} - \bar{\lambda} Id_V)v = 0.$$

This shows when f is normal, f^{ad} and f have the same eigenvectors and their corresponding eigenvalues are complex conjugate. $\qquad\square$

In the following proposition, we see that for normal operators, the eigenvectors corresponding to different eigenvalues are linearly independent and orthogonal.

Proposition 10.14 *Eigenvalues of distinct eigenvalues are orthogonal. Let V be an inner product space over $\mathbb{K} \in \{\mathbb{R}, \mathbb{C}\}$ and let $f \in \text{End}(V)$ be normal. If (λ_1, v_1) and (λ_2, v_2) are eigenelements so that $\lambda_1 \neq \lambda_2$, then $\langle v_1 | v_2 \rangle = 0$.*

Proof For $(\lambda_2 - \lambda_1) \neq 0$ we have

$$(\lambda_2 - \lambda_1)\langle v_2 | v_1 \rangle = \lambda_2 \langle v_2 | v_1 \rangle - \lambda_1 \langle v_2 | v_1 \rangle$$
$$= \langle \bar{\lambda}_2 v_2 | v_1 \rangle - \langle v_2 | \lambda_1 v_1 \rangle.$$

Using that f is normal and Proposition 10.13, we obtain:

$$(\lambda_2 - \lambda_1)\langle v_2 | v_1 \rangle = \langle f^{ad} v_2 | v_1 \rangle - \langle v_2 | f v_1 \rangle$$
$$= \langle v_2 | f v_1 \rangle - \langle v_2 | f v_1 \rangle = 0.$$

$\qquad\square$

10.6 The Spectral Theorems

As already pointed out, whether a matrix is diagonalizable depends on the field $\mathbb{R}$ or $\mathbb{C}$ we are working over. In general, the situation for a complex vector space is simpler to deal with. Therefore, the mathematical literature will usually refer to complex and real spectral theorems separately. We follow this course for instructional reasons, but we show in a corollary that if we have a slightly different perspective, it is possible to consider only one spectral theorem for the general field $\mathbb{K}$.

As discussed in Sect. 9.4, the diagonalizability of an operator F on an n-dimensional vector space V implies both the existence of a diagonal representation $f_B = \mathrm{diag}(\lambda_1, \ldots, \lambda_n)$ and the existence of a basis $B = (b_1, \ldots, b_n)$ consisting of eigenvectors b_s, correspond to the eigenvalues λ_s for every $s \in I(n)$. It is reasonable to call B an eigenbasis of f. We already know that not every operator f has the privilege to be diagonalizable or equivalently to have an eigenbasis. In a complex vector space, the normal operators are precisely those operators which have the privilege not only to be diagonalizable but in addition to be diagonalizable with an orthonormal eigenbasis!

This is essentially the content of the spectral theorems:

Theorem 10.1 *The complex spectral theorem ($\mathbb{K} = \mathbb{C}$).*

Let V be an n-dimensional complex vector space and $f \in \mathrm{End}(V)$. Then f is normal if and only if V has an orthonormal basis of eigenvectors of f.

Proof Suppose that an orthonormal basis eigenbasis $C = (c_1, \ldots, c_n)$ exists. Then f is diagonalizable and the matrix $F := f_C$ can be written as $F = \mathrm{diag}(\lambda_1, \ldots, \lambda_n)$. The adjoint $f_C^{ad} = F^\dagger$ is also diagonal and so $F^\dagger F = F F^\dagger$. Thus $f^{ad} \circ f = f \circ f^{ad}$ and f is normal.

Suppose that f is normal. Then, by Schur's theorem (see Corollary 10.1), f is triangularizable so there exists an orthonormal basis $C = (c_1, \ldots, c_n)$ and $f_C \equiv F = (\varphi_s^i)$ is an upper triangular matrix: $\varphi_s^i = 0$ for $i > s$, $\quad i, s \in I(n)$.

Then the matrix F is given by

$$
F = \begin{bmatrix}
\varphi_1^1 & \varphi_2^1 & \cdots & \cdots & \cdots & \varphi_n^1 \\
0 & \varphi_2^2 & \cdots & \cdots & \cdots & \vdots \\
\vdots & 0 & \varphi_3^3 & \cdots & \cdots & \vdots \\
\vdots & \vdots & \cdots & 0 & \varphi_{n-1}^{n-1} & \varphi_n^{n-1} \\
0 & 0 & \cdots & & 0 & \varphi_n^n
\end{bmatrix}.
$$

The adjoint of f is given by

$$
f_C^{ad} = F^\dagger = ((\varphi^{ad})_s^i) \quad \text{with} \quad (\varphi^{ad})_s^i = \bar{\varphi}_i^s. \tag{10.1}
$$

Equation (10.1) tells that $F^\dagger$ is lower triangular and we can write:

$$
F = \begin{bmatrix} * & * & * \\ 0 & * & * \\ 0 & 0 & * \end{bmatrix} \quad \text{and} \quad F^{ad} = \begin{bmatrix} * & 0 & 0 \\ * & * & 0 \\ * & * & * \end{bmatrix}. \tag{10.2}
$$

F and F^{ad} represent f and f^{ad}. This means:

$$f c_s = c_i \varphi_s^i \quad \text{and} \quad f_s^{ad} = c_i (\varphi^{ad})_s^i = \sum_{i=1}^{n} c_i \bar{\varphi}_i^s. \tag{10.3}$$

Since f is normal, we have in addition:

$$\| f c_s \|^2 = \| f^{ad} c_s \| \quad \text{or equivalently} \quad \sum_{i=1}^{n} |\varphi_s^i|^2 = \sum_{i=1}^{n} |\varphi_i^s|^2. \tag{10.4}$$

This means that the norms of the corresponding columns of F and $F^\dagger$ are equal. This leads by induction of Eq. (10.2) directly to the result that only the diagonal elements of F are nonzero. So F is diagonal and f is diagonalizable. This proves the Theorem.   □

Theorem 10.2 *The real spectral theorems* ($\mathbb{K} = \mathbb{R}$).

Let V be an n-dimensional real vector space and $f \in \mathrm{End}(V)$. Then f is self-adjoint if and only if f has an orthonormal basis of eigenvectors of f.

Proof By Schur's theorem, there is a basis $B = (b_1, \ldots, B_n)$ such that f_B is upper triangular. There is

$$f(b_i) = \sum_{j=1}^{i} \varphi_i^j b_j \quad \text{for some} \quad \varphi_i^j \in \mathbb{R}.$$

Through the Gram-Schmidt algorithm, one can construct an orthogonal basis $C = (c_1, \ldots, c_n)$ such that for each $i \in I(n)$,

$$b_i \in \mathrm{span}\{c_1, \ldots, c_i\}$$

and

$$c_i \in \mathrm{span}\{b_1, \ldots, b_i\},$$

so that the change of basis matrix $T = T_{BC}$ is triangular. Hence $f_C = T f_B T^{-1}$ is upper triangular. But since f_C is an orthogonal representation of f, we see $(f_C)^\dagger = f_C$. Thus, f_C must be diagonal, and C is an orthonormal eigenbasis for f.   □

Remark 10.1 Since a self-adjoint operator is a normal operator, the complex, self-adjoint case was already proven in the previous theorem. We give here a second proof for the case $\mathbb{K} = \mathbb{C}$.

Corollary 10.3 *Real and complex spectral theorems.*
If the characteristic polynomial of $f \in \text{End}(V)$ decomposes into linear factors of $\mathbb{K} \in \{\mathbb{R}, \mathbb{C}\}$, then f is normal if and only if f has an orthonormal basis of eigenvectors of f.

Proof The proof goes through as in the complex spectral theorem since Schur's theorem can also be applied here!   □

There exists a more direct formulation of the spectral theorem if we combine it with Theorem 9.1. We thus obtain a spectral decomposition of every normal operator f parametrized by the set of its eigenvalues:

Theorem 10.3 *Spectral decomposition theorem.*

Let f be a normal operator on an inner product space V with the distinct eigenvalues $\lambda_1, \ldots, \lambda_r$ and a characteristic polynomial which decomposes into linear factors. If V_j is the eigenspace corresponding to the eigenvalue λ_j, $j \in I(r)$ and P_j is the orthogonal projection of V on V_j, then the following statements hold:

(i) $V = V_1 \ominus V_2 \ominus \cdots \ominus V_r$.
(ii) $P_i P_j = \delta_{ij} P_i$ *for* $i, s \in I(r)$.
(iii) $id_V = P_1 + P_2 + \cdots + P_r$.
(iv) $f = \lambda_1 P_1 + \lambda_2 P_2 + \cdots + \lambda_r P_r$.

Proof This theorem is simply Theorem 9.1 with the additional information that the V_i are orthogonal. The orthogonality part follows the spectral theorem of $\mathbb{K}$.   □

Comment 10.5 Normal operators revisited.

Now that we clarified the structure of all normal operators, we would like to discuss their content. Returning to the relation $f^{ad} \circ f = f \circ f^{ad}$, we observe the two "trivial" realizations as discussed in Sect. 10.5: $f^{ad} \circ f = Id_V = f \circ f^{ad}$ and $f^{ad} = f$. It turns out that the first relation corresponds to isometries or equivalently to unitary or orthogonal operators and the second relation to self-adjoint operators. At the matrix level, one writes $A^\dagger A = \mathbb{1}_n = A A^\dagger$ which corresponds to the unitary or orthogonal matrices $A \in U(n)$ or $A \in O(n)$ if $A \in \mathbb{C}^{n \times n}$ or $A \in \mathbb{R}^{n \times n}$ respectively. Similarly, the condition $A^\dagger A = A A^\dagger$ follows trivially when $A^\dagger = A$ which means that if A is complex, then A belongs to the Hermitian matrices $H(n)$, or if A is real, then A belongs to $S(n)$, the symmetric matrices. It is interesting to notice that $U(n)$ is a group, we may consider $O(n)$ as a subgroup of $U(n)$ and we have $O(n) \leq U(n) \leq Gl(n, \mathbb{C})$, and that $H(n)$ is a vector space and we have

$$S(n) \leq H(n) < \mathbb{C}^{n,n}.$$

As usual, the notation "$\leq$", "$<$" indicates a subspace with a similar structure.

It turns out that, as we shall see in Chap. 11, isometries and self-adjoint operators are indeed the essential part of the normal operators.

Summary

In physics, when referring to a vector space, it almost always implies an inner product vector space. In this chapter, we covered everything related to an inner product space. This mainly includes concepts associated with orthogonality. Previously known results were reiterated, summarized, and supplemented.

The normal operators, those endomorphisms relating to the inner product, were extensively motivated and discussed here with emphasis on their analogies to complex and real numbers. It was noted that precisely these operators are the most commonly used in physics.

The spectral theorem applies to normal operators. Here, it was shown for complex vector spaces that normal operators, such as isometries, self-adjoint operators, and nonnegative operators, possess the best properties regarding their eigenvalues and eigenvectors.

It was demonstrated that normal operators are the only ones that have orthogonal and orthonormal eigenbases. Moreover, self-adjoint operators even have real eigenvalues. This allows them to act as observables in quantum mechanics.

Finally, it was pointed out that most operators describing symmetries in physics are elements of unitary or orthogonal groups, and they also belong to the set of normal operators.

Exercises with Hints

In the first five exercises, you learn how to express a covector with the help of a corresponding vector as scalar product. Furthermore, you learn to distinguish a basis dependent isomorphism from a basis free one (canonical isomorphism) with the example of a vector space V and its dual V^*.

Exercise 10.1 *Riesz representation theorem.* Let V be an inner product vector space and ξ a covector (linear function, linear form, $\xi \in V^* := \mathrm{Hom}(V, \mathbb{K})$). Show that there exists a unique vector u such that we can express $\xi(v)$ with all $v \in V$ as a scalar product:

$$\xi(v) = \langle v | u \rangle.$$

Exercise 10.2 Let $(a_1, \ldots, a_r)$ be a linearly independent list of vectors in V not necessarily orthonormal. Show that there exists a vector $u \in V$ such that $\langle a_s | u \rangle$ is positive for all $s \in I(r)$.

Exercise 10.3 *An isomorphism* $V \underset{B}{\cong} V^*$.

Let V be a vector space with $B = (b_1, \ldots, b_n)$ a basis of V and $V^* = \mathrm{Hom}(V, \mathbb{R})$ its dual with basis $B^* = (\beta^1, \ldots, \beta^n)$, such that $\beta^i(b_s) = \delta_s^i$, $i, s \in I(n)$. Show that there exists a map $\psi_B : V \to V^*$, that is, an isomorphism between V and V^*.

Exercise 10.4 Let V be an inner product vector space, $f \in \mathrm{Hom}(V, V)$, and $u, w \in V$, such that

$$f(v) = w \langle u | v \rangle \quad \text{for all} \quad v \in V.$$

Show that f is normal if and only if u and w are linearly dependent ($w = \lambda u$ for some $\lambda \in \mathbb{K}$).

Exercise 10.5 *Canonical isomorphism* $V \underset{can}{\cong} V^*$.

Let (V, s) be an n-dimensional Euclidean vector space with $s \equiv \langle | \rangle$ a symmetric positive definite bilinear form. Show that the map

$$\hat{s} : \quad V \longrightarrow V^*$$
$$u \longmapsto \hat{s}(u) := s(u, \cdot) \equiv \langle u | \cdot \rangle \equiv \hat{u} \in V^*,$$

is a canonical isomorphism between V and V^*.

Exercise 10.6 *Transitive action* (*see Definition* 1.6) of reflections on spheres.
Given two distinct vectors u and w with $\|u\| = \|w\|$, show that there exists a third vector $a \in V$ and a map

$$s_a(v) = v - 2\frac{\langle a | v \rangle}{\langle a | a \rangle} a \quad \text{for all} \quad v \in V,$$

such that $s_a(u) = w$ and $s_a(w) = u$.

Exercise 10.7 Use the Cauchy-Schwarz inequality to show that

$$(\alpha_i \beta^i)^2 \leq (i \, \alpha_i \, \alpha^i)(k \, \beta_k \, \beta^k),$$

with $\alpha_i = \alpha^i$, $\beta_i = \beta^i \in \mathbb{R}$, and $i, k \in I(n)$.

Exercise 10.8 Let V be a vector space and $u, v \in V$ with $\|u\| = \|v\|$. Show that $\|\alpha u + \beta v\| = \|\beta u + \alpha v\|$ for all $\alpha, \beta \in \mathbb{R}$.

Exercise 10.9 Let V_1 and V_2 be inner product vector spaces and $(V_1 \times V_2, s \equiv \langle \cdot, \cdot | \cdot, \cdot \rangle)$ given by

$$\langle u_1, u_2 | v_1, v_2 \rangle := \langle u_1 | v_1 \rangle + \langle u_2 | v_2 \rangle.$$

Show that s is an inner product in $V_1 \times V_2$.

> The next two exercises refer to Proposition 10.7 and to Comment 10.2.

Exercise 10.10 Let U be a subspace of an inner product space V and P_U an orthogonal projection. Show the following assertions:

(i) P_U is a linear operator;
(ii) $P_U|_U = id_U$;
(iii) $P_U|_{U^\perp} = \hat{0}$, the zero operator;
(iv) $\ker P_U = U^\perp$;
(v) $\operatorname{im} P_U = U$.

Exercise 10.11 Let U be a subspace of an inner product space V and P_U an orthogonal projection. Show the following assertions:

(i) $P_U^2 = P_U$;
(ii) For all $v \in V$, $v - P_U(v) \in U^\perp$;
(iii) $id_V - P_U = P_{U^\perp}$.

Exercise 10.12 Let V be a vector space and P an operator with $P^2 = P$ such that $\ker P$ is orthogonal to $\operatorname{im} P$. Check that $V = \ker P \oplus \operatorname{im} P$ and show that there exists a subspace U such that $P = P_U$.

Exercise 10.13 Let U be a subspace of a vector space V and $f \in \operatorname{Hom}(V, V)$. Show that if

$$P_U \circ f \circ P_U = f \circ P_U,$$

then U is an f-invariant subspace of f.

Exercise 10.14 Let U be a subspace of V with basis $A = (a_1, \ldots, a_r)$ and $B = (a_1, \ldots, a_r, b_1, \ldots, b_l)$ a basis of V. The Gram-Schmidt procedure applied to B produces an orthonormal basis $E = (c_1, \ldots, c_r, d_1, \ldots, d_l) = (C, D)$. Show that C and D are orthonormal bases of U and $U^\perp$ respectively.

Exercise 10.15 Let

$$f \in \text{Hom}(V, V')$$

be given by

$$f(v) = w\langle u|v\rangle,$$

with $w \in V'$ and $u, v \in V$. Determine the adjoint operator f^{ad}. Write the result in the Dirac formalism.

Exercise 10.16 Let

$$F \in \text{Hom}(\mathbb{K}^n, \mathbb{K}^n)$$

be given by

$$F[e_1 \cdots e_n] = [0\, e_1 \cdots e_{n-1}].$$

Determine $F^{ad} \in \text{Hom}(\mathbb{K}^n, \mathbb{K}^n)$.

Exercise 10.17 Let $f \in \text{Hom}(V, V)$. Show that $\lambda \in \mathbb{K}$ is an eigenvalue of f if and only if $\bar\lambda$ is an eigenvalue of f^{ad}.

Exercise 10.18 Let $f \in \text{Hom}(V, V')$. Show that the following assertions hold:

(i) $\dim \ker f^{ad} - \dim \ker f = \dim V' - \dim V$;
(ii) $\text{rank } f^{ad} = \text{rank } f$.

Exercise 10.19 Let V be a complex inner product vector space. Show that the set of self-adjoint operators is not a complex vector space.

Exercise 10.20 Let $f, g \in \text{Hom}(V, V)$ be self-adjoint operators. Show that $g \circ f$ is self-adjoint if and only if $f \circ g = g \circ f$.

Exercise 10.21 Let V be an inner product vector space and P an operator with $P^2 = P$. Show that P is an orthogonal projection if and only if $P^{ad} = P$.

Exercise 10.22 Let $f \in \text{Hom}(V, V)$ be a normal operator. Show that

$$\ker f^{ad} = \ker f \quad \text{and}$$
$$\text{im } f^{ad} = \text{im } f.$$

Exercise 10.23 Let $f \in \text{Hom}(V, V)$ be a normal operator. Show that $\ker f^m = \ker f$ for every $m \in \mathbb{N}$.

Exercise 10.24 Let V be a real inner product vector space and $f \in \text{Hom}(V, V)$. Show that f is self-adjoint if and only if all pairs of eigenvectors corresponding to distinct eigenvalues $\lambda_1, \ldots, \lambda_r$ of f are orthogonal and

$$V = E(\lambda_1, f) \oplus \cdots \oplus E(\lambda_r, f).$$

References and Further Reading

1. S. Axler, *Linear Algebra Done Right* (Springer Nature, 2024)
2. G. Fischer, B. Springborn, *Lineare Algebra. Eine Einführung für Studienanfänger*. Grundkurs Mathematik (Springer, 2020)
3. S.H. Friedberg, A.J. Insel, L.E. Spence, *Linear Algebra* (Pearson, 2013)
4. K. Jänich, *Mathematik 1. Geschrieben für Physiker* (Springer, 2006)
5. N. Johnston, *Advanced Linear and Matrix Algebra* (Springer, 2021)
6. M. Koecher, *Lineare Algebra und analytische Geometrie* (Springer, 2013)
7. G. Landi, A. Zampini, *Linear Algebra and Analytic Geometry for Physical Sciences* (Springer, 2018)
8. J. Liesen, V. Mehrmann, *Linear Algebra* (Springer, 2015)
9. P. Petersen, *Linear Algebra* (Springer, 2012)
10. S. Roman, *Advanced Linear Algebra* (Springer, 2005)
11. B. Said-Houari, *Linear Algebra* (Birkhäuser, 2017)
12. A.J. Schramm, *Mathematical Methods and Physical Insights: An Integrated Approach* (Cambridge University Press, 2022)
13. G. Strang, *Introduction to Linear Algebra* (SIAM, 2022)

Chapter 11
Positive Operators–Isometries–Real Inner Product Spaces

In this chapter, we proceed with special normal operators such as the nonnegative operators. The analogical comparison of normal operators with the complex numbers continues in that we now introduce nonnegative operators which correspond to nonnegative real numbers.

Subsequently, we discuss isometries. These are closely related to symmetries in physics, in particular to symmetries in quantum mechanics.

We then use properties of operators in complex vector spaces to derive properties of operators in real vector spaces. An instrument for this method is complexification which we explain in detail. In this way, we obtain the spectral theorem for real normal operators which have not been accessible so far.

11.1 Positive and Nonnegative Operators

As discussed in Sect. 10.5, the normal operators are precisely those operators that have a remarkable analogy to the complex numbers. This analogy goes one step further and extends to positive and nonnegative numbers. This analogy also allows us to speak of the root of a nonnegative operator. It is useful to remember first the situation with the complex numbers. A complex number z is positive if z is real and positive. This is equivalent to the existence of some $w \neq 0$ such that $z = \bar{w}w$, or to having a positive square root $\sqrt{z} \equiv \underset{+}{\sqrt{z}}$ (positive). Likewise, z is nonnegative if z is positive or zero. Coming now to the operators, the expected analogy with the complex numbers is the following: The self-adjoint operators correspond to real numbers; the positive operators which are always self-adjoint, correspond to positive numbers, and the nonnegative operators correspond, of course, to nonnegative numbers. This leads to this definition:

© The Author(s), under exclusive license to Springer Nature Switzerland AG 2024
N. A. Papadopoulos and F. Scheck, *Linear Algebra for Physics*,
https://doi.org/10.1007/978-3-031-64908-0_11

> **Definition 11.1** Positive and nonnegative operators.
> A self-adjoint operator f on an inner product space V is positive if $\langle v|fv\rangle > 0$
> and is nonnegative if $\langle v|fr\rangle \geqslant 0$ for all $v \in V\backslash\{0\}$.

Warning: Confusingly, some authors use the term "positive" to mean "nonnegative".

> **Definition 11.2** Square root.
> An operator g is a square root of an other operator f if $g^2 = f$.

Proposition 11.1 below gives some properties of nonnegative operators.

> **Proposition 11.1** *Nonnegative operator.*
> *The following statements are equivalent:*
>
> *(i)*   *An operator f is nonnegative.*
> *(ii)* *The eigenvalues of f are nonnegative.*
> *(iii)* *f has a nonnegative root g.*
> *(iv)* *There exists an operator g so that $f = g^{ad}g$.*

Proof (i) $\Rightarrow$ (ii)
Suppose (i) holds. Then f is self-adjoint and $\langle v|fv\rangle \geqslant 0$ for all $v \in V$. By the spectral theorem, there is an orthonormal eigenbasis $(c_1, \ldots, c_n)$ of f with eigenvalues $(\lambda_1, \ldots, \lambda_n)$. Then for each $j \in I(n)$, $\langle c_j|fc_j\rangle = \langle c_j|\lambda_j c_j\rangle = \lambda\langle c_j|c_j\rangle = \lambda_j \geqslant 0$. This proves (ii).

Proof (ii) $\Rightarrow$ (iii)
Suppose (ii) holds. We can define a linear operator g on the basis $C = (c_1, \ldots, c_n)$ so that $gc_j = \sqrt{\lambda_j}c_j$. Then $\sqrt{\lambda_j} \geqslant 0$ and $g^2c_j = \sqrt{\lambda_j}\sqrt{\lambda_j}c_j = \lambda_j c_j$ for all $j \in I(n)$ so that $g^2 = f$. This g is a nonnegative square root of f.

Proof (iii) $\Rightarrow$ (iv)
Suppose (iii) holds. Then it follows that (iv) is also valid since g is self-adjoint and $f = gg = g^{ad}g$ (and of course $f^{ad} = f$).

Proof (iv) $\Rightarrow$ (i)
If $f = g^{ad}g$ holds, then $\langle v|fv\rangle = \langle v|g^{ad}gv\rangle = \langle gv|gv\rangle \geqslant 0$ by the definition of $\langle |\rangle$. So f is nonnegative. This completes the proof of the Proposition. $\square$

In the proof of (iii), we have $\lambda_j \geqslant 0$ and we chose for g: $\sqrt{\lambda_j} \equiv \underset{+}{\sqrt{\lambda_j}} \geqslant 0$. This means that the nonnegative operator g is uniquely defined. This is in perfect analogy to complex numbers: $z \neq 0$ has a unique positive square root $\underset{+}{\sqrt{z}} > 0$.

11.2 Isometries

We repeat the geometric definition of an isometry valid for every $\mathbb{K}$-vector space with an inner product.

Definition 11.3 Isometry.
An operator f on an inner product space $(V, \langle | \rangle)$ is an isometry if it preserves the norm

$$\|fv\| = \|v\| \text{ for all } v \in V.$$

In addition, if $\mathbb{K} = \mathbb{C}$, an isometry f is also called a unitary operator f and if $\mathbb{K} = \mathbb{R}$, an isometry f is called an orthogonal operator f.

The term "orthogonal" is a tradition and actually means "orthonormal"!

The following proposition gives a series of equivalent descriptions of what an isometry is.

Proposition 11.2 *Equivalent formulations of an isometry.*
If f is an operator on V, the following statements are equivalent:

(i) *f is an isometry ($\|fv\| = \|v\|$).*
(ii) *$\langle fv | fv \rangle = \langle u | v \rangle$ for all $u, v \in V$.*
(iii) *For every orthonormal list $(c_1, \ldots, c_k)$ of vectors in V, $(fc_1, \ldots, fc_k)$ is also orthonormal.*
(iv) *There exists an orthonormal basis $(c_1, \ldots, c_n)$ of V so that $(fc_1, \ldots, fc_n)$ is orthonormal.*
(v) *$f^{ad} \circ f = Id_V = f \circ f^{ad}$.*
(vi) *f^{ad} is an isometry.*
(vii) *f is invertible and $f^{-1} = f^{ad}$.*

This proposition specifies the following. (ii) shows that an isometry preserves the inner product. It is compatible with the inner product and is therefore exactly what we expect of an operator associated with the inner product.

(iii) and (iv) show that an isometry transforms an orthonormal basis into an orthonormal basis.

(v), (vi), and (vii) show not only that an isometry is invertible, but also that we can express this inverse simply by the adjoint ($f^{-1} = f^{ad}$). This leads to the important fact that the isometries form a group.

Proof (i) $\Rightarrow$ (ii) We prove (ii) expressing the inner product by the norm, as given by the polarization identity in Sect. 10.2.

Proof (ii) $\Rightarrow$ (iii) For a given orthonormal list $(c_1, \ldots, c_k)$ of the vectors $\langle c_i | c_j \rangle = \delta_{ij}$ $i, j \in I(k)$, the preservation of the inner product gives

$$\langle fc_i | fc_j \rangle = \langle c_i | c_j \rangle = \delta_{ij} \ \forall \, i, j \in I(k).$$

Proof (iii) $\Rightarrow$ (iv): If $(c_1, \ldots, c_n)$ is an orthonormal basis in V, we have as above for $k = n$

$$\langle fc_i | fc_j \rangle = \langle c_i | c_j \rangle = \delta_{ij} \quad i, j \in I(n).$$

Proof (iv) $\Rightarrow$ (v) Consider

$$\langle v | f^{ad} \circ f v \rangle = \langle fv | fv \rangle = \langle v | v \rangle = \langle v | id_V v \rangle$$

for all $v \in V$ and so we have $\langle v | (f^{ad} f - id_V) v \rangle = 0$. As $f^{ad} f$ is self-adjoint and Proposition 10.11, valid for complex and real vector spaces, we get $f^{ad} \circ f - id_V = 0 \Leftrightarrow f^{ad} \circ f = id_V$. It follows that $f^{ad} \circ f$ is invertible and hence, so are f^{ad} and f which means

$$f^{ad} \circ f = f \circ f^{ad} = id_V.$$

Proof (v) $\Rightarrow$ (vi)

$$\|f^{ad} v\|^2 = \langle f^{ad} v | f^{ad} v \rangle = \langle v | f \circ f^{ad} v \rangle = \langle v | id_V v \rangle =$$
$$= \langle v | v \rangle = \|v\|^2.$$

So f^{ad} is also an isometry.

Proof (vi) $\Rightarrow$ (vii) If f^{ad} is an isometry, then

$$(f^{ad})^{ad} = f \quad \text{and} \quad f^{ad} \circ f = id_V$$

since

$$(i) \text{ implies } (v),$$

so

$$f^{ad} \circ f \circ f^{-1} = f^{-1} \Rightarrow f^{ad} = f^{-1}.$$

Proof (vii) $\Rightarrow$ (i) If $f^{-1} = f^{ad}$, then

$$id_V = f^{ad} \circ f,$$

with

$$\langle v | f^{ad} fv \rangle = \langle fv | fv \rangle$$

gives

$$\|fv\|^2 = \langle fv | fv \rangle = \langle v | f^{ad} \circ fv \rangle = \langle v | id_V v \rangle = \langle v | v \rangle = \|v\|^2.$$

The sequence of proofs (i) $\Rightarrow$ (ii) $\Rightarrow \cdots \Rightarrow$ (vii) $\Rightarrow$ (i) is complete and so is the proof of the Proposition. $\square$

The structure of isometries was given essentially by the spectral theorems in Sect. 10.6 since isometries are a subset of normal operators. However, here we have to distinguish between $\mathbb{C}$- and $\mathbb{R}$-vector spaces. This is also clear with the following two corollaries.

Corollary 11.1 *Complex spectral theorems for isometries.*
An operator f is an isometry on a complex vector space V if and only if V has an orthonormal basis of eigenvectors of f and all eigenvalues have the absolute value 1 (i.e., $|\lambda| = 1$).

Proof We have only to prove

$$|\lambda_i| = 1 \text{ for every } i \in I(n).$$

For an orthonormal eigenbasis $(c_1, \ldots, c_n)$ of f, we have $fc_i = \lambda_i c_i$ and $\|fc_i\| = \|c_i\| = 1$ so that

$$1 = \|c_i\| = \|fc_i\| = \|\lambda_i c_i\| = |\lambda_i| \|c_i\| = |\lambda_i|.$$

Conversely, if f has such an eigenbasis, then for each eigenvector c_i,

$$\|fc_i\| = \|\lambda_i c_i\| = |\lambda_i| \, \|c_i\| = \|c_i\|$$

so that by linearity and Pythagoras,

$$\|fv\| = \|v\| \ \forall \, v \in U,$$

so f is an isometry. $\square$

Corollary 11.2 *Real spectral theorem for isometries.*
Let f be an operator on a real inner product vector space V, then f has an orthonormal eigenbasis with eigenvalues of absolute value 1 if and only if it is both, orthogonal (isometry) and self-adjoint.

Proof $\Rightarrow$
If V has an orthonormal basis $(c_1, \ldots, c_n)$ of eigenvectors $fc_j = \lambda c_j$ and

$|\lambda_j| = 1$ $j \in I(n)$, then f is self-adjoint ($f^{ad} = f$) by the real spectral theorem (Theorem 10.2) since the eigenvalues are real ($\lambda_j \in \mathbb{R}$). It follows for every $j \in I(n)$:

$$(f^{ad} \circ f)(c_j) = (f \circ f)(c_j) = f(fc_j) = f(\lambda_j c_j) = \lambda_j^2 c_j = 1 c_j.$$

So we have $f^{ad} \circ f = Id_V$ which means that f is orthogonal. So f is both orthogonal and self-adjoint. This was one direction of the proof.   □

Proof $\Leftarrow$
If f is self-adjoint and orthogonal at the same time, then, by the real spectral theorem, an orthonormal eigenbasis $(c_1, \ldots, c_n)$ exists with $f c_i = \lambda_i c_i$. By the fact that f is also orthogonal, we have $\|f c_i\| = \|c_i\|$ and so

$$1 = \|c_i\| = \|f c_i\| = \|\lambda_i c_i\| = |\lambda_i| \|c_i\| = |\lambda_i| 1 = |\lambda_i|.$$

This is the other direction of the proof.   □

This corollary shows that an orthogonal operator must be self-adjoint to be diagonalizable. This means that, in general, an orthogonal operator will be nondiagonalizable (see Examples 9.10 and 9.17). Hence, we have to discuss separately normal operators and particularly isometries in a real inner product space.

11.3   Operators in Real Vector Spaces

It is easier to deal with operators in complex vector spaces than in real vector spaces. As we already saw, over real vector spaces, only in the very special case where an operator whose characteristic polynomial splits into linear factors, is it possible to proceed analogously to the case of complex vector spaces. Therefore, we generally expect the structure of normal operators in a real vector space to be quite different from the corresponding case in a complex vector space. In contrast to the complex spectral theorem, a rotation on a two-dimensional real vector space is generally not diagonalizable. However, as we know, when going from $\mathbb{R}$ to $\mathbb{C}$ and even from $\mathbb{R}^n$ to $\mathbb{C}^n$, there are connections between real and complex vector spaces. These connections also affect the behavior of the corresponding operator. Therefore, it is reasonable to examine how the results in the complex case affect the operators in the real case.

The instrument for such a procedure is complexification. As $\mathbb{R}$ is embedded naturally in $\mathbb{C}$ and $\mathbb{R}^n$ in $\mathbb{C}^n$, a real vector space U can be embedded naturally in a complex vector space U_c, the complexification of U.

The best way to get a feeling for this rather abstract situation with operators, is to consider what happens with matrices. How do we obtain information about real matrices from what we know about the structure of complex matrices? And yet, this is not the first time we are faced with complexification!

The introduction of complex numbers was already a case of complexification, though on the lowest possible level. It is instructive to recall the steps to construct

$\mathbb{C} = \mathbb{R}_c$ from $\mathbb{R}$.

$$\mathbb{R}_c \cong \mathbb{C} \cong \mathbb{R} \times \mathbb{R} = \{z = (\zeta_0, \zeta_1) : \zeta_0, \zeta_1 \in \mathbb{R}\}.$$

What is completely new about $\mathbb{R}_c \cong \mathbb{C}$ with respect to $\mathbb{R} \times \mathbb{R}$, is that we have an additional notion of complex multiplication. Complex multiplication on $\mathbb{R}^2$ makes it a field like $\mathbb{R}$. This gives us an action of $\mathbb{C}$ on the vector space $\mathbb{R} \times \mathbb{R}$. Equivalently, starting with the following procedure, taking

$$z := (\zeta_0, \zeta_1) = \zeta_0 + i\zeta_1, \quad x := (\xi_0, \xi_1) = \xi_0 + i\xi_1,$$

with

$$\zeta_0, \zeta_1, \xi_0, \xi_1 \in \mathbb{R} \quad \text{and} \quad i := (0, 1),$$

we obtain:

$$\mathbb{C} \times \mathbb{R}_c \longrightarrow \mathbb{R}_c$$
$$(z, x) \longmapsto zx := (\zeta_0 \xi_0 - \zeta_1 \xi_1, \zeta_0 \xi_1 + \zeta_1 \xi_0).$$

The above definition implies that

$$ii = (0, 1)(0, 1) = (-1, 0)$$

which justifies the identification

$$\mathbb{C} = \mathbb{R}_c = \mathbb{R} \times \mathbb{R} = \mathbb{R} + i\mathbb{R}.$$

We are ready to generalize the above formalism to a real vector space U. We hope that the attentive reader will also accept our choice of the index $(0, 1)$ instead of $(1, 2)$, we write (ζ_0, ζ_1) instead of (ζ_1, ζ_2)!

Definition 11.4 Complexification of U.
The complexification U_c of U is given by $U_c := U \times U = U + iU$ and the complex scalar multiplication by

$$\mathbb{C} \times U_c \longrightarrow U_c,$$
$$(\zeta_0 + i\zeta_1, u_0 + iu_1) \longmapsto (\zeta_0 + i\zeta_1)(u_0 + iu_1) := \zeta_0 u_0 - \zeta_1 u_1 + i(\zeta_0 u_1 + \zeta_1 u_0),$$
$$\zeta_0, \zeta_1 \in \mathbb{R} \text{ and } u_1, u_2 \in U.$$

It is important to distinguish between the three vector spaces U, $U \times U$, and U_c. As U and $U \times U$ are real vector spaces with real dimensions dim $U = n$, dim $U \times U = 2n$. The vector space U_c, however, is a complex vector space with the definition given above. Still, nobody can hinder us from also considering U_c as a real vector space given by $U_c = U \times U$ with dimension $\dim_{\mathbb{R}} U_c = 2n$ of course! But what is the dimension of U_c as a complex vector space?

For the dimension of U_c (where we denote dim $U_c \equiv \dim_{\mathbb{C}} U_c$), a basis of U plays again a central role:

Proposition 11.3 *Basis of U_c.*
If U is a real vector space (dim $U = n$) with a basis $B = (b_1, \ldots, b_n)$, then this same B is a basis of the complex vector space U_c.

Proof We have only to test that the list B is linearly independent and spanning. Setting $b_s \lambda^s = 0$, with $\lambda^s := \lambda_0^s + i\lambda_1^s$, $\lambda_0^s, \lambda_1^s \in \mathbb{R}$, $s \in I(n)$, we obtain $b_s(\lambda_0^s + i\lambda_1^s) = b_s\lambda_0^s + ib_s\lambda_1^s = 0$. This means that $b_s\lambda_0^s = 0$ and $b_s\lambda_1^s = 0$.

From the linear independence of B in U, we obtain $\lambda_0^s = 0$, $\lambda_1^s = 0$ and so $\lambda^s = 0$ for all $s \in I(n)$, and B is linearly independent in U_c.

B is also spanning in U_c: If $v \in U_c$, $v = (v_0, v_1) \in U \times U$, we have $v_0 = b_s\lambda_0^s$ and $v_1 = b_s\lambda_1^s$ since B is a basis in U. So we obtain

$$v = (v_0, v_1) = (b_s\lambda_0^s, b_s\lambda_1^s) = b_s(\lambda_0^s, \lambda_1^s) = b_s(\lambda_0^s + i\lambda_1^s) = b_s\lambda^s$$

with $\lambda^s \in \mathbb{C}$. So B is indeed spanning in U_c and B is a basis in U_c. $\square$

Corollary 11.3 *The dimension of U_c is* dim $U_c \equiv \dim_{\mathbb{C}} U_c = n = \dim_{\mathbb{R}} U \equiv$ dim U.

$\square$

The extension of U to $U_c = U + iU$ leads as expected to the question of whether one can extend maps $f : U \to V$ on real vector spaces to maps $f_c : U_c \to V_c$ on their complexifications. Indeed, one can, with the following definition.

Definition 11.5 Complexification of f.
If U is a real vector space, and $f : U \to V$ a real linear map, the complexification f_c of f, $f_c \in \text{Hom}(U_c, V_c)$ is given with $u_0, u_1 \in U$ by

$$f_c(u_0 + iu_1) := fu_0 + ifu_1.$$

This means equivalently that we have

$$f_c \equiv f \times f : \qquad U \times U \longrightarrow V \times V$$
$$(u_0, u_1) \longmapsto (fu_0, fu_1).$$

Remark 11.1 f_c is $\mathbb{C}$-linear.
Indeed, f_c is a $\mathbb{C}$-linear operator, as $v = u_0 + iu_1$ and $\lambda = \lambda_0 + i\lambda_1$, $u_0, u_1 \in U$, $\lambda_0, \lambda_1 \in \mathbb{R}$, we obtain

$$
\begin{aligned}
f_c(\lambda v) &= f_c((\lambda_0 + i\lambda_1)(u_0 + iu_1)) = f_c(\lambda_0 u_0 - \lambda_1 u_1 + i(\lambda_0 u_1 + \lambda_1 u_0)) \\
&= f(\lambda_0 u_0 - \lambda_1 u_1) + if(\lambda_0 u_1 + \lambda_1 u_0) \\
&= \lambda_0 f u_0 - \lambda_1 f u_1 + i(\lambda_0 f u_1 + \lambda_1 f u_0) \\
&= (\lambda_0 + i\lambda_1)(f u_0 + if u_1) \\
&= \lambda f(v).
\end{aligned}
$$

$\square$

We can of course apply the notion of complexification immediately to matrices $\mathbb{R}^{n \times n} \subseteq \mathbb{C}^{n \times n}$ because a matrix $A \in \mathbb{R}^{n \times n}$ can be understood as an operator, $A \in \text{End}(\mathbb{R}^n)$. So for a real matrix A

$$
\begin{aligned}
A : \qquad & \mathbb{R}^n \to \mathbb{R}^n, \\
& \vec{u} \mapsto A\vec{u}. \\
A_c : \qquad & \mathbb{C}^n \longrightarrow \mathbb{C}^n \\
& \vec{u}_0 + i\vec{u}_1 \longmapsto A_c(\vec{u}_0 + i\vec{u}_1) := A\vec{u}_0 + iA\vec{u}_1.
\end{aligned}
$$

This is obviously what we would do intuitively: Science is nothing but intuition that became rational.

In this context, the question of how to represent f_c can easily be answered: If we have $f_{BB} = F \in \mathbb{R}^{n \times n}$, then $(f_c)_{BB} = F$ again. This follows from the fact that, as the above proposition shows, any basis of U is also a basis of U_c. With the above notation we have:

$$f b_s = b_t \varphi_s^t \quad \varphi_s^t \in \mathbb{R} \quad s, t \in I(n) \quad \text{and}$$
$$f_c b_s = b_t \varphi_s^t.$$

This is consistent with $F = (\varphi_s^t) \in \mathbb{R}^{n \times n}$ and $F_c = (\varphi_s^t) \in \mathbb{R}^{n \times n} \subseteq \mathbb{C}^{n \times n}$.

The last but not trivial question is what happens with the eigenvalues and eigenvectors of f and f_c. This leads us back to our original discussion of what can be learned from the complexification f_c of f for the f itself. The best way to understand this problem is to consider a simple example at the level of matrices in low dimensions.

We start with a real matrix in two dimensions, a rotation about the angle $\frac{\pi}{2}$ which we know to be normal but nondiagonalizable:

$$A = \begin{bmatrix} 0 & -1 \\ 1 & 0 \end{bmatrix} \quad \text{and} \quad A_c = \begin{bmatrix} 0 & -1 \\ 1 & 0 \end{bmatrix}.$$

The characteristic polynomial is given by

$$\chi_A(x) = \det \begin{bmatrix} 0 & -1 \\ 1 & 0 \end{bmatrix} = x^2 + 1$$

and it has no real solution. This means that the operator

$$A : \mathbb{R}^2 \longrightarrow \mathbb{R}^2$$
$$\vec{u} \longmapsto A\vec{u}$$

has no eigenvalues and the spectrum (i.e., the set of eigenvalues) of A is empty, given by $\sigma(A) = \emptyset$. The complexification operator is given by

$$A_c : \mathbb{C}^2 \longrightarrow \mathbb{C}^2$$
$$\vec{v} \longmapsto A\vec{v} := A\vec{v}$$

with $\vec{v} = \vec{u}_0 + i\vec{u}_1 \equiv (\vec{u}_0, \vec{u}_1), \vec{u}_0, \vec{u}_1 \in \mathbb{R}^2$. The eigenvalues of A_c are the solutions of

$$\chi_A(x) = x^2 + 1 = 0, \quad \lambda_1 = i \quad \text{and} \quad \lambda_2 = -i.$$

So the spectrum of A_c is $\sigma(A_c) = \{i, -i\}$. For the corresponding eigenvectors, we may choose the two orthogonal vectors $b_1 = \begin{bmatrix} 1 \\ -i \end{bmatrix}$ and $b_2 = \begin{bmatrix} 1 \\ i \end{bmatrix}$. We therefore have an eigenbasis given by $B = (b_1, b_2)$ with $< b_1 | b_2 > = 0$.

After this short excursion, we want to find out what we can generally learn about an operator on a real vector space from an operator on a complex vector space. As we already saw, every operator in a complex vector space can be triangularized. This is possible since every operator on a complex vector space has at least one eigenvalue.

This is not the case for every operator on a real vector space. But the existence of an eigenvalue in the complex setting guarantees at least the presence of an invariant, one or two-dimensional invariant subspace corresponding to the subspace of the real operator. This is the content of the following proposition.

Proposition 11.4   *f-invariant spaces in real vector spaces.*
Every operator in a real vector space has an invariant subspace of dimension 1 or 2.

Proof If V is a real vector space, $f \in \mathrm{End}(V)$ and $f_c \in \mathrm{End}(V_c)$, then there exists an eigenvalue $\lambda \in \mathbb{C}$ of f_c:

$$f_c v = \lambda v.$$

We can write $\lambda = \lambda_0 + i\lambda_1$, $\lambda_0, \lambda_1 \in \mathbb{R}$ and $v = u_0 + iu_1$, $u_0, u_1 \in V$, so we have $f_c v = fu_0 + ifu_1$ and the eigenvalue equation $fu_0 + ifu_1 = (\lambda_0 + i\lambda_1)(u_0 + iu_1)$. This leads to

$$f u_0 = \lambda_0 u_0 - \lambda_1 u_1 \quad \text{and} \quad f u_1 = \lambda_0 u_1 + \lambda_1 u_0.$$

We may define $U := \mathrm{span}_{\mathbb{R}}(u_0, u_1)$ and, as seen above, U is an f-invariant subspace of V with dimension 1 or 2. $\qquad\square$

11.4   Normal Operators on Real Vector Spaces

We can now give a complete description of normal operators on real vector spaces. The prerequisite for the notion of the normal operator is, of course, the existence of an inner product vector space. So here, we are dealing with normal operators on a Euclidean vector space. We also know from the previous section that every operator on a real or a complex vector space has an invariant subspace of dimension 1 or 2. Therefore, we must first describe the complete prescription of normal operators in dimensions 1 and 2. For that purpose and to prepare the following procedure, it is advantageous to primarily discuss the very pleasant properties of normal operators in real and complex vector spaces, in the context of their restrictions on invariant subspaces. The following two propositions illustrate this.

Lemma 11.1 *Normal matrices.*
A block matrix $F = \begin{bmatrix} A & C \\ 0 & D \end{bmatrix}$ with the blocks $A, C,$ and D is normal if and only if the matrix C is zero ($C = 0$) and the matrix A and D are normal.

Proof Given

$$F = \begin{bmatrix} A & C \\ 0 & D \end{bmatrix} \text{ and } F^\dagger = \begin{bmatrix} A^\dagger & 0 \\ C^\dagger & D^\dagger \end{bmatrix},$$

we have

$$F^\dagger F = \begin{bmatrix} A^\dagger & 0 \\ C^\dagger & D^\dagger \end{bmatrix} \begin{bmatrix} A & C \\ 0 & D \end{bmatrix} = \begin{bmatrix} A^\dagger A & A^\dagger C \\ C^\dagger A & C^\dagger C + D^\dagger D \end{bmatrix}$$

and

$$F F^\dagger = \begin{bmatrix} A & C \\ 0 & D \end{bmatrix} \begin{bmatrix} A^\dagger & 0 \\ C^\dagger & D^\dagger \end{bmatrix} = \begin{bmatrix} AA^\dagger + CC^\dagger & CD^\dagger \\ DC^\dagger & DD^\dagger \end{bmatrix}.$$

If F is normal, then

$$F^\dagger F = F F^\dagger.$$

This leads to the condition

$$A^\dagger A = AA^\dagger + CC^\dagger. \tag{11.1}$$

Since, in general, we have $A^\dagger A \neq AA^\dagger$, we have to proceed to take the trace of the above equation. We need first the following definition:

> **Definition 11.6** Trace of a matrix.
> The trace $\mathrm{tr}(A)$ of a square matrix $A = (\alpha_s^i) \in \mathbb{K}^{n \times n}$ is the sum of the diagonal entries of A:
> $$\mathrm{tr}(A) := \alpha_s^s \equiv \sum_{s=1}^{n} \alpha_s^s.$$

The expression $\mathrm{tr}(A^\dagger A)$ is very interesting: firstly, the symmetry equation $\mathrm{tr}(A^\dagger A) = \mathrm{tr}(AA^\dagger)$ and secondly $\mathrm{tr}(A^\dagger A)$ is a sum of squares so we have

$$\mathrm{tr}(A^\dagger A) = \bar{\alpha}_\mu^s \alpha_s^\mu = \alpha_\mu^s \bar{\alpha}_s^\mu = \mathrm{tr}(AA^\dagger) \tag{11.2}$$

and

$$\mathrm{tr}(A^\dagger A) = \sum_{s,i}^{n} |\alpha_s^i|^2. \tag{11.3}$$

The trace in Eq. (11.1) leads to

$$\mathrm{tr}(A^\dagger A) = \mathrm{tr}(AA^\dagger) + \mathrm{tr}(CC^\dagger). \tag{11.4}$$

With the result of Eq. (11.2), we obtain

$$\text{tr}(CC^{\dagger}) = 0. \tag{11.5}$$

From the result in Eq. (11.3) there follows that also $C = 0$. □

Proposition 11.5 *Restriction of a normal operator.*
Let f be a normal operator f on an inner product vector space V, and U an f-invariant subspace of V. Then

(i) The subspace $U^{\perp}$ is also f-invariant;
(ii) U and $U^{\perp}$ are f^{ad}-invariant;
(iii) The restrictions to U satisfy $(f|_U)^{ad} = (f^{ad})|_U$;
(iv) $f|_U$ and $f^{ad}|_U$ are normal.

Remark 11.2 This means that, if we use the notation $f_U := f|_U$, we have $f_U, f_U^{ad} \in \text{End } V$ and $f_U, f_U^{\perp}, f_U^{ad}, f^{ad}{}_U^{\perp}$ are normal.

Proof (i)

Let $C_r := (c_1, \ldots, c_r)$ be an orthonormal basis of U. We extend C_r to an orthonormal basis C on V, $C = (C_r, B_s) = (c_1, \ldots, c_r, b_1, \ldots, b_s)$, so that $r + s = n = \dim V$. Then B_s, being orthogonal to U, is a basis of $U^{\perp}$. Since U is f-invariant, the representation of f, with respect to the basis C, is given by the block matrix $f_C \equiv F$:

$$F = \begin{bmatrix} F_1 & F_2 \\ 0 & F_3 \end{bmatrix}.$$

Since f is normal, F is a normal matrix. By Lemma 11.1, $F_2 = 0$ and F is a block diagonal matrix:

$$F = \begin{bmatrix} F_1 & 0 \\ 0 & F_3 \end{bmatrix}.$$

This shows instantly that $U^{\perp}$ is f-invariant. □

Proof (ii)

The matrix of f^{ad}, with respect to the basis C, is given by

$$f_C^{ad} \equiv F^{\dagger} = \begin{bmatrix} F_1^{\dagger} & 0 \\ 0 & F_3^{\dagger} \end{bmatrix}$$

and this shows again that U and $U^{\perp}$ are f^{ad}-invariant. □

Proof (iii) First proof.

For $u_1, u_2 \in U$ and the definitions $f|_U(u_1) = fu_1$ and $f^{ad}|_U(u_2) = f^{ad}(u_2)$ using $f^{ad}(U) \leqslant U$ from (ii), we obtain

$$\langle (f|_U)^{ad} u_2 | u_1 \rangle = \langle u_2 | f|_U u_1 \rangle = \langle u_2 | f u_1 \rangle = \langle f^{ad} u_2 | u_1 \rangle = \langle f^{ad}|_U(u_2) | u_1 \rangle$$

which shows that $(f|_U)^{ad} = f^{ad}|_U$. This also justifies the notation $f^{ad}{}_U$:

$$f^{ad}{}_U := f^{ad}|_U!$$

$\square$

Proof (iii) Second proof.

The representation matrix $f_C^{ad} \equiv F^\dagger$ shows this result directly: We have $f_{C_r} = F_1$ and $f_{C_r}^{ad} = F_1^\dagger$. $\square$

Proof (vi) First proof.

Observe:

$$\begin{aligned} f_U^{ad} f_U &= (f^{ad})_U f_U \\ &= (f^{ad} f)_U \\ &= (f f^{ad})_U f_U \\ &= f^{ad}{}_U f_U. \end{aligned}$$

This shows that f_U is normal. $\square$

Proof (iv) Second proof.

It follows from Proposition 11.1 that F normal is equivalent to F_1 and F_3 being normal. $\square$

This completes our preparation for normal operators. As we know from the last section, every normal operator in a Euclidean vector space has a one-dimensional or a two-dimensional invariant subspace. We will now determine the structure, starting with this low-dimensional normal operator.

Nothing needs to be said in the one-dimensional case because, in this case, every operator is a normal operator and invariant subspace here means eigenspace. The two-dimensional case is not quite trivial. It is clarified in the following proposition.

> **Proposition 11.6** *Normal operators in a two-dimensional real vector space.*
>
> *Suppose a normal operator f on a two-dimensional Euclidean vector space V, has no one-dimensional f-invariant subspace (i.e., no real eigenvalues). In that case, every orthonormal basis gives a representation of f by a matrix of the form:*
>
> $$F = \begin{bmatrix} \alpha & -\beta \\ \beta & \alpha \end{bmatrix}$$
>
> *with $\alpha, \beta \in \mathbb{R}$, $\beta \neq 0$. Without loss of generality, we may also assume that $\beta > 0$.*

Proof For any orthonormal basis $C = (c_1, c_2)$, the matrices $f_C \equiv F$ and $f_C^{ad} \equiv F^\mathsf{T}$ are given by

$$F = \begin{bmatrix} \alpha & \gamma \\ \beta & \delta \end{bmatrix} \text{ and } F^\mathsf{T} = \begin{bmatrix} \alpha & \beta \\ \gamma & \delta \end{bmatrix}.$$

Note that f is normal precisely when F is normal: $F^\mathsf{T} F = F F^\mathsf{T}$. This leads to the two independent equations

$$\alpha^2 + \beta^2 = \alpha^2 + \gamma^2 \tag{11.6}$$

and

$$\alpha\beta + \gamma\delta = \alpha\gamma + \beta\delta. \tag{11.7}$$

From Eq. (11.6), we obtain

$$\gamma^2 = \beta^2 \text{ and } \gamma = \pm\beta.$$

We exclude the case $\gamma = \beta$ (and $\beta = 0$) because it leads to $F^\mathsf{T} = F$. But a symmetric matrix has a real eigenvalue and a one-dimensional eigenspace which is f-invariant. So we have to take $\beta \neq 0$ and $\gamma = -\beta$. From Eq. (11.7), we obtain $\alpha\beta - \beta\delta = -\alpha\beta + \beta\delta$ and $\alpha - \delta = -\alpha + \delta$ which leads to $\alpha = \delta$ and to

$$F = \begin{bmatrix} \alpha & -\beta \\ \beta & \alpha \end{bmatrix}.$$

$\square$

After having obtained all the above results, we expect that a normal operator f on a real inner product vector space will have an orthogonal decomposition, consisting of normal operators restricted to one-dimensional or two-dimensional Euclidean vector spaces. This means that we will have an orthogonal decomposition of V, we use the symbol "$\ominus$" for it:

$$V = U_1 \ominus \cdots \ominus U_i \ominus \cdots \ominus U_s \quad s \in \mathbb{N}, i \in I(s),$$

with dim $U_i = 1$ or dim $U_i = 2$ and the corresponding decomposition of f:

$$f = f_1 \ominus f_2 \ominus \cdots \ominus f_i \ominus \cdots f_s,$$

with normal $f_i \in \text{End}(U_i)$ and if dim $U_i = 2$ with f_i of the type given in the previous proposition. This is given in the next theorem.

Theorem 11.1 *Normal operators in real vector spaces.*
If f is a normal operator on a real inner product vector space V, then the following are equivalent:

   (i) f is normal;
   (ii) There is an orthogonal decomposition of V into one-dimensional or two-dimensional f-invariant subspaces of V, with a corresponding orthonormal basis of V with respect to which f has a block diagonal representation such that each block is a 1×1 matrix (with $F_i = \alpha_i$) or a 2×2 matrix of the form

$$F_i = \begin{bmatrix} \alpha_i & -\beta_i \\ \beta_i & \alpha_i \end{bmatrix},$$

with $\alpha_i, \beta_i \in \mathbb{R}, s \in \mathbb{N}, i \in I(s)$ and $\beta_i > 0$.

Proof Proof of the Theorem.
We prove (ii) by induction on $\dim(V)$. We assume that dim $V \geqslant 3$ (if dim $V = 1$ or dim $V = 2$ there is nothing to be proven). Let U be a one-dimensional f-invariant or, if no such one-dimensional subspace exists, let U be a two-dimensional f-invariant subspace of V (without a one-dimensional subspace of V) according to Proposition 11.6.

If dim $U = 1$, any nonzero vector of U is an eigenvector of f in U. We normalize this vector to norm 1 and we obtain an orthonormal basis in U. In this case, the matrix of $f|_U$ is given by $F_1 = (\alpha_1)$, $\alpha_1 \in \mathbb{R}$. If dim $U = 2$, then the matrix of $f|_U$ is given by Proposition 11.6. This matrix has, with respect to an orthonormal basis of U, the form:

$$F_1 = \begin{bmatrix} \alpha_i & -\beta_i \\ \beta_1 & \alpha_1 \end{bmatrix},$$

with $\alpha_1, \beta_1 \in \mathbb{R}$ and $\beta_1 > 0$.

The subspace $U^\perp$ of V has fewer dimensions than V. According to Proposition 11.5, $U^\perp$ is also an f-invariant subspace of V and $f|_{U^\perp}$ is a normal operator in $U^\perp$. Therefore we can apply the hypothesis of induction on $U^\perp$: There exists an orthonormal basis with respect to which the matrix of $f|_{U^\perp}$ has the expected block diagonal form. This basis of $U^\perp$, together with the basis of U, gives an orthonormal basis of V with respect to which the matrix of f has the form given in the above theorem. □

Modulo reordering the basis vectors of an orthonormal basis C, we obtain directly the following corollary.

Corollary 11.4 *Spectral theorem for orthogonal operators.*
Let f be an orthogonal operator in a Euclidean vector space V with $\dim V = n$. Then there exists an orthonormal basis C such that the matrix F representing f can be given in the following block diagram form with dimensions $k + l + 2r = n$:

$$F = 1 \ominus \cdots \ominus 1 \ominus (-1) \ominus \cdots \ominus (-1) \ominus F_1 \ominus \cdots \ominus F_j \ominus \cdots \ominus F_r .$$

That is, there are k trivial eigenvalues with $\lambda = 1$, l trivial eigenvalues with $\lambda = -1$ and orthogonal matrices F_j, $j \in I(r)$ given by the angles $\varphi_j \in [0, 2\pi]$, $\varphi_j \neq 0, \pi, 2\pi$ and

$$F_j = \begin{bmatrix} \cos \varphi_j & -\sin \varphi_j \\ \cos \varphi_j & \sin \varphi_j \end{bmatrix}.$$

Summary

In this chapter, we first briefly discussed the nicest operators in mathematics: non-negative operators and isometries. Both are special, normal operators and therefore subjects of the spectral theorem.

Our main concern was to examine operators in real vector spaces, and especially normal operators in real inner product spaces.

Here, the question of diagonalization was much more challenging than in complex vector spaces.

The process of complexification was extensively discussed. This approach allows us to relate the question of diagonalization to the known results in complex vector spaces. In this way, the spectral theorem was also applied to real vector spaces.

Exercises with Hints

Exercise 11.1 Show that the sum of two positive operators is a positive operator.

Exercise 11.2 Show that a nonnegative operator is positive if and only if it is invertible.

Exercise 11.3 Show that a nonnegative operator in a vector space V has a unique nonnegative square root.

> In the next exercises, we consider symmetric matrices S (symmetric opera-
> tors). We define the vector space of symmetric matrices (see Exercise 2.10) by
> $\text{Sym}(n) := \{S \in \mathbb{R}^{n \times n} : S^{\mathsf{T}} = S\}$.

Exercise 11.4 Show that $\dim \text{Sym}(n) = \frac{1}{2}n(n+1)$.

Exercise 11.5 Let F be a matrix $F \in \mathbb{R}^{n \times n}$ and $\phi_F : \text{Sym}(n) \to \mathbb{R}^{n \times n}$ be the linear map given by $\phi_F(S) := F^{\mathsf{T}} S F$. Prove the following assertions:

(i) $\phi_F \in \text{End}(S(n))$;
(ii) ϕ_F is bijective if and only if F is invertible.

Exercise 11.6 Consider the function $f(x) = \alpha x^{\mathsf{T}} S x$ to show that the matrix $S = \begin{bmatrix} \alpha & \beta \\ \beta & \delta \end{bmatrix} \in \text{Sym}(2)$ is positive definite if and only if

$$\alpha > 0 \quad \text{and} \quad \alpha\delta - \beta^2 > 0.$$

Exercise 11.7 Let $S = (\sigma_{is}) \in \text{Sym}(n)$ with $i, s \in I(n)$. Show that the following conditions are necessary for S to be positive definite.

(i) $\sigma_{ii} > 0$;
(ii) $\sigma_{ii}\sigma_{ss} - \sigma_{is}^2 > 0$ for every $i < s$.

Exercise 11.8 Let $S \in \text{Sym}(n)$ and $F \in Gl(n)$. Show that the following assertions are equivalent.

(i) S is positive definite;
(ii) $F^{\mathsf{T}} S F$ is positive definite.

Exercise 11.9 *Criterion for positive definite matrices.*
For $S = (\sigma_{is})$ with $i, s \in I(n)$, show that the following assertions are equivalent.

(i) S is positive definite;
(ii) There exists $F \in Gl(n)$ such that $S = F^{\mathsf{T}} F$;

(iii) All main minors $\delta_k = \delta_k(S) := \det \begin{bmatrix} \sigma_{11} & \cdots & \sigma_{1k} \\ \vdots & & \vdots \\ \sigma_{k1} & \cdots & \sigma_{kk} \end{bmatrix}$ are positive.

Exercise 11.10 For $S = (\sigma_{is})$ with $i, s \in I(n)$ and $S \in \mathrm{Sym}(n)$, show that for any $m \in \mathbb{N}$ the following assertions hold.

 (i) $\ker S^m = \ker S$;
(ii) $\operatorname{im} S^m = \operatorname{im} S$;
(iii) $\operatorname{rank} S^m = \operatorname{rank} S$.

Exercise 11.11 Let S, T be positive definite matrices and suppose $ST = TS$. Show that TS is also positive definite.

Exercise 11.12 Let $S, T \in \mathrm{Sym}(n)$ and S be positive definite. Show that there exists some $B \in Gl(n)$ with $B^\mathsf{T} S B = \mathbb{1}_n$ and $B^\mathsf{T} T B = D$ with D diagonal.

Exercise 11.13 Check that the statements of Exercises 11.6, 11.7, and 11.8, also hold for positive semidefinite (nonnegative) matrices if we interchange the symbols $>$ and $\geq$.

Exercise 11.14 *Criterion for positive semidefinite matrices.*
For $S = (\sigma_{is}) \in \mathrm{Sym}(n)$, $i, s \in I(n)$, show that the following assertions are equivalent:

 (i) S is positive semidefinite (nonnegative);
(ii) There exists $F \in \mathbb{R}^{n \times n}$ such that $S = F^\mathsf{T} F$.

Exercise 11.15 Let $S \in S(n)$ be positive definite. Show that then S is positive semidefinite if and only if $\det S \neq 0$.

Exercise 11.16 Let $A \in \mathbb{R}^{2 \times 2}$ be given by $A = \begin{bmatrix} \alpha & \beta \\ \gamma & \delta \end{bmatrix}$. Show that A has real eigenvalues if and only if
$$(\alpha - \delta)^2 + 4\beta\gamma > 0.$$

The next exercise shows again the analogy between real numbers and self-adjoint operators. We have $x^2 + 2\beta + \gamma = (x - \beta)^2 + \gamma - \beta^2 > 0$ if $\gamma - \beta^2 > 0$. We may expect a similar relation with f a self-adjoint operator to x a real number.

Exercise 11.17 If $f \in \mathrm{Hom}(V, V)$ is self-adjoint and $\beta, \gamma \in \mathbb{R}$ such that $(\gamma - \beta^2) > 0$, then show that
$$f^2 + 2\beta f + \gamma id$$

is invertible.

References and Further Reading

1. S. Axler, *Linear Algebra Done Right.* (Springer Nature, 2024)
2. G. Fischer, B. Springborn, *Lineare Algebra, Eine Einführung für Studienanfänger. Grundkurs Mathematik.* (Springer, 2020)
3. S.H. Friedberg, A.J. Insel, L.E. Spence, *Linear Algebra* (Pearson, 2013)
4. K. Jänich, *Mathematik 1. Geschrieben für Physiker* (Springer, 2006)
5. N. Johnston, *Advanced Linear and Matrix Algebra.* (Springer, 2021)
6. M. Koecher, *Lineare Algebra und analytische Geometrie.* (Springer, 2013)
7. G. Landi, A. Zampini, *Linear Algebra and Analytic Geometry for Physical Sciences.* (Springer, 2018)
8. P. Petersen, *Linear Algebra.* (Springer, 2012)
9. S. Roman, *Advanced Linear Algebra.* (Springer, 2005)
10. B. Said-Houari, *Linear Algebra.* (Birkhäuser, 2017)
11. A.J. Schramm, *Mathematical Methods and Physical Insights: An Integrated Approach.* (Cambridge University Press, 2022)
12. G. Strang, *Introduction to Linear Algebra.* (SIAM, 2022)

Chapter 12
Applications

In the following three subsections, we will discuss some of the most important special cases and applications of standard operators and linear maps, generally. We start with orthogonal operators, isometries in real vector spaces, including reflections.

Next, we return to linear maps, $\mathrm{Hom}(V, V')$, on inner product vector spaces and explain in some detail the role of singular value decomposition (SVD). This leads smoothly to the polar decomposition operators of square matrices.

We finally discuss the Sylvester's law of inertia and investigate shortly its connection with special relativity.

12.1   Orthogonal Operators–Geometric Aspects

Prime examples of operators contain what we usually call rotations and reflections. Since we have described rotations in the previous Chap. 11, we now broaden our study to reflections too. In two dimensions, the reader ought to be familiar with both. In higher dimensions, the story becomes a little less straightforward. Some important observables in physics are connected mainly with reflections, such as in quantum mechanics and elementary particle physics. It is interesting to notice that some observables which are also described by reflections, are involved in some of the still unsolved problems in physics. As example, we mention observables which are connected with the CPT theorem.

We start, as in the previous sections, with the two-dimensional case since this shows all the essential geometric properties that also appear in higher dimensions.

Isometries are special cases of normal operators. Nontrivial normal operators in two dimensions and real vector spaces have the form

$$\begin{bmatrix} \alpha & -\beta \\ \beta & \alpha \end{bmatrix}, \quad \text{with} \quad \alpha, \beta \in \mathbb{R},$$

© The Author(s), under exclusive license to Springer Nature Switzerland AG 2024
N. A. Papadopoulos and F. Scheck, *Linear Algebra for Physics*,
https://doi.org/10.1007/978-3-031-64908-0_12

as it was given by the Theorem 11.1. Isometries are operators and have only one additional condition: $\alpha^2 + \beta^2 = 1$ which means we can find some $\varphi \in \mathbb{R}$ such that $\alpha = \sin\varphi$ and $\beta = \cos\varphi$, and so the matrix takes the form:

$$A \equiv A(\varphi) = \begin{bmatrix} \cos\varphi & -\sin\varphi \\ \sin\varphi & \cos\varphi \end{bmatrix} \quad \varphi \in \mathbb{R}.$$

This may also be expressed differently: A is an operator which transforms an orthonormal basis and particularly the canonical orthonormal basis in $\mathbb{R}^2$ to another orthonormal basis into $\mathbb{R}^2$:

$$A : \mathbb{R}^2 \longrightarrow \mathbb{R}^2$$
$$E = [e_1 e_2] \longmapsto [a_1 a_2] = A.$$

If we consider the unit circle $S^1 := \{e(\varphi) := \begin{bmatrix} \cos\varphi \\ \sin\varphi \end{bmatrix} : \varphi \in \mathbb{R}\}$, we may choose $a_1 = e(\varphi)$ and $a_2 = e(\varphi + \frac{\pi}{2})$ (which corresponds to $a_2 = \begin{bmatrix} -\sin\varphi \\ \cos\varphi \end{bmatrix}$ so that $a_2 \perp a_1$) and obtain A as above. Of course, if $\{a_1, a_2\}$ is orthonormal, then so is $\{a_1, a_2\}$ and thus we obtain a second possibility, parametrized again by the angle φ giving

$$B \equiv B(\varphi) = \begin{bmatrix} \cos\varphi & \sin\varphi \\ \sin\varphi & -\cos\varphi \end{bmatrix}.$$

It is immediately apparent that A corresponds to a rotation around the angle φ and it turns out that B is a reflection over the line $\mathrm{span}(e(\frac{\varphi}{2}))$. For both A and B, we have $A^\top A = \mathbb{1}_2$ and $B^\top B = \mathbb{1}_2$ but $\det A = 1$ and $\det B = -1$. So we may define:

$$O(2) := \{A \in \mathbb{R}^{2\times 2} : A^\top A = \mathbb{1}_2\} \quad \text{and}$$
$$SO(2) := \{A \in \mathbb{R}^2 \times \mathbb{R}^2 : A^\top A = \mathbb{1}_2 \quad \text{and} \quad \det A = 1\}.$$

These sets, $O(2)$ and $SO(2)$, are subgroups of $Gl(2, \mathbb{R}) \equiv Gl(2)$ and we have the relation $SO(2) < O(2) < Gl(2)$. If we consider the elements of the group $SO(2)$ as points in $\mathbb{R}^2$, we see the equivalence of $SO(2) \cong S^1$. As for the rest of $O(2)$, since $A \in O(2)$ implies

$$1 = \det(A^\top A)$$
$$= \det(A^\top)\det(A)$$
$$= \det(A)^2,$$

we see that

$$O(2) - SO(2) = \{B \in O(2) : \det B = -1\}.$$

The main difference between A and B is that B is a symmetric matrix and is therefore diagonalizable.

It is easy to check that the eigenvectors of B are given by $(c_1, c_2) = (e(\frac{\varphi}{2}), e(\frac{\varphi}{2} + \frac{\pi}{2}))$ with the corresponding eigenvalues $\lambda_1 = 1$ and $\lambda_2 = -1$. So we have $Bc_1 = c_1$ and $Bc_2 = -c_2$. If we define the lines $U = \text{span}(c_2) = \mathbb{R}c_2$ and $H = U^\perp$, we obtain the orthogonal decomposition $\mathbb{R}^2 = U \ominus H$. Then B describes a reflection of the vectors $v \in \mathbb{R}^2$ in the line H so that we have:

$$B|_U = -id_U \quad \text{and} \quad B|_H = id_H.$$

The points of H are the fixed points of B. It is interesting to see that B factorizes!

$$B = \begin{bmatrix} \cos\varphi & \sin\varphi \\ \sin\varphi & -\cos\varphi \end{bmatrix} = \begin{bmatrix} \cos\varphi & -\sin\varphi \\ \sin\varphi & \cos\varphi \end{bmatrix} \begin{bmatrix} 1 & 0 \\ 0 & -1 \end{bmatrix}.$$

The matrix $S := \begin{bmatrix} 1 & 0 \\ 0 & -1 \end{bmatrix}$ is a reflection with the eigenvectors e_1 and e_2 and of course $\det S = -1$. So the group $O(2)$ can be described as:

$$O(2) = \{A, AS : A \in SO(2)\} \text{ or}$$
$$O(2) = SO(2) \cup SO(2)S \quad \text{with} \quad SO(2) \cap SO(2)S = \varnothing.$$

A reflection can be expressed in terms of a vector a and a corresponding projection,

$$P_a = \frac{|a\rangle\langle a|}{\langle a|a\rangle} \equiv \frac{aa^\mathsf{T}}{a^\mathsf{T}a},$$

as such

$$S_a := id - 2\frac{|a\rangle\langle a|}{\langle a|a\rangle} \equiv id - 2P_a.$$

So we have

$$S = id - 2|e_2\rangle\langle e_2|$$

and

$$B = id - 2|e(\varphi/2 + \pi/2)\rangle\langle e(\varphi/2 + \pi/2)| \quad \text{and} \quad id = |e_1\rangle\langle e_1| + |e_2\rangle\langle e_2|.$$

At this point, it is also interesting to ask what happens when we compose reflections $S_b S_a \{b \neq \pm a\}$. It turns out that $S_b S_a$ is again a rotation. The fixed points of $S_b S_a$ are given by $H_a \cap H_b = \{0\}$! Since $S_b S_a$ is an orthogonal operator and the only fixed point is zero, $S_b S_a$ must be a rotation. An explicit calculation can also show this:

$$B(\varphi)B(\psi) = \begin{bmatrix} \cos\varphi & +\sin\varphi \\ \sin\varphi & -\cos\varphi \end{bmatrix} \begin{bmatrix} \cos\psi & \sin\psi \\ \sin\psi & -\cos\psi \end{bmatrix} =$$
$$\begin{bmatrix} \cos(\varphi - \psi) & -\sin(\varphi - \psi) \\ \sin(\varphi - \psi) & +\cos(\varphi - \psi) \end{bmatrix} = A(\varphi - \psi).$$

As we now see, we can express every rotation in $\mathbb{R}^2$ by a composition of two reflections. As shown in the above equation, a possibly less pleasant result is that the group $O(2)$ is a non-abelian group (noncommutative). Nevertheless, its subgroup $SO(2)$ is a commutative group: $A(\varphi)A(\psi) = A(\varphi + \psi)$.

All the above results can be generalized to any n-dimension, in particular the Theorem 11.1 of the previous section. The following formulation is slightly different from Corollary 11.4:

Corollary 12.1 *Orthogonal operators.*
If f is an orthogonal operator on a Euclidean vector space, then the following statements are equivalent:

- *f is orthogonal.*
- *There is an orthogonal decomposition of V into one-dimensional or two-dimensional f-invariant subspaces of V: $V = U_1 \ominus U_2 \ominus \cdots \ominus U_s$, $s \in \mathbb{N}$, $i \in I(n)$ and a corresponding orthonormal basis of V. With respect to this basis, f has a block diagonal representation such that each block is a 1×1-matrix ($F_i = \pm 1$) or a 2×2-matrix of the form*

$$F_i = \begin{bmatrix} \cos \varphi_i & -\sin \varphi_i \\ \sin \varphi_i & \cos \varphi_i \end{bmatrix}$$

with $\varphi_i \in \mathbb{R}$, $i \in I(s)$. $\qquad\qquad\square$

We will now discuss the role of reflections as special orthogonal operators on an n-dimensional Euclidean vector space V. The definition is the same as in the $\mathbb{R}^2$ case above. It is characterized by the orthogonal decomposition of V into a one-dimensional subspace U and its orthogonal complement $H := U^\perp$ so that dim $H = n - 1$ and $V = U \ominus H$. This is explained in the following proposition.

12.1.1   The Role of Reflections

Proposition 12.1 *Reflections in n dimensions, reflections in V.*
Let V be an n-dimensional real inner product space, $a \in V \setminus \{0\}$ and $s_a : V \to V$ be the map:

$$s_a(v) = v - 2 \frac{\langle a | v \rangle}{\langle a | a \rangle} a.$$

Then the following statements are valid:

> (i) $s_a(a) = -a$.
> (ii) $\langle w|s_a(v)\rangle = \langle s_a(w)|v\rangle$, $w, v \in V$.
> (iii) $s_a \circ s_a = id$.
> (iv) s_a is an orthogonal operator on V (i.e., $s_a \in O(V)$).

Some additional explanation may be quite useful. If we define

$$U \equiv U(a) := \mathbb{R}a \equiv \text{span}(a) \text{ and } H = H(a) := U^\perp = a^\perp,$$

then

$$V = U \ominus H \text{ and } s_a i_H = id_H \text{ with } s_a i_U = -id_U.$$

This means that the vectors of all $w \in H$ are fixed points of $s_a : s_a(w) = w$. The map s_a describes a reflection of all vectors $u \in U$ over zero and a reflection of the vectors $v \in V - H$ over the hyperplane H. Hence s_a is a symmetric, involutive, and orthogonal operator on V. We are coming now to the proof.

Proof For the four different points of Proposition 12.1, an explicit calculation leads to

(i): $s_a(a) = a - 2\frac{\langle a|a\rangle}{\langle a|a\rangle}a = a - 2a = -a$.

(ii): $\langle w|s_a(v)\rangle = \langle w|v\rangle - 2\frac{\langle a|v\rangle\langle a|w\rangle}{\langle a|a\rangle}$. This is explicitly symmetric in v, w.

(iii):

$$s_a(s_a(v)) = s_a(v) - 2\frac{\langle a|s_a v\rangle}{\langle a|a\rangle}a$$

$$= v - 2\frac{\langle a|v\rangle}{\langle a|a\rangle}a - 2\frac{1}{\langle a|a\rangle}\left[\langle a|v\rangle - 2\frac{\langle a|v\rangle\langle a|a\rangle}{\langle a|a\rangle}\right]a$$

$$= v - 2\frac{\langle a|v\rangle}{\langle a|a\rangle}a - 2\frac{\langle a|v\rangle}{\langle a|a\rangle}a + 4\frac{\langle a|v\rangle}{\langle a|a\rangle}a = v, \text{ utilizing (ii)}.$$

(iv): $\langle s_a w|s_a v\rangle = \langle w|s_a s_a v\rangle = \langle w|v\rangle$, utilizing (iii). $\qquad\square$

We will now discuss additional properties of orthogonal operators, particularly a surprising connection of reflections to all other orthogonal transformations.

An obvious fact is that orthogonal transformations act on the sphere of radius r given by

$$S^{(n-1)}(r) := \{v \in V : \|v\| = r\}.$$

In other words, whenever $f \in O(V)$:

$$f : S^{(n-1)}(r) \longrightarrow S^{(n-1)}(r)$$
$$v \longmapsto f(v)$$
$$\text{since } \|f(v)\| = \|v\| = r.$$

The next proposition shows a key property of reflections that determines the role of reflections within the orthogonal operators. A reflection can connect two distinct vectors on a sphere:

Proposition 12.2 *Transitive action of reflections on spheres.*
Let V be a real inner product space. Suppose two distinct vectors u and v have $\|u\| = \|v\|$, then there exists a third vector $a \in V$ such that

$$s_a(u) = v \quad and \quad s_a(v) = u.$$

Proof We choose the reflection with $a = u - v$ and we observe that $\langle u - v|u\rangle = 1/2\|u - v\|^2 = 1/2\langle u - v|u - v\rangle$. So we have

$$s_{u-v}(u) = u - 2\frac{\langle u - v|u\rangle}{\langle u - v|u - v\rangle}(u - v) = u - 2\frac{1}{2}\frac{\langle u - v|u - v\rangle}{\langle u - v|u - v\rangle}(u - v)$$

$$= u - (u - v) = v.$$

Finally, $s_a(v) = u$ follows simply as s_a is an involution. $\square$

This result shows that the whole group $O(V)$ acts on the sphere $S^{(n-1)}(r)$ transitively (see Definition 1.6 and also generally Sect. 1.3 on group actions). The following theorem is an observation made by Élie Cartan. When it was published, it was indeed a surprise because of the delay with which it appeared and because of the statement itself. Cartan's point was simply that the orthogonal group $O(V)$ is generated by reflections:

Theorem 12.1 *Reflections and orthogonal operators. (Élie Cartan)*
Let V be an n-dimensional real inner product space. Every orthogonal operator $f \in O(V)$ is a product of at most $n = \dim V$ reflections.

Proof We will prove this theorem by induction on $n = \dim V$. For $n = 1$, we have $V = \mathbb{R}$ and $\langle y|x\rangle = yx$ so that $s_a(x) = -x$ $\forall x \in \mathbb{R}$, so that in the notation of Proposition 12.1, $U = \mathbb{R}$, $H = \{0\}$ and $O(V) = \{\pm id\}$. Suppose now, for all dimensions smaller than n, that the theorem holds. Since $f \in O(V)$ and $f \neq Id$, we consider $v_0 \in V$ with $f(v_0) = v_0' \neq v_0$ and of course $\|f(v_0)\| = \|v_0\|$. So we can apply the above Proposition 12.1 for v_0 and v_0'. There exists a nonzero vector a and a reflection s_a so that $s_a(v_0') = v_0$ or $s_a(f v_0) = v_0$ and $s_a \circ f(v_0) = v_0$. We set $Z := \mathbb{R}v_0$ and $W := (\mathbb{R}_{v_0})^\perp \equiv v_0^\perp$. Since $s_a \circ f$ is an orthogonal transformation, we see immediately that Z and W are $s_a \circ f$-invariant Euclidean subspaces of V with

$$s_a \circ f|_Z = id_Z \quad and \quad g := s_a \circ f|_W \in O(W).$$

Since dim $W = n - 1$, we can apply the induction hypothesis and we have at most $n - 1$ reflections on W:

$$s_{b_1}, \ldots, s_{b_r} \text{ where } b_i \in W \quad i \in I(r), \quad r \leqslant n - 1$$

so that

$$s_{b_r} \circ \cdots \circ s_{b_1}|_W \in O(W) \quad \text{and} \quad g = s_a \circ f|_W = s_{b_r} \circ \cdots \circ s_{b_1}|_W.$$

Since $b_i \in W \subseteq V$ and $z \in Z = W^\perp$, we have $s_{b_i} z = z$ for all $i \in I(r)$ and so $s_{b_r} \circ \cdots \circ s_{b_1}(z) = z$ or $s_{b_r} \circ \cdots \circ s_{b_1}|_Z = Id_Z$.

This means that we can continue consistently the map g from W to $V = W \ominus Z$:

$$g = s_a \circ f = s_{b_r} \circ \cdots \circ s_{b_1} \in O(V).$$

This leads to

$$f = s_a \circ g = s_a \circ s_{b_r} \circ \cdots \circ s_{b_1}.$$

This proves the factorization of f to at least n reflections. □

12.2 Singular Value Decomposition (SVD)

In this section, we return to general linear maps $f \in \mathrm{Hom}(V, V')$ on inner product vectors spaces. We discover that if we take not one but two special orthonormal bases, we obtain extremely pleasant results.

Up to this point, we have had frequent opportunities to see the importance of bases for understanding the algebraic and geometric structure of linear maps. In this section, we realize this fact once more. Considering a linear map $f : V \to V'$, the choice of a tailor-made basis $B = [b_1 \ldots b_n]$ and $B' = [b'_1 \ldots b'_m]$ for V and V' (of dimension n and m) respectively, led us in Chap. 3 to the equation

$$fB = B'\Sigma_1$$

and to a matrix representation of f of the form $f_{B'B} \equiv \Sigma_1 = \begin{bmatrix} 1_r & 0 \\ 0 & 0 \end{bmatrix}$ which is the normal form.

The matrix Σ_1 is as simple as possible. The number r, the rank of f, is an important geometric property. But by choosing such perfect tailor-made bases, we lost some other important geometric properties of f. In the case of the endomorphisms $f \in \mathrm{Hom}(V, V) \equiv \mathrm{End}(V)$, since we consider only one vector space V, it seems unnatural to consider more than one basis B, and this makes such a classification much more complicated. It is more difficult to determine the corresponding normal form of f. For endomorphisms, the theory leads in the end to what is called a Jordan normal form of f. But, within End(V), there are "privileged" endomorphisms that allow a most simple matrix representation while at the same time keeping the important geometric properties of f. These are the diagonalizable endomorphisms. The corresponding tailor-made bases, here $B^\circ = [b_1^\circ, \ldots, b_n^\circ]$, are given by the eigenvectors b_s° with eigenvalues λ_s, $s \in I(n)$:

$$f b_s^\circ = \lambda_s b_s^\circ$$

which we may express also by the equation

$$F B^\circ = B^\circ \Lambda \quad \text{with} \quad \Lambda = \begin{bmatrix} \lambda_1 & & 0 \\ & \ddots & \\ 0 & & \lambda_n \end{bmatrix}.$$

Apart from this, there are even more privileged endomorphisms, the normal operators. As we know from the spectral theorems, the tailor-made bases are given by the orthonormal bases $C = [c_1, \ldots, c_n]$. And one of the results was that these are the only linear maps that are orthogonally diagonalizable:

$$f C = C \Lambda.$$

There is no doubt that normal operators are the "nicest" operators there are! It is remarkable that in physical theories like quantum mechanics, they are essentially the only ones we need. But from a mathematical point of view, there is the question of universality. We would like to have a normal form for all endomorphisms! To achieve this, we have to go one step back and try to use two bases for one vector space! This will lead us to what we call singular value decomposition (SVD) which is applicable to all $f \in$ End(V) over real or complex inner product spaces and by construction to all $f \in$ Hom(V, V').

The appropriate tailor-made bases for f are two orthonormal bases that are connected to the two diagonalizable self-adjoint operators $f^{ad} f$ and $f f^{ad}$. The use of orthonormal bases instead of general bases preserves the information about eigenvalues even when considering general linear maps $f \in$ Hom(V, V').

This is why we discuss the SVD for $f \in$ Hom(V, V'), the special case $f \in$ End(V) is completely trivially included in the general situation. As we shall see, a kind of miracle makes the whole procedure possible: The map f essentially transforms the eigenvectors, the positive part of the eigenbasis of $f^{ad} f$ into the eigenvectors of $f f^{ad}$, and we are led to the following theorem:

Theorem 12.2 *Singular value decomposition (SVD).*

Let V, V' be inner product spaces of dimensions n, m respectively and $f : V \rightarrow V'$ a linear map of rank r. Then there are orthonormal bases $U = [u_1 \ldots u_n]$ of V and $W = [w_1 \ldots w_m]$ of V' and positive scalars $\sigma_1 \geqslant \sigma_2 \geqslant \cdots \geqslant \sigma_r$, the so-called singular values of f, such that

$$f(u_s) = \sigma_s w_s \qquad \text{if } s \in I(r) \quad \text{and}$$
$$f(u_s) = 0 \qquad \text{if } s > r.$$

Here, u_s are eigenvectors of $f^{ad} f$ with eigenvalues $\lambda_s = \sigma_s^2$ whenever $s \in I(r)$, and $\lambda_s = \sigma_s^2 = 0$ if $s > r$.

This may be expressed by the compact notation:

$$fU = W\Sigma \tag{12.1}$$

or equivalently by

$$f = W\Sigma U^{ad} \tag{12.2}$$

with the matrix

$$\Sigma = \begin{bmatrix} \sigma_1 & & & & 0 \\ & \ddots & & & \\ & & \sigma_r & & \\ & & & 0 & \\ & & & & \ddots \\ 0 & & & & 0 \end{bmatrix} \quad \text{uniquely defined.} \tag{12.3}$$

If we use the same letters U and W for corresponding bases in $\mathbb{K}^n$ and $\mathbb{K}^m$, we have the unitary (orthogonal) matrices

$$U = [\vec{u}_1 \ldots \vec{u}_n] \text{ and } W = [\vec{w}_1 \ldots \vec{w}_m].$$

For the representation $F := f_{WU}$ of f, we may write the above equation in the standard matrix form with entries consisting of scalars only:

$$FU = \Sigma W \tag{12.4}$$
$$F = W\Sigma U^\dagger. \tag{12.5}$$

Remark 12.1 If we denote by $U^* = (u_s^*)_n$ the dual basis of $U = (u_s)_n$ (i.e., $U_s^*(u_r = \delta_s)$, we may express all the above equations as a sum of rank 1 operators (matrices):

$$f = \sigma_1 w_1 u_1^\dagger + \dots \qquad \sigma_r w_r u_r^\dagger , \qquad (12.6)$$

$$F = \sigma_1 \vec{w}_1 \vec{u}_1^\dagger + \dots \qquad \sigma_r \vec{w}_r \vec{u}^\dagger \ \text{ or equivalently} \qquad (12.7)$$

$$F = \sigma_1 |w_1\rangle\langle u_1| + \dots \qquad \sigma_r |w_r\rangle\langle u_r|. \qquad (12.8)$$

Proof Theorem 12.2

By Proposition 11.1, $f^{ad} f \in \text{End}(V)$ is a nonnegative operator (positive semidefinite). An orthonormal eigenbasis $(u_1, \dots, u_n)$ of $f^{ad} f$ is guaranteed by the spectral theorems with the corresponding eigenvalues λ which we can order so that:

$$\lambda_s \neq 0 \text{ if } s \in I(r) \text{ and } \lambda_s = 0 \text{ if } s > r ,$$

$$\text{with } \lambda_1 \geqslant \dots \geqslant \lambda_r > 0 \text{ and } \lambda_s = \sigma_s^2. \qquad (12.9)$$

Since $\ker f^{ad} f = \ker f$ ($\langle v | f^{ad} f v \rangle = \langle f v | f v \rangle$), we can uniquely define $\sigma_s = \sqrt{\lambda_s}$ when $s \in I(r)$ and $\sigma_s = 0$ when $s > r$, so we have

$$f u_s = \sigma_s w_s \qquad\qquad \text{if } s \leqslant r \qquad (12.10)$$

$$\text{and } f u_s = 0 \qquad\qquad \text{if } s > r. \qquad (12.11)$$

It turns out that $(w_s)_r$ is an orthonormal basis of $\text{im } f = \text{span}(w_1, \dots, w_r) \leqslant V'$: When $s, t \in I(r)$, we have

$$\langle \sigma_t w_t | \sigma_s w_s \rangle = \langle f u_t | f u_s \rangle = \langle f^{ad} f u_t | u_s \rangle. \qquad (12.12)$$

This leads to

$$\bar{\sigma}_t \sigma_s \langle w_t | w_s \rangle = \langle \sigma_t^2 u_t | u_s \rangle = \sigma_t^2 \langle u_t | u_s \rangle = \sigma_t^2 \delta_{ts} \qquad (12.13)$$

and so

$$\langle w_t | w_s \rangle = \delta_{ts}. \qquad (12.14)$$

We extend $(w_1, \dots, w_r)$ to an orthonormal basis in V': $(w_1, \dots, w_r, w_{r+1}, \dots, w_m)$ and so we have

$$f u_s = \sigma_s w_s \qquad (12.15)$$

if $s \leqslant r$, as above.

If $s > r$, since $f u_s = 0 \Leftrightarrow f^{ad} f u_s = 0$, we have from $f^{ad} f u_s = \lambda_s u_s = 0, \lambda_s = 0$ and $\sigma_s = 0$, also $f u_s = 0$. $\square$

Comment 12.1 *SVD direct.*

It is interesting to notice that if $f \in \mathrm{Hom}(V, V')$ is given by the data $f u_s = \sigma_s w_s$ with $s \leqslant r$, and $f u_s = 0$ with $s > r$ where $U = (u_1, \ldots, u_n)$ and where $W = (w_1, \ldots, w_m)$ are orthonormal bases in V and V', then it follows that U is an eigenbasis of $f^{ad} f$ with eigenvalues $\lambda_s = \sigma_s^2$ with $s \leqslant r$ and $\lambda_s = 0$ with $s > r$.

This means that

$$f^{ad} f u_s = \sigma_s^2 u_s \qquad\qquad \text{if } s \leqslant r \quad \text{and} \qquad (12.16)$$

$$f^{ad} f u_s = 0 \qquad\qquad \text{if } s > r. \qquad (12.17)$$

Proof A direct calculation leads to the above result: We expand $f^{ad} w_t \in V$ according to the orthonormal basis U:

$$
\begin{aligned}
f^{ad} w_t &= u_s \langle u_s | f^{ad} w_t \rangle = u_s \langle f u_s | w_t \rangle \\
&= u_s \langle \sigma_s w_s | w_t \rangle \\
&= \sigma_s u_s \langle w_s | w_t \rangle = \sigma_s u_s \delta_{st} \qquad (12.18)
\end{aligned}
$$

so that

$$f^{ad} w_s = \sigma_s u_s. \qquad (12.19)$$

Starting again with

$$f u_s = \sigma_s w_s, \qquad (12.20)$$

we test

$$f^{ad} f u_s = \sigma_s (f^{ad} w_s) = \delta_s (\delta_s u_s) = \sigma_s^2 u_s. \qquad (12.21)$$

For $s > r$ we have $f^{ad} f u_s = 0 \Leftrightarrow f u_s = 0$. □

The singular value decomposition (SVD) can be used to give direct proof of the polar decomposition (see below) of an endomorphism of a square matrix. The polar decomposition gives a factorization of every square matrix as a product of a unitary and a nonnegative matrix. This means that we can express every square matrix by two special normal operators which we can completely describe and understand by the spectral theorems. In addition, the analogy to complex numbers which was explained in Sect. 10.5 appears again: Just as we have for every $z \in \mathbb{C}$ $z = e^{i\varphi}|z|$ with $\varphi \in \mathbb{R}$, so we have for every $A \in \mathbb{K}^{n \times n}$ $A = QP$ with Q an isomorphism and P a nonnegative matrix. The analogy $e^{i\varphi} \sim Q$ and $|z| \sim P$ should be clear. This leads to the following result.

> **Proposition 12.3** *Polar decomposition.*
> *For any square matrix $A \in \mathbb{K}^{n \times n}$, there exists a unitary matrix Q and a non-negative matrix P such that*
>
> *(a) $A = QP$ and*
> *(b) if A is invertible, this decomposition is unique.*

Proof Using the SVD as in the above proposition for matrices, we can express A by the equation

$$A = W \Sigma U^{\dagger}.$$

U and W are unitary matrices and Σ is a diagonal matrix with nonnegative entries. Using $U^{\dagger}U = \mathbb{1}_n$, we obtain

$$A = W \mathbb{1}_n \Sigma U^{\dagger} = WU^{\dagger}(U \Sigma U^{\dagger}).$$

We set $Q := WU^{\dagger}$ and $P := U \Sigma U^{\dagger}$. It is obvious that Q is unitary and P nonnegative with the same rank as Σ. So the above equation leads to

$$A = Q P$$

which is assertion a.

To prove assertion b of Proposition 12.3, we assume that

$$A = QP = Q_0 P_0$$

again with Q_0 unitary and P_0 nonnegative. Now, since A is invertible, P and P_0 are also invertible and therefore positive.

$$\text{Consider} \quad A^{\dagger}A = (QP)^{\dagger}QP = P^{\dagger}Q^{\dagger}QP = PQ^{\dagger}QP = P^2$$
$$\text{and} \quad A^{\dagger}A = (Q_0 P_0)^{\dagger}Q_0 P = P_0^2 .$$

We obtain $P^2 = P_0^2$ and since $P_1 P_0$ are positive, $P_0 = P$. With $A = Q_0 P_0 = QP$ we obtain $Q_0 = AP^{-1}$. This shows that the decomposition $A = QP$ is unique. $\square$

12.3   The Scalar Product and Spacetime

If you did not want to understand the structure of spacetime, you would probably not need, as a physicist, to read this section. If Einstein had not discovered the theory of relativity, the problem discussed in this section would not be so relevant for physical investigations. In particular, as we have special relativity, we may ask how many kinds of "special relativities" could be possible, in principle. Special relativity is known to be characterized by three space dimensions and one time dimension. This ratio could be (theoretically) different, especially if we want to consider models of spacetime with larger spacetime dimension. In other words, we would like to classify all possible special relativities. Today, this is useful for instructional reasons and also in connection, for example, with cosmological models. Mathematicians have already solved this mathematical problem. It is the classification of all symmetric bilinear forms which James Joseph Sylvester found in the middle of the 19th century. The discussion of this problem here allows us to examine the role of matrices. They are used to represent, for example, linear maps or bilinear forms. We expect that this additional and, in a way, technical point will be particularly useful for physicists.

We consider symmetric bilinear forms in a real vector space for the sake of simplicity and their relevance in physics. We start with representing a symmetric bilinear form on a real vector space, here also called scalar product, and the corresponding transformation formula by changing the bases.

We consider an n-dimensional vector space V with a bilinear, symmetric scalar product s:

$$s: \quad V \times V \longrightarrow \mathbb{R}$$
$$(u, v) \longmapsto s(u, v).$$

We choose a basis $B = (b_1, \ldots, b_n)$. We define a matrix $S_B = (\sigma^B_{\mu\nu})$ with $\sigma^B_{\mu\nu} := s(b_\mu, b_\nu)$ $\mu, \nu \in I(n)$. Since s is a symmetric scalar product, it is clear that S_B is a symmetric matrix. If $u = b_\mu u^\mu$ and $v = b_\mu v^\mu \in V$, we obtain

$$s(v, u) = s(b_\mu v^\mu, b_\nu u^\nu) = v^\mu \sigma^B_{\mu\nu} u^\nu = v_B^\mathsf{T} S_B u_B. \tag{12.22}$$

If we choose a second basis $C = (c_1, \ldots, c_n)$, we have $\sigma^C_{\mu\nu} := s(c_\mu, c_\nu)$ $S_C := (\sigma^C_{\mu\nu})$ and $s(v, u) = v_C^\mathsf{T} S_C u_C$. The change of the bases from B to C is given by $b_s = c_\mu \tau_s^\mu$, $\tau_s^\mu, \in \mathbb{R}$ and we may define the transition matrix

$$T \equiv T_{CB} := (\tau_t^\mu).$$

We also have $B = CT$, using matrix notation with entries vectors in V, $B = [b_1, \ldots, b_n]$ and $C = [c_1, \ldots, c_n]$ for B and C. So we have

$$\sigma^B_{\mu,t} = s(b_\mu, b_t) = s(c_\nu \tau_\mu^\nu, c_r \tau_t^r) \tag{12.23}$$

and

$$\sigma^B_{\mu,t} = \tau_\mu^\nu s(c_\nu, c_r) \tau_t^r = \tau_\mu^\nu \sigma^C_{\nu r} \tau_t^r \equiv \tau_{\nu\mu} \sigma^C_{\nu r} \tau_{rt} , \quad \text{with } r, t \in I(n). \tag{12.24}$$

We set $\tau_\mu^\nu \equiv \tau_{\nu\mu}$ for comparison with the usual notation in the mathematical literature. At the level of matrices, we find

$$S_B = T_{CB}^\mathsf{T} S_C \, T_{CB} \equiv T^\mathsf{T} S_C \, T \qquad \text{or equivalently} \qquad (12.25)$$

$$S_C = T_{BC}^\mathsf{T} \, S_B \, T_{BC} \equiv (T^{-1})^\mathsf{T} S_B \, T^{-1}. \qquad (12.26)$$

This is the transition formula for bilinear forms.

It is important to compare this with the transformation formula for endomorphisms. If $f \in \text{End}(V)$ with corresponding representations f_{BB} and f_{CB}, with the bases B and C, then we have

$$f_{BB} = T_{BC} \, f_{CC} \, T_{CB} = T^{-1} f_{CC} \, T \qquad \text{or equivalently} \qquad (12.27)$$

$$f_{CC} = T_{CB} \, f_{BB} \, T_{BC} = T \quad f_{BB} \, T^{-1}. \qquad (12.28)$$

As one can see, the difference is the T^{-1} that appears for endomorphism transformations. This means that if we know precisely what a given matrix represents, we know its transformation law. This is extremely important to know, not only in relativity but across physics. It is useful to give or recall some further definitions to proceed.

We observed above an equivalence relation corresponding to symmetric bilinear forms. This is usually called congruence: The matrices A and B are congruent if an invertible matrix $G \in Gl(m)$ exists so that $A = G^\mathsf{T} B G$ holds. In this sense, S_B and S_C are congruent. The rank is mainly defined for a linear map, but it may also be defined for a symmetric bilinear form:

Definition 12.1 Rank of a bilinear form.
Suppose $s = S$ is a bilinear form on V. The rank of s, rank(s), is the rank of any representing matrix, that is, for any basis B of V

$$\text{rank}(s) = \text{rank}(S_B).$$

We say s is degenerate if there is some $v \in V | \{0\}$ such that $s(v, u) = 0$ for all $u \in V$. Otherwise, we say s is nondegenerate.

The subspace U_0 of V given by

$$U_0 := \{v \in V : s(v, u) = 0 \quad \text{for all} \quad u \in V\},$$

what may also be called degenerate.

If s is nondegenerate, then we have of course $U_0 = \{0\}$ and rank(s) = dim V. We further recall the following definition which is valid for symmetric and hermitian forms.

- s is positive definite if $s(v, v) > 0$ for all $v \neq 0$, $v \in V$,
- s is negative definite if $s(v, v) < 0$ for all $v \neq 0$, $v \in V$,
- s is indefinite if there exist u and v in V,

so that $s(v, v) > 0$ and $s(u, u) < 0$.

The analog definition is valid for a symmetric or a Hermitian matrix and in an obvious notation we have to replace $s(v, v)$ or $s(u, u)$ by $\vec{v}^\dagger S \vec{v}$ and $\vec{u}^\dagger S \vec{u}$.

We are now returning to our classification problem for symmetric bilinear forms which corresponds, in the case of spacetime, to the classification of all theoretically possible models for a flat (linear) spacetime, that is, for special relativity. Mathematically speaking, we have to consider all possible pairs (V, s) with s not only not positive definite but also semidefinite and even s degenerate. This means we have to consider also non-Euclidean geometries. As is well-known, it took science more than 2000 years to get to that point!

Here we restrict ourselves, of course, to the mathematical problem and, without loss of generalization and for the sake of simplicity, to the level of matrices.

The first step is to diagonalize a representation, a matrix like S_B. The spectral theorem tells us that we can even use an orthogonal matrix $Q \in O(n)$. This fits very nicely with the transformation formula for scalar products since whenever $Q \in O(n)$, $Q^{-1} = Q^\mathsf{T}$! So the eigenvalue equations at the level matrices give us

$$S_B Q = Q \Sigma_B \quad \text{and} \quad \Sigma_B = Q^\mathsf{T} S_B Q \qquad (12.29)$$

with

$$\Sigma_B = \begin{bmatrix} \sigma_1^B & & 0 \\ & \ddots & \\ 0 & & \sigma_n^B \end{bmatrix} \quad \text{and} \quad \sigma_\mu^B \in \mathbb{R}. \qquad (12.30)$$

The scalars (numbers) σ_μ^B are the eigenvalues of S_B if we consider S_B as an endomorphism in $\mathbb{R}^n$. Here, S_B instead represents the bilinear form s given in the basis B.

The process of diagonalization leads mathematically to the result in Eq. (12.30). But the notion of eigenvalue is not relevant for a bilinear form, because the scalars $\sigma_1^B, \ldots, \sigma_n^B$ are not invariant. We show this fact explicitly by the index B. We can see the reason for the non-invariance in the transformation formula for scalar products above (Eq. (12.29)). We have in $S_B = Q^\mathsf{T} S_C Q$ on the left hand side of S_C the matrix Q^T instead of the matrix Q^{-1}. This scales the coefficients in S_C, and we obtain

$$(\sigma_1^C, \ldots, \sigma_n^C) \neq (\sigma_1^B, \ldots, \sigma_n^B).$$

This is clear in the case dim $V = 1$. The basis is given by $b \in V$ and $b \neq 0$. So we have $\sigma^B = s(b, b)$. A second basis given by $c \in V$ and $c \neq 0$ with $c = \lambda b, \lambda \in \mathbb{R}, \lambda \neq 0$ leads to $\sigma^C = s(c, c) = s(\lambda b, \lambda b) = \lambda^2 s(b, b) = \lambda^2 \sigma^B$. It is interesting to notice that λ^2 being positive, the sign of σ^B is conserved. The so-called Sylvester's law of inertia determines exactly the invariance we are looking for: The matrices A and B have the same invariants if an invertible matrix $G \in Gl(n)$ exists, so that $A = G^\mathsf{T} B G$. It turns out that the invariants of the congruence are three numbers: The number of positive (n_+), negative (n_-), and zero (n_0) entries in the diagonal matrix Σ_B. This means that we can write this Σ_B as a block diagonal matrix:

$$\Sigma_B = \begin{bmatrix} \Sigma^{(+)} & & 0 \\ & \Sigma^{(-1)} & \\ 0 & & \Sigma^{(0)} \end{bmatrix} = \mathrm{diag}(\sigma_1^{(+)}, \ldots, \sigma_{n_+}^{(+)}, \sigma_1^{(-)}, \ldots, \sigma_{n_-}^{(-)}, \sigma_1^{(0)}, \ldots, \sigma_{n_0}^{(0)})$$

with $\sigma_1^{(+)}, \ldots, \sigma_{n_+}^{(+)}$ positive, $\sigma_1^{(-)}, \ldots, \sigma_{n_-}^{(-)}$ negative, and $\sigma_1^{(0)} = 0, \ldots, \sigma_{n_0}^{(0)} = 0$. For a compact formulation of the above, we propose this definition.

Definition 12.2 The inertia of s are the numbers (n_+, n_-, n_0) $(n = n_+, n_-, n_0)$.

All the above discussion leads to the following theorem.

Theorem 12.3 *Sylvester's law of inertia.*
The inertia (n_+, n_-, n_0) of a symmetric bilinear form s or of a symmetric matrix S is invariant under the congruence. This means that we have $n_+(S_B) = n_+(S_C)$, $n_-(S_B) = n_-(S_C)$, and $n_0(S_B) = n_0(S_C)$.

Proof As we see from the above discussion, we may, according to the spectral theorem, diagonalize the matrix $S = S_B$. We obtain the diagonal matrix Σ using the orthogonal matrix $Q \in O(n)$. This corresponds to an ordered orthonormal basis.

$$Q = (q_i^{(+)}, q_j^{(-)}, q_k^{(0)}) \quad \text{with} \quad i \in I(n_+), j \in I(n_-), k \in I(n_0), \tag{12.31}$$

with

$$\Sigma_B = \mathrm{diag}\left(\sigma_i^{(+)}, \sigma_j^{(-)}, \sigma_k^{(0)}\right)_B \equiv \mathrm{diag}\left(\sigma(B)_i^{(+)}, \sigma(B)_j^{(-)}, \sigma(B)_k^{(0)}\right) \tag{12.32}$$

and $\sigma_i^{(+)}$ positive, $\sigma_j^{(-)}$ negative, and $\sigma_k^{(0)} = 0$ for all $i \in I(n_+)$, $j \in I(n_-)$, $k \in I(n_0)$. Now we can scale the basis vectors in Q and we obtain $\bar{Q} = (\bar{q}_i^{(n_+)}, \bar{q}_j^{(n_-)}, \bar{q}_k^{(n_0)})$ with

$$\vec{q}_i^{(+)} = \frac{1}{\sqrt{\sigma_i}} q_i^{(+)}, \quad \vec{q}_j^{(-)} = \frac{1}{\sqrt{|\sigma_j|}} q_j^{(-)} \quad \text{and} \quad \vec{q}_k^{(0)} = \vec{q}_k^{(0)}. \tag{12.33}$$

This leads to

$$\Sigma_B^0 = \begin{bmatrix} \mathbb{1}_{n_+} & & 0 \\ & \mathbb{1}_{n_-} & \\ 0 & & 0_{n_0} \end{bmatrix}. \tag{12.34}$$

We consider the subspaces

$$U_B^{(+)} := \mathrm{span}(\vec{q}_1^{(+)}, \ldots, \vec{q}_{n_+}^{(+)}), \quad U_B^{(-)} := \mathrm{span}(\vec{q}_1^{(-)}, \ldots, \vec{q}_{n_-}^{(-)}) \quad \text{and} \tag{12.35}$$

$$U_B^{(0)} := \mathrm{span}(\vec{q}_1^{(0)}, \ldots, \vec{q}_{n_0}^{(0)}). \tag{12.36}$$

It is clear that for $u_+^B \in U_B^{(+)}$, $u_-^B \in U_B^{(-)}$ and $u_0^B \in U_B^{(0)}$, we have $s(u_+, u_+) > 0$, $s(u_-, u_-) < 0$ and $s(u_0, u_0) = 0$.

At this point, we may have realized that the numbers (n_+^B, n_-^B, n_0^B) are well defined since $n_0 = \dim V - \mathrm{rank}(s)$ and $n_+^B(n_-^B)$ correspond to the maximum dimension of a subspace with positive (negative) definite restriction of s. Apparently, we here have a direct sum:

$$V = U_B^{(+)} \oplus U_B^{(-)} \oplus U_B^{(0)}. \tag{12.37}$$

We can of course follow the same procedure for any other basis C and S_C and we come to similar expressions:

$$n_+^C = \dim U_C^{(+)}, \quad n_-^C = \dim U_C^{(-)}, \quad n_0^C = \dim U_C^{(0)}. \tag{12.38}$$

This means that if a subspace Z has the property that for all $z \in Z$, $s(z, z) > 0$ holds, then its dimension is not bigger than n_+^B so that $\dim Z \leqslant n_+^B = \dim U_B^{(+)}$. This can also be proven by contradiction:

$$\text{If} \quad \dim Z > n_+^B,$$

$$\text{then there exists a} \quad \tilde{z} \in Z,$$

so that $\tilde{z} \notin U_B^{(+)}$ and $\tilde{z} \in U_B^{(-)} + U_B^{(0)}$ which signifies $s(\tilde{z}, \tilde{z}) \leqslant 0$ or, equivalently, that $\dim(Z \cap U^{(0)} \oplus U^{(-1)}) \geqslant 1$. This is in contradiction to the hypothesis $s(z, z) > 0$ for all $z \in Z$. So $\dim Z \leqslant n_+^B$ holds.

Therefore, if we take $Z = U_C^{(+)}$ and $\dim U_C^{(+)} = n_+^C$, we have $n_+^C \leqslant n_+^B$. Interchanging B and C, we get $n_+^B \leqslant n_+^C$. This leads to $n_+^B = n_+^C$ and shows that

$$(n_+^B, n_-^B, n_0^B) = (n_+^C, n_-^C, n_0^C). \tag{12.39}$$

We see that the numbers (n_+, n_-, n_0) are basis-independent and depend only on the scalar product s. This means further that they are invariant under the congruence. This proves the theorem. $\qquad\square$

Summary

Three important applications of the last three chapters to operators were discussed.

Firstly, the orthogonal group in two dimensions was thoroughly examined. Then, the structure of orthogonal operators was presented for arbitrary dimensions, corresponding to the spectral theorem for the orthogonal group. Reflections were also presented as a multiplicatively generating system of the orthogonal group.

Next, the singular value decomposition (SVD) was introduced, a highly relevant method in the field of modern numerical linear algebra. It allows for a universal representation of any operator through the use of two specially tailored orthonormal bases.

Finally, in terms of the structure of spacetime in special relativity, the classification of symmetric, nondegenerate, bilinear forms was discussed. This was a well-known mathematical problem that was addressed as early as the 19th century by the mathematician James Joseph Sylvester.

References and Further Reading

1. G. Fischer, B. Springborn, *Lineare Algebra* (Eine Einführung für Studienanfänger, Grundkurs Mathematik (Springer, 2020)
2. S.H. Friedberg, A.J. Insel, L.E. Spence, *Linear Algebra* (Pearson, 2013)
3. N. Johnston, *Advanced Linear and Matrix Algebra*. (Springer, 2021)
4. M. Koecher, *Lineare Algebra und analytische Geometrie*. (Springer, 2013)
5. G. Strang, *Introduction to Linear Algebra*. (SIAM, 2022)

Chapter 13
Duality

In physics, there is a tendency to believe that dual vector spaces $V^* = \text{Hom}(V, \mathbb{K})$ are unnecessary and superfluous. This can be fallacious and cause unnecessary obstacles to understanding crucial aspects of linear algebra related to theoretical physics. Duality is also essential for understanding tensor formalism, the subject of the following chapter. A particular difficulty with the dual space is that we have to know precisely in what situations one must and in what situations one need not pay close attention to it. Even in this case, the elements of V^*, which appear as elements of V, have a different transformation behavior from the standard elements in V. This is why our intention in this section is also to help avoid such confusion.

To clarify this situation, we have to proceed slowly and repeat, when necessary, some well-known facts. So we will distinguish precisely the formalism corresponding to an abstract vector space (without an inner product) from a vector space with a scalar product. We believe that this section is necessary to understand tensor formalism. This makes it indispensable for a good understanding of special and general relativity and relativistic field theory.

13.1 Duality on an Abstract Vector Space

We focus our attention on the different transformation behavior of vectors and covectors by changing bases, and briefly reviewing what we already know. This includes our conventions and notations, which should not be underestimated here. We try to use as clear a notation as possible, especially for the desiderata in physics.

We start with an abstract n-dimensional vector space V, $\dim V = n$ and its dual V^*. We choose two bases

$$B := (b_1, \ldots, b_n) \equiv (b_s)_n \text{ and } C := (c_1, \ldots, c_n) \equiv (c_i)_n \,,$$

© The Author(s), under exclusive license to Springer Nature Switzerland AG 2024
N. A. Papadopoulos and F. Scheck, *Linear Algebra for Physics*,
https://doi.org/10.1007/978-3-031-64908-0_13

with the corresponding dual bases

$$B^* := (\beta^1, \ldots, \beta^n) \equiv (\beta^s)_n \text{ and } C^* := (\gamma^1, \ldots, \gamma^n) \equiv (\gamma^i)_n$$

given by

$$\beta^s(b_r) = \delta_r^s \text{ and } \gamma^i(c_j) = \delta_j^i \qquad (13.1)$$

where $r, s, i, j \in I(n)$.

> **Remark 13.1** Cobasis notation.
> A consistent notation within our sign convention is to write $\mathcal{B} := [\beta^1 \ldots \beta^n]^\mathsf{T}$ and $\mathcal{C} := [\gamma^1 \ldots \gamma^n]^\mathsf{T}$ vertically. Yet, if it is useful for our demonstration, we identify the corresponding lists, colists, and matrices.

For a vector $v \in V$ and a covector $\xi \in V^*$, we have the following expression, using the Einstein convention:

$$v = b_s v_B^s = c_i v_C^i \text{ and } \xi = \xi_s^B \beta^s = \xi_i^C \gamma^i. \qquad (13.2)$$

The coefficients (components) $v_B^s, v_C^i, \xi_s^B, \xi_i^C \in \mathbb{K}$ can be expressed by

$$v_B^s = \beta^s(v), \quad v_C^i = \gamma^i(v) \text{ and } \xi_s^B = \xi(b_s), \quad \xi_i^C = \xi(c_i). \qquad (13.3)$$

The change of basis transformation matrix $T = (\tau_s^i)$ with the transformation coefficient τ_s^i is given by:

$$b_s = c_i \tau_s^i \text{ or } c_i = b_s \bar{\tau}_i^s \text{ and } \bar{T} = (\bar{\tau}_i^s) = T^{-1}. \qquad (13.4)$$

The coefficients $\tau_s^i, \bar{\tau}_i^s$ can also be expressed by

$$\tau_s^i = \gamma^i(b_s) \text{ and } \bar{\tau}_i^s = \beta^s(c_i). \qquad (13.5)$$

The matrix form of the change of basis is given by

$$\mathcal{B} = C T \text{ or } C = \mathcal{B} T^{-1}. \qquad (13.6)$$

Note the correspondence of Eqs. (13.6) to (13.4). Using the above duality relation (Eq. 13.1), we obtain

$$\beta^s = \bar{\tau}_i^s \gamma^i \text{ or } \gamma^i = \tau_s^i \beta^s. \qquad (13.7)$$

The corresponding matrix form is given by

$$\mathcal{B} = T^{-i} \mathcal{C} \text{ or } \mathcal{C} = T \mathcal{B}. \qquad (13.8)$$

For the coefficients of v and ξ, we write

$$v_B := \begin{bmatrix} v_B^1 \\ \vdots \\ v_B^n \end{bmatrix}, \quad v_C := \begin{bmatrix} v_C^i \\ \vdots \\ v_?^n \end{bmatrix}, \quad \xi^B := [\xi_1^B \ldots \xi_n^B], \quad \xi^C := [\xi_1^C \ldots \xi_n^C], \qquad (13.9)$$

and we obtain

$$v_B^s = \bar{\tau}_i^s v_C^i \quad \text{or} \quad v_C^i = \tau_s^i v_B^s \quad \text{and} \quad \xi_s^B = \xi_i^C \tau_s^i \quad \text{or} \quad \xi_i^C = \xi_s^B \bar{\tau}_i^s. \qquad (13.10)$$

Note the correspondence of Eq. (13.10) to Eqs. (13.7) and (13.4). The matrix form of the above equation is expressed as

$$v_B = T^{-1} v_C \quad \text{or} \quad v_C = T v_B \quad \text{and} \quad \xi^B = \xi^C T \quad \text{or} \quad \xi^C = \xi^B T^{-1}. \qquad (13.11)$$

Hence,

> vector coefficients transform like T, covector coefficients like $(T^{-1})^\mathsf{T}$,
>
> cobasis elements transform like T, basis elements like $(T^{-1})^\mathsf{T}$.

We may also express this fact slightly differently:

> v^s transforms like β^s, ξ_s transforms like b_s,
>
> v^i transforms like γ^i, ξ_i transforms like c_i. $\qquad (13.12)$

Note that $v^s, \xi_s, v^i, \xi_i \in \mathbb{K}$, $b_s, c_i \in V$ and $\beta^s, \gamma^i \in V^*$.

We therefore see explicitly the importance of the position of the indices (upstairs and downstairs). This is a great advantage of our tensor notation within linear algebra. It allows to see immediately why and how $v \in V$ and $\xi \in V^*$ are coordinate invariant:

$$v = b_s v_B^s = c_i v_C^i \quad \text{and} \quad \xi = \xi_s^B b^s = \xi_i^C \gamma^i, \qquad (13.13)$$

with

$$v_B^s, v_C^i, \xi_s^B, \xi_i^C \in \mathbb{K}.$$

In our discussion, the distinction between V and V^* is at least implicitly always present. If we use the basis B and its dual $\mathcal{B}$, we can determine an isomorphism:

$$g : V \longrightarrow V^*$$
$$b_s \longmapsto g(b_s) = \beta^s. \qquad (13.14)$$

Needless to say that we have here a basis dependent isomorphism $g \equiv g^B$ and that the bases B and $\mathcal{B}$ are the tailor-made bases of g. So we have the representation

$$g^B_{BB} = \mathbb{1}_n.$$ (13.15)

The same can be done with the second bases C and $\mathcal{C}$ and we again get the representation

$$g^C_{CC} = \mathbb{1}_n.$$ (13.16)

Since no other structure is present in V, the explicit use of bases is needed for any isomorphism. So we may conclude this "obvious" isomorphism is not canonical as it depends on extra information. In fact, one can prove that no "canonical" isomorphism exists for general vector spaces.

Before proceeding, let us have a short break. There is a point about duality that we have to think over. The question arises of what will happen if we consider further duals. What happens if we consider $(V^*)^*$ and $((V^*)^*)^*$, and so forth. Here, we are lucky when V is finite-dimensional. In contrast to what happens in an infinite-dimensional vector space, the duality operation stops. We set $(V^*)^* \equiv V^{**} := \mathrm{Hom}(V^*, \mathbb{K})$. The reason for this break is that between V^{**} and V, there exists a basis-independent (canonical) isomorphism.:

$$ev : V \longrightarrow V^{**} = \mathrm{Hom}(V^*, \mathbb{K})$$
$$v \longmapsto ev(v) \equiv v^\# : v^\#(\xi) := \xi(v).$$ (13.17)

A direct inspection shows that the map ev is linear, injective, and surjective and we have

$$V \underset{can}{\cong} V^{**}.$$ (13.18)

We therefore can identify V^{**} with V and write $v^\# = v$ and define

$$v(\xi) := \xi(v)!$$ (13.19)

This is a fundamental relation used implicitly in tensor formalism and which justifies the dual nomenclature.

13.2 Duality and Orthogonality

We can now come to duality with the additional structure of an inner product in a vector space and so bring together duality and orthogonality. The scalar (inner) product changes the connection between V and V^* drastically, as it obtains for us, without using a basis, not only a canonical isomorphism between V and V^*, but even more a canonical isometry between V and V^*. If we want to, we can identify V and V^* as vector spaces with a scalar product, here meaning a "nondegenerate bilinear form".

In physics, without discussing this point, we make this identification from academic infancy, for example, in Newtonian mechanics, in electrodynamics, and in special relativity. Since the most critical applications of duality in physics, in special and general relativity and relativistic field theory, concern real vector spaces, we restrict ourselves in what follows to real vector spaces. For simplicity's sake, we go one step further and discuss the case of a positive definite scalar product, that is, we consider Euclidean vector spaces. The formalism is precisely the same for the more general case of a nondegenerate scalar product. In addition, it is comfortable to always have in mind our three-dimensional Euclidean vector space.

We consider an n-dimensional Euclidean vector space (V, s) with a symmetric positive definite bilinear form s:

$$s : V \times V \longrightarrow \mathbb{R}$$
$$(u, v) \longmapsto s(u, v) \equiv \langle u | v \rangle.$$

The first pleasant achievement is that we get from s a canonical isomorphism between V and V^*:

Proposition 13.1 *Canonical isomorphism.*
The map

$$\hat{s} : V \longrightarrow V^*$$
$$u \longmapsto \hat{s}(u) := s(u, \cdot) = \langle u | \cdot \rangle \equiv \hat{u} \in V^* \qquad (13.20)$$

is a canonical isomorphism between V and V^*:

$$V \cong_s V^*. \qquad (13.21)$$

Proof. The map $\hat{u}$ is a linear form since $\hat{u}(v) = s(u, v)$ and s is a bilinear form. Here, V is a real vector space and s is linear in both arguments.

- $\hat{s}$ is injective since $\ker \hat{s} = \{0\}$: If $\hat{s}(u) = \hat{u} = 0^* \in V^*$, then $\hat{u}(v) = 0^*(v) = 0 \, \forall \, v \in V$. This means that $\hat{u}(v) = \langle u | v \rangle = 0 \, \forall \, v \in V$. Setting $v = u$, we also have $\langle u | u \rangle = 0$ so that $u = 0$ and $\hat{s}$ is injective.
- $\hat{s}$ is also surjective: Since $\dim V^* = \dim V$ and $\hat{s}(V) \leq V^*$, we have $\hat{s}(V) = V$.
- $\hat{s}$ is surjective and injective and so $\hat{s}$ is a vector space isomorphism.

□

Remark 13.2 Notation for the canonical isomorphism (isometry).
We believe that the following notation is quite suggestive and useful: For the
inverse of $\hat{s}$ we use the symbol $\check{s} \equiv (\hat{s})^{-1}$.

$$\check{s} : V^* \longrightarrow V$$

$$\xi \longmapsto \check{s}(\xi) \equiv \check{\xi} \equiv u(\xi) \in V$$

so that

$$\langle u(\xi)|v\rangle \equiv \langle \check{\xi}|v\rangle := \xi(v) \in \mathbb{R}. \qquad (13.22)$$

The second achievement of the inner product, also called a metric in physics, is that
s canonically induces in V^* a metric which we denote by s^*. Therefore both, (V, s)
and (V^*, s^*), are Euclidean vector spaces. The previously defined map $\hat{s} : V \to V^*$
is now an isometry and not "only" an isomorphism (see Comment 13.1 below).

Definition 13.1 Induced inner product on V^*.
Given an inner product space (V, s), the inner product on V^* is given by

$$s^* : V^* \times V^* \longrightarrow \mathbb{R}$$

$$(\xi, \zeta) \longmapsto s^*(\xi, \zeta) \equiv \langle \xi|\zeta\rangle_* \equiv \langle \xi|\zeta\rangle, \qquad (13.23)$$

where $s(\check{\xi}, \check{\zeta}) = s^*(\xi, \zeta)$ or equivalently by

$$(\hat{u}, \hat{v}) \longmapsto s^*(\hat{u}, \hat{v}) := s(u, v). \qquad (13.24)$$

With this metric, (V^*, s^*) is also a Euclidean vector space.

Comment 13.1 *The canonical isometry.*

(i) *Equations (13.23) and (13.24) mean that $\langle \hat{s}(u)|\hat{s}(v)\rangle \equiv \langle \hat{u}|\hat{v}\rangle_* = \langle u|v\rangle$ so
that $\hat{s} : V \to V^*$ is a canonical isometry $V \underset{s}{\cong} V^*$.*
(ii) *We have*

$$\|\hat{u}\| = \|u\|, \|\check{\xi}\| = \|\xi\| \qquad (13.25)$$

and from Eqs. (13.23) and (13.24),

$$\langle \xi|\hat{v}\rangle_* \equiv \langle \xi|\hat{v}\rangle = \xi(v). \qquad (13.26)$$

Notice the correspondence between Eqs. (13.26) and (13.22).

If we choose the bases $B = (b_1, \ldots, b_n)$ and $\mathcal{B} = (\beta^1, \ldots, \beta^n)$ with $\beta^i(b_j) = \delta^i_j$, $i, j \in I(n)$, the representation of s^* is given by the matrix S^*:

$$S^* = (\sigma^{ij}) \tag{13.27}$$

with $\sigma^{ij} := s^*(\beta^i, \beta^j)$.

It is interesting that the matrices S^* and $S = (\sigma_{ij})$ with $\sigma_{ij} = s(b_i, b_j)$ represent the isometry between V and V^*.

The map $\hat{s}$ and its inverse $\check{s} \equiv (\hat{s})^{-1}$ produce a new basis in V^* and V:

$$\hat{B} := (\hat{b}_1, \ldots, \hat{b}_n) \quad \text{and} \quad \check{\mathcal{B}} := (\check{\beta}^1, \ldots, \check{\beta}^n).$$

There are now two bases in V^*: $\mathcal{B}$ and $\hat{B}$.

To keep with our convention of using latin letters for vectors in V and greek letters for covectors in V^*, we have to redefine the elements $\hat{b}_i$ and $\check{\beta}^i$ and to write:

- $\beta_i := \hat{b}_i$ and $b^i := \check{\beta}^i$, so we write
- $\hat{B} = (\beta_1, \ldots, \beta_n)$ compared to $\mathcal{B} = (\beta^1, \ldots, \beta^n)$, both in V^* ;
- $\check{B} = (b^1, \ldots, b^n)$ compared to $B = (b_1, \ldots, b_n)$, both in V.

We show and proof all the above assertions in the next proposition.

> **Proposition 13.2** *Representation of the isomorphism $\hat{s} : V \to V^*$.*
> *Let V be a real inner product space. With all notation defined as above, the following statements hold true.*
>
> (i) $\beta_i = \sigma_{ij}\beta^{ij}$ and $S = (\sigma_{ij})$ invertible;
> (ii) $S^* = S^{-1}$, $S^*S = \mathbb{1}_n$ and $\sigma^{ik}\sigma_{kj} = \delta^i_j$;
> (iii) $b^i = \sigma^{ij}b_j$.

Proof. (i)

Since $\hat{s}(b_i) \in V^*$, we may write $\hat{s}(b_i) = \lambda_{ij}\beta^{ij}$ with $\lambda_j \in \mathbb{R}$. On the other hand, we have $\hat{s}(b_i) = \hat{b}_i \equiv \beta_i$. This leads to $\lambda_{ij}\beta^i(b_k) = \beta_i(b_k)$, giving

$$\lambda_{ij}\delta^i_k = \beta_i(b_k) \equiv \hat{b}_i(b_k) = \langle b_i | b_k \rangle = \sigma_{ik}$$

and so $\lambda_{ik} = \sigma_{ik}$. S is invertible since $\hat{s}$ is bijective.

Proof. (ii)

As $\hat{s}$ is an isometry, we have

$$\sigma_{ij} = s(b_i, b_j) = s^*(\hat{b}_i, \hat{b}_j) \equiv s^*(\beta_i, \beta_j) \stackrel{(i)}{=}$$
$$= s^*(\sigma_{ik}\beta^k, \sigma_{jl}\beta^l) = \sigma_{ik}s^*(\beta^k, \beta^l)\sigma_{jl} = \sigma_{ik}\sigma^{kl}\sigma_{lj}$$

or equivalently in matrix form s

$$S = S\,S^*\,S \quad \text{since} \quad \mathbb{1}_n = S^*\,S \quad \text{since} \quad (S^* = S^{-1}).$$

This means that $\sigma^{ik}\sigma_{kj} = \delta^i_j$. Here, we do not write the index B in S since it is not necessary.

Proof. (iii)
Analogously to (i), since $\check{\beta}^i \in V$, we have $\check{\beta}^i = \mu^{ik} b_k$ with $\mu^{ik} \in \mathbb{R}$. Taking the scalar product on both sides, we obtain $\mu^{ik}\langle b_k | b_j \rangle = \langle \beta^i | b_j \rangle$. This leads to

$$\mu^{ik}\sigma_{kj} = \beta^i(b_j)$$

$$\text{and to} \quad \mu^{ik}\sigma_{kj} = \delta^i_j,$$

$$\text{which means} \quad \mu^{ik} = \sigma^{ik}.$$

$$\square$$

All this shows that, mathematically, the two Euclidean vector spaces (V, s) and (V^*, s^*) are indistinguishable and can be identified. This means we are left with only one vector space V but with two bases, $(b_1, \ldots, b_n)$ and $(b^1, \ldots, b^n)$ in V, and the following relations:

$$\langle b_i | b_j \rangle = \sigma_{ij}, \quad \langle b^i | b^j \rangle = \sigma^{is} \quad \text{and} \quad \langle \beta^i | b_j \rangle = \delta^i_j.$$

This can be shown by a direct calculation:

$$\langle b^i | b^j \rangle \equiv \langle \check{\beta}^i | \check{\beta}^j \rangle \underset{isometric}{=} \langle \beta^i | \beta^j \rangle = \sigma^{ij},$$

$$\langle b^i | b_j \rangle \underset{(a)}{=} \langle b^i | \sigma_{jk} b^k \rangle = \langle b^i | b^k \rangle \sigma_{jk} = \sigma^{ik}\sigma_{jk} = \delta^i_j.$$

These two bases, bases $(b_i)_n$ and $(b^i)_n$, are used in many fields in physics and are often called reciprocal. The following discussion clarifies this situation even more.

The representations S and S^* of the isometry between V and V^* are symmetric but not diagonal. We expect that if we find a tailor-made basis, we can diagonalize S and S^*. We can indeed do this if we choose an orthonormal basis $C = (c_1, \ldots, c_n)$, with its dual $\mathcal{C} = (\gamma^1, \ldots, \gamma^n)$ and $\gamma^i(c_j) = \delta^i_j$. The matrices S_C and S^*_C also represent the scalar products s and s^*, as they were introduced before. It is evident that with the basis C we have

$$\sigma^C_{ij} = s(c_i, c_j) = \delta_{ij} \quad \text{and so} \tag{13.28}$$

$$S = \mathbb{1}_n \quad \text{and} \quad S^* = \mathbb{1}_n. \tag{13.29}$$

Furthermore, if we consider C and $\check{C}$, the two bases in V, and $\mathcal{C}$ and $\hat{C}$, the two bases in V^*, we have

$$c_i = \sigma_{ij}^C c^j = \delta_{ij} c^j = c^i \quad \text{and} \qquad (13.30)$$

$$\gamma_i = \sigma_{ij}^C \gamma^j = \gamma^i \quad \text{which means} \qquad (13.31)$$

$$C = \check{C} \quad \text{and} \quad C = \hat{C}. \qquad (13.32)$$

For this reason, we often use expressions (see Eq. 13.30) like

$$v = c_s v^s = c_s \langle c^s | v \rangle. \qquad (13.33)$$

So, in this case, with an orthonormal basis in V, we need only to consider one basis in V and one basis in its dual V^*. In most cases in physics and geometry, it is reasonable to work with an orthonormal basis. Nevertheless, there are cases where choosing an orthonormal basis is not possible, for example , when selecting coordinate bases on a manifold with nonzero curvature. This is one of the main reasons why we considered the general situation in the discussions of this section.

Summary

This chapter was introduced to dispel certain biases in physics related to the dual space of a vector space and its necessity.

In physics, vector spaces with an inner product are initially used. As demonstrated here, this renders the dual space practically obsolete and therefore leads to a series of misunderstandings. However, a residual form remains known as the reciprocal basis, which finds broad application in certain areas of physics. Reciprocal bases are clearly visible and distinguishable when nonorthogonal bases are used.

In the context of tensors, it is also important to differentiate whether the tensors are defined on an abstract vector space or on an inner product space. For example, the tensors we have discussed so far have all been tensors in an abstract vector space. It is only in the next chapter that tensors defined on an inner product space will appear for the first time. These are precisely the tensors that are preferred in physics.

Chapter 14
Tensor Formalism

Without any exaggeration, we can say that in physics tensor formalism is needed and used as much as linear algebra itself. For example, it is impossible to understand electrodynamics, relativity, and many aspects of classical mechanics without tensors. Therefore, there is no doubt that a better understanding of tensor formalism leads to a better understanding of physics.

Engineers and physicists first came across tensors in terms of indices to describe certain states of solids. They were first realized as very complicated, unusual objects with many indices, and their mathematical significance was even questionable. Later, mathematicians found out that these objects correspond to a very precise and exciting mathematical structure. It turned out that this structure is a generalization of linear algebra. This is multilinear algebra and can be considered the modern name for what physicists usually refer to as tensor calculus.

This chapter will discuss tensor formalism, also known as multilinear algebra, in a basis-independent way. But of course, as we know from linear algebra, we cannot do without basis-dependent representations of tensors. From Sect. 3.5 and Chap. 8 we already know what tensors are, and we know at least one possibility to arrive at tensor spaces. Before, this was obtained by explicitly utilizing bases of vector spaces. Now, we would like to achieve this differently, which will allow us to further expand and consolidate the theory of tensors.

14.1 Covariant Tensors and Tensor Products

We start with the most general definition of a multilinear map and consider a vector-valued multilinear map. Dealing with various special cases later will allow a better understanding of the topic.

© The Author(s), under exclusive license to Springer Nature Switzerland AG 2024
N. A. Papadopoulos and F. Scheck, *Linear Algebra for Physics*,
https://doi.org/10.1007/978-3-031-64908-0_14

Definition 14.1 Multilinear vector valued map.
Suppose $V_1, \ldots, V_k$ and Z are vector spaces. A map

$$f : V_1 \times \cdots \times V_k \longrightarrow Z$$
$$(v_1, \ldots, v_k) \longmapsto f(v_1, \ldots, v_k) \in Z$$

is said to be multilinear if it is linear in every variable individually, so with $\lambda, \mu \in \mathbb{R}$:

$$F(v_1, \ldots, \lambda u_i + \mu v_i, \ldots, v_k) = \lambda f(v_1, \ldots, u_i \ldots v_k) + \mu f(v_1, \ldots, v_i, \ldots v_k).$$

We denote the space of such multilinear maps by

$$T(V_1, \ldots V_k; Z).$$

It is clear that if we take $k = 1$, we have a linear map $f \in T(V; Z) \equiv \mathrm{Hom}(V, Z)$. If we take $k = 2$, $f \in T(V_1, V_2; Z)$ is a vector valued bilinear form and for $k = 3$ we may talk about trilinear forms. Since Z is a vector space, it is evident that Z also induces a vector structure on $T(V_1, \ldots, V_k; Z)$:

$$(f + g)(v_1, \ldots, v_k) := f(v_1, \ldots, v_k) + g(v_1, \ldots, v_k)$$

and

$$(\lambda f)(v_1, \ldots, v_k) := \lambda f(v_1, \ldots, v_k).$$

If we take $Z = \mathbb{R}$, we may also write

$$T(V_1, \ldots V_k; \mathbb{R}) \equiv T(V_1, \ldots V_k).$$

and in this case, we call f a multilinear $\mathbb{R}$ valued function or a multilinear form. If we take $V_1 = \cdots = V_k =: V$, we may also write

$$T^k(V; Z) \quad \text{or} \quad T^k(V, \mathbb{R}) \equiv T^k(V).$$

Definition 14.2 A covariant k-tensor.
A covariant tensor on V is a real valued multilinear function of k vectors of V:

$$\varphi: \underbrace{V \times \cdots \times V}_{k\text{-times}} \longrightarrow \mathbb{R}$$

$$(v_1, \ldots, v_k) \longmapsto \varphi(v_1, \ldots, v_k).$$

Such φ are also called k-linear forms of k-forms or k-tensors or more precisely covariant k-tensors.

When we talk about multilinear maps, we can ask what differences there are between the properties of multilinear and linear maps. This leads us to the following comment and remark.

Comment 14.1 The image of multilinear maps.

Although the image of a linear map, as we know, is a vector space, the image of a multilinear map is not, in general. A simple way to verify this fact is to consider a bilinear map given by the product of two polynomials. Consider the $(n + 1)$-dimensional and $(2n + 1)$-dimensional spaces of polynomials;

$$\mathbb{R}[x]_n := \mathrm{span}(1, x, \ldots, x^n) \quad \text{and} \quad \mathbb{R}[x]_{2n} := \mathrm{span}(1, x, \ldots, x^{2n}).$$

We set

$$V = \mathbb{R}[x]_1 = \{\alpha_0 + \alpha_1 x : \alpha_0, \alpha_1 \in \mathbb{R}\}$$

and

$$W = \mathbb{R}[x]_2 = \{x_0 + \alpha_1 x + \alpha_2 x^2 : \alpha_0, \alpha_1, \alpha_2 \in \mathbb{R}\}.$$

If we consider $f \in T^2(V; W)$, given by

$$f : V \times V \longrightarrow W$$

$$(\varphi, \chi) \longmapsto f(\varphi, \chi) = \varphi \cdot \chi,$$

and take $\varphi, \chi \in \{1, x\}$, we see that $1, x^2 \in \mathrm{im}\, f$ since $1 \cdot 1 = 1$ and $x \cdot x = x^2$. Yet, we observe that $1 + x^2 \notin \mathrm{im}\, f$ because otherwise $1 + x^2$ would have real roots, but as we know, both roots of $1 + x^2 = (x - i)(x + i)$ are strictly imaginary.

Our next remark is more pleasant.

Remark 14.1 The role of bases in multilinear maps.

Just as in the case of linear maps, any multilinear map $f \in T(V_1, \ldots, V_k; Z)$ is uniquely determined by its values on the basis vectors of the V_s.

Proof To avoid risking confusion due to burdensome notation, we will only prove this in the case of a bilinear map $f \in T(U, V; Z)$: We take $B = (b_1, \ldots, b_k)$, a basis of U, and $C = (c_1, \ldots, c_l)$, a basis of V. We set

$$f(b_s, c_i) = z_{si} \in Z,$$
$$s \in I(k), \ i \in I(l),$$

and for $u = b_s u_B^s \in U$ and $v = c_i v_C^i \in V$, we define:

$$f(u, v) := f(b_s, c_i) u_B^s v_C^i = z_{si} u_B^s v_C^i.$$

This shows the uniqueness since every single one of z_{si}, u_B^s, v_C^i is uniquely given.

This f is bilinear as the partial maps $f_u : V \to Z$ and $f_v : U \to Z$ are linear. For f_u we have for example,

$$f_u(v) = f(u, v) = z_{si} u_B^s v_C^i = (z_{si} \cdot u_B^s) v_C^i$$

which is explicitly linear in V.   □

Remark 14.2 Maps of covariant tensors, the pullback.

A fact that is valid only for covariant tensors, is the connection of covariant tensors on different vector spaces: Any linear map of vector spaces,

$$F : \ V \longrightarrow W$$
$$v \longmapsto F(v) = w$$

induces a linear map F^*

$$F^* : T^k(W) \longrightarrow T^k(V)$$
$$\psi \longmapsto F^*\psi,$$

given by $(F^*\psi)(v_1, \ldots, v_k) := \psi(Fv_1, \ldots, Fv_k).$

This generalizes the dual map of covectors:

$$F^* : W^* \longrightarrow V^*$$
$$\eta \longmapsto F^*\eta, \quad \text{given by}$$
$$(F^*\eta)(v) := \eta(Fv).$$

We may call F^* pullback of covariant tensors.

We now present several examples which illuminate various aspects of multilinear maps and thus also proceed towards the definition of a tensor. This should also show how natural the notion of tensor product is, even if it looks very complicated.

14.1.1 Examples of Covariant Tensors

Example 14.1 Bilinear Maps
As we have already seen in Comment 14.1 and Remark 14.1, multiplication provides an example of a bilinear map.

Example 14.2 Bilinear Forms
The inner product in (V, s) and the dot product in $(\mathbb{R}, <|>)$ are, as we know, symmetric bilinear forms.

Example 14.3 Constant (Scalars)
When $k = 0$, we define $T^0(V) = \mathbb{R}$ and the elements of $T^0(V)$ are usually called scalars.

Example 14.4 Linear Forms
When $k = 1$ we have

$$T^1(V, \mathbb{R}) \equiv \mathrm{Hom}(V, \mathbb{R}) \equiv V^*$$

and any element of V^* is called a covector, a linear form, a linear functional, or a 1-form. The different names for $\xi \in V^*$ demonstrate appropriately the importance of V^*. Indeed, this V^* is very special in the context of multilinear forms because, as we shall see, it generates multiplicatively $T^k(V)$ for every $k \in \{2, 3, \dots\}$.

The following examples also give a good idea of the corresponding multiplication, the tensor product.

Example 14.5 On the Way to Tensor Products
This example shows a basic covariant tensor that is nontrivial and created by two covectors leading to an elementary tensor product. Given the covectors ξ and η, we can obtain in a natural way from the Cartesian product $V^* \times V^*$, the bilinear form $\xi\eta$ on V:

$$\begin{aligned} \xi\eta : \quad V \times V \quad &\longrightarrow \quad \mathbb{R} \\ (u, v) \quad &\longmapsto \quad \xi\eta(u, v) := \xi(u)\eta(v). \end{aligned} \tag{14.1}$$

This is obviously a nontrivial covariant tensor $\xi\eta \in T^2(V)$.
At the same time, we obtain a bilinear product P from the Cartesian product $V^* \times V^*$ to the bilinear forms in V:

$$\begin{aligned} P : \quad V^* \times V^* \quad &\longrightarrow \quad T^2(V) \\ (\xi, \eta) \quad &\longmapsto \quad P(\xi, \eta) := \xi\eta \in T^2(V). \end{aligned} \tag{14.2}$$

Comment 14.2 The bilinear form $\xi\eta$ and analysis.

The bilinear map P can also be defined on all functions, not only on linear maps. Given two sets X and Y and the two functions f and g correspondingly on X and Y (we write $f \in \mathcal{F}(X) \equiv \mathrm{Map}(X, \mathbb{R})$ and $g \in \mathcal{F}(Y)$), we may obtain a new function on the Cartesian product $X \times Y$:

$$
\begin{aligned}
fg: \quad X \times Y &\longrightarrow \mathbb{R} \\
(f, g) &\longmapsto fg(x, y) := f(x)g(y).
\end{aligned}
\tag{14.3}
$$

At the same time, we obtain the product P:

$$
\begin{aligned}
P: \quad \mathcal{F}(X) \times \mathcal{F}(Y) &\longrightarrow \mathcal{F}(X \times Y) \\
(f, g) &\longmapsto fg.
\end{aligned}
\tag{14.4}
$$

This product can also be called a tensor product. The product fg is naturally induced since it is based on the multiplication on their common codomain $\mathbb{R}$.

There is also a complementary view concerning the above product P given in Eq. (14.2). Instead of $T^2(V)$, we may introduce a second vector space $V^* \otimes V^*$ which is directly algebraically generated from the Cartesian product $V^* \times V^*$ by a new product which we denote here by Θ. In this way, we obtain a second bilinear map given by

$$
\begin{aligned}
\Theta: \quad V^* \times V^* &\longrightarrow V^* \otimes V^* \\
(\xi, \eta) &\longmapsto \Theta(\xi, \eta) := \xi \otimes \eta.
\end{aligned}
\tag{14.5}
$$

If we use a basis $B^* = (\beta^1, \ldots, \beta^n) = (\beta^i)_n \subseteq V^*$, we can make this more concrete with $i, j \in I(n)$:

$$
(\beta^i, \beta^j) \longmapsto \Theta(\beta^i, \beta^j) := \beta^i \otimes \beta^j.
\tag{14.6}
$$

So we obtain

$$
V^* \otimes V^* = \mathrm{span}\{\beta^i \otimes \beta^j : i, j \in I(n)\}.
\tag{14.7}
$$

In addition, we can see that $V^* \otimes V^*$ is a free vector space (see Definition 8.1 since every set can be used as a basis of some vector space) over $B^* \times B^* = \{(\beta^i, \beta^j) : i, j \in I(n)\}$ or equivalently over the set $I(n) \times I(n) =$

$\{(i, j)$ where $i, j \in I(n)\}$ since every set can be used as a basis of a vector space!

$$V^* \otimes V^* \cong \mathbb{R}(I(n) \times I(n)). \tag{14.8}$$

As expected, Θ is the tensor product and $V^* \otimes V^*$ is a tensor space. It turns out that this tensor product Θ has a more general character. This means that we also have a tensor product for V. We denote it by the same symbol:

$$\Theta : V \times V \longrightarrow V \otimes V$$
$$(u, v) \longmapsto \Theta(u, v) := u \otimes v. \tag{14.9}$$

As we see, via the same kind of product Θ, we also obtain a tensor space $V \otimes V$ over V. For this reason, we see various symbols in literature which are used instead of our character Θ, such as the symbol $\otimes$.

To fully understand what is happening, we have to compare $\xi\eta$ with $\xi \otimes \eta$. We do not expect, of course, $\xi\eta$ and $\xi \otimes \eta$ to be equal, and for the same reason, we do not expect $T^2(V)$ and $V^* \otimes V^*$ to be the same.

What we expect is the two vector spaces above to be equivalent, that is, we expect a canonical isomorphism

$$V^* \otimes V^* \cong T^2(V). \tag{14.10}$$

Indeed, both have the same dimension and the exact relationship between $\xi\eta$ and $\xi \otimes \eta$ is given by the following equation:

$$\xi \otimes \eta(u \otimes v) := \xi\eta(u, v) = \xi(u)\eta(v). \tag{14.11}$$

This can also be expressed by the commutative diagram:

So we have:

$$\tilde{P}(\xi \otimes \eta)(u, v) = P(\xi, \eta)(u, v) = \xi\eta(u, v) = \xi(u)\eta(v). \tag{14.12}$$

We observe that P is a bilinear and $\tilde{P}$ a linear map. Because of the above isomorphism $V^* \otimes V^* \cong T^2(V)$, we may identify $\xi \otimes \eta$ with $\xi\eta$ and write $\xi \otimes \eta(u, v) = \xi\eta(u, v)$, and we use the unambiguous notation $\xi \otimes \eta$, as is often done in the literature.

Our intention is to give an informal introduction to tensor products with these examples. It might be instructive to see the same results for $k = 3$, as in the following example.

Example 14.6 Tensor Product
Starting with the covectors $\xi, \eta, \theta \in V^*$ and proceeding similarly as in example 5, we may write for their tensor product,

$$\xi \otimes \eta \otimes \theta \in V^* \otimes V^* \otimes V^*$$

given by

$$(\xi \otimes \eta \otimes \theta)(u, v, w) = \xi(u)\eta(v)\theta(w) \tag{14.13}$$

and we expect a canonical isomorphism given by

$$T^3(V) \cong V^* \otimes V^* \otimes V^*. \tag{14.14}$$

Comment 14.3 Product of polynomials as tensor product.

Taking into account the above considerations and coming back to Comment 14.1, we can additionally give a new interpretation of the product of polynomials: Taking $f = \Theta$, we have $\Theta(\varphi, \chi) = \varphi \otimes \chi$. This means that we can also consider the product of polynomials as example of a tensor product.

We are now ready for the general definition of a tensor product.

Definition 14.3 Tensor product on V.
For two given covariant tensors, $T \in T^k(V)$ and $S \in T^l(V)$, the tensor product $\otimes$ is defined by

$$\Theta : T^k(V) \times T^l(V) \longrightarrow T^{k+l}(V)$$
$$(T, S) \longmapsto T \otimes S(v_1, \dots, v_k, v_{k+1}, \dots, v_{k+l}) := T(v_1, \dots, v_k)S(v_{k+1\dots, v_{k+l}}).$$

We say $T \otimes S$ is a covariant tensor of rank or order $k + l$.

From the multilinearity of T and S follows the multilinearity of their product since the expression

$$T(v_1, \ldots, v_k) \, S(v_{k+1}, \ldots, v_{k+l})$$

is linear in each argument v_i, $i \in I(k + l)$ separately. It is further clear that the operation

$$(T, S) \overset{\Theta}{\mapsto} T \otimes S$$

is bilinear; since when $T, T_1, T_2 \in T^k(V)$, $S, S_1, S_2 \in T^l(V)$ and $\lambda_1, \lambda_2, \mu_1, \mu_2 \in \mathbb{R}$:

$$(\lambda_1 T_1 + \lambda_2 T_2) \otimes S = \lambda_1 (T_1 \otimes S) + \lambda_2 (T_2 \otimes S)$$

and

$$T \otimes (\mu_1 S_1 + \mu_2 S_2) = \mu_1 (T \otimes S_1) + \mu_2 (T \otimes S_2).$$

As we see, tensor product is bilinear, just as every product should be.

In addition, it is associative:

$$L \otimes (T \otimes S) = (L \otimes T) \otimes S$$

which is also easy to verify. This means that we can write tensor products of several tensors without parentheses.

Since $T^k(V)$ is a vector space, we can choose a basis and determine the coefficients of $T \in T^k(V)$.

Proposition 14.1 *Basis in $T^k(V)$.*
Let V be a vector space. If $B = (b_1, \ldots, b_n)$ is a basis of V and $B^* = (\beta^1, \ldots, \beta^n)$ a basis of V^* with $\beta^j(b_i) = \delta_i^j$ $i, j \in I(n)$, then there is a basis of $T^k(V)$ given by

$$B^{(k)} = \{\beta^{i_1} \otimes \ldots \otimes \beta^{i_k} : i_1, \ldots, i_k \in I(n)\}. \qquad (14.15)$$

It follows that when $T \in T^k(V)$, the decomposition

$$T = \tau_{i_1 \ldots i_k} \beta^{i_1} \otimes \ldots \otimes \beta^{i_k},$$

with $\tau_{i_1 \ldots i_k} \in \mathbb{R}$ and $\dim T^k(V) = n^k$.

Proof For a basis we have to show that

(i) $\mathrm{span}(B^{(k)}) = T^k(V)$,
(ii) $B^{(k)}$ is linearly independent .

We first show (i).

The multilinear form T is uniquely determined by its values on certain lists of basis vectors like $(b_{i_1}, \ldots, b_{i_k})$ by Remark 14.1. Therefore, T is defined by the values

$$T(b_{i_1}, \ldots, b_{i_k}) = \tau_{i_1 \ldots i_k}. \tag{14.16}$$

That $B^{(k)}$ spans $T^k(V)$ means that we have T as a linear combination

$$T = \lambda_{i_1 \ldots i_k} \beta^{i_1} \otimes \ldots \otimes \beta^{i_k} \quad \text{with} \quad \lambda_{i_1 \ldots i_k} \in \mathbb{R}.$$

Using the multilinearity of T, we may determine $\lambda_{i_1 \ldots i_k}$, taking

$$\begin{aligned}
T(b_{j_1}, \ldots, b_{j_k}) &= \lambda_{i_1 \ldots i_k} (\beta^{i_1} \otimes \ldots \otimes \beta^{i_k})(b_{j_1}, \ldots, b_{j_k}) \\
&= \lambda_{i_1 \ldots i_k} (\beta^{i_1}(b_{j_1})) \ldots \beta^{i_k}(b_{j_k}) \\
&= \lambda_{i_1 \ldots i_k} \delta^{i_1}_{j_1} \ldots \delta^{i_k}_{j_k} \\
&= \lambda_{j_1 \ldots j_k}.
\end{aligned}$$

This gives, based on Eq. (14.16), $\lambda_{i_1 \ldots i_k} = T_{i_1 \ldots i_k}$ and (a) holds.

To show (ii), we set $\lambda_{i_1} \ldots i_k \beta^{i_1} \otimes \ldots \otimes \beta^{i_k} = 0$ and by the same computation as above, we obtain $\lambda_{i_1} \ldots i_k = 0$, so (b) is also valid and the proposition is proven. $\qquad\square$

Comment 14.4 Determination of the coefficients of tensors.

Notice that with this proof we once again showed that the coefficients of T are given by $T(b_{j_1}, \ldots, b_{j_k})$. Additionally, this result makes the expression

$$T^k(V) \cong \underbrace{V^* \otimes \cdots \otimes V^*}_{k\text{-times}}$$

and the corresponding identification

$$T^k(V) = \underbrace{V^* \otimes \cdots \otimes V^*}_{k\text{-times}}$$

plausible.

As we see, every covariant tensor can be written as a linear combination of tensor products of covectors. Further more, we may always assume that every $T \in T^k(V)$ can be written as $T = \mu_{i_1 \ldots i_k} \xi^{i_1} \otimes \ldots \otimes \xi^{i_k}$ with $\xi^{i_1}, \ldots, \xi^{i_k} \in V^*$ not necessarily basis vectors of V^*. We see this if we take

$$\xi^{i_1} = \xi_j^{i_1} \beta^j, \quad \ldots, \quad \xi^{i_k} = \xi_j^{i_k} \beta^j, \tag{14.17}$$

with

$$\xi_j^{i_1}, \ldots, \xi_j^{i_k} \in \mathbb{R} \tag{14.18}$$

and do the same calculation as above.

Definition 14.4 Decomposable tensors.
A tensor of the form $\xi^1 \otimes \ldots \otimes \xi^k$ is called decomposable, $\xi^1, \ldots, \xi^* \in V^*$.

We stated above that the set of decomposable tensors spans $T^k(V)$.

Comment 14.5 Tensor algebra on a vector space.

We may ask: Well, we see the product, the so-called tensor product. But where is the algebra? Because in Definition 14.3, we leave both, $T^k(V)$ and $T^l(V)$.

The answer is that we have to take all the tensor spaces together and so to obtain the tensor algebra over V:

$$T^*(V) := T^0(V) \oplus T^1(V) \oplus \cdots \oplus T^k(V) \oplus \cdots .$$

14.2 Contravariant Tensors and the Role of Duality

In Example 14.5 in the last section, we also saw, in addition to the covariant tensor space $V^* \otimes V^*$, the tensor space of contravariant tensors $V \otimes V$. It is reasonably possible that as a student of physics, the first tensor space we meet is $V \otimes V$. We first started with covariant tensors because, coming from analysis, it seems quite natural to introduce and discuss the tensor product within covariant tensors. This was also demonstrated in Example 14.5.

It is, therefore, necessary to explain the connection between covariant and contravariant tensors. As the reader has probably already realized, this connection has a name: duality. To clarify the role of duality thoroughly, it is helpful to revue what we know and to consider all the relevant possibilities which appear. If we start by comparing V with its dual V^*, we also have to compare V^* with its dual $(V^*)^* \equiv V^{**}$, the so-called double dual of V. We cannot stop here since we have also to compare V^{**} with $(V^{**})^*$ and so on. But when the dimension of V is finite, this procedure stops.

As we saw in Eq. 13.17 in Sect. 13.1, it turns out that V^{**} is canonically isomorphic to $V : V^{**} \underset{can}{\cong} V$. As in Sect. 14.1, V is an abstract vector space without further structure. We know that in this case V and V^* are isomorphic, but this isomorphism is basis dependent: $V \underset{B}{\cong} V^*$ (noncanonical). So we have, for example,

$$\psi_B : V \longrightarrow V^*$$
$$b_i \longmapsto \beta^i \tag{14.19}$$

with $\beta^i(b_j) = \delta^i_j$ $i, j \in I(n)$. The identification between V and V^* in the case of an abstract vector space V, is not possible. Again, if we want to compare V^* with $(V^*)^* \equiv V^{**} \equiv \mathrm{Hom}(V^*, \mathbb{R})$, the same is true for the same reason.

Yet, if we want to compare V with V^{**}, the situation changes drastically. Here, there exists a canonical isomorphism and an identification is possible:

$$ev : V \underset{can}{\overset{\cong}{\to}} V^{**}$$
$$v \longmapsto v^{\#} := ev(v) \in \mathrm{Hom}(V^*, \mathbb{R}), \tag{14.20}$$

given by

$$v^{\#} : V^* \longrightarrow \mathbb{R}$$
$$\xi \longmapsto v^{\#}(\xi) := \xi(v). \tag{14.21}$$

This is the reason why we may identify V with $V^{**} : V^{**} = V$, and we do not make a distinction between the elements $v^{\#}$ and v. This identification is beneficial since it simplifies a lot the tensor formalism. On the other hand, if we want to profit from this simplification, we have to get a good understanding of this identification.

Proceeding similarly as in Sect. 14.1, we consider contravariant tensors essentially by exchanging V and V^*. We now consider multilinear functions on V^* instead of considering them on V. So the contravariant tensor of order k, which we denote by $T^k(V^*)$,

$$T^k(V^*) \cong \underbrace{V \otimes \cdots \otimes V}_{k\text{-times}}.$$

Taking into account Proposition 14.1 and using the notation $B^{\#} = (b^{\#}_1, \ldots, b^{\#}_n)$ for a basis in V^{**} and the identification $b^{\#}_i = b_i$, we obtain for $T \in T^k(V^*)$

$$T = b^{\#}_{i_1} \otimes \cdots \otimes b^{\#}_{i_k} \, T^{i_1 \cdots i_k} = b_{i_1} \otimes \cdots \otimes b_{i_k} \, T^{i_1 \cdots i_k}.$$

That should be compared to the actual result in Proposition 14.1 for a covariant tensor.

$$T \in T^k(V):$$

$$T = \tau_{i_1 \cdots i_k} \; \beta^{i_1} \otimes \cdots \otimes \beta^{i_k}.$$

For $T^k(V^*)$ we find in the literature also the notation $T_k(V) := T^k(V^*)$. So we have, including the corresponding identifications, the expressions

$$T_k(V) = \underbrace{V \otimes \cdots \otimes V}_{k\text{-times}} \quad \text{and} \quad T^k(V) = \underbrace{V^* \otimes \cdots \otimes V^*}_{k\text{-times}}.$$

14.3 Mixed Tensors

We start with an example of a mixed tensor, the following canonical bilinear form given by V^* and V which is the evaluation of covectors on vectors:

$$\phi : V^* \times V \longrightarrow \mathbb{R}$$
$$(\xi, v) \longmapsto \phi(\xi, v) := \xi(v). \tag{14.22}$$

This bilinear form ϕ is nondegenerate and inherently utilizes the canonical isomorphism between V and V^{**}, given by the partial map ψ_v.

$$\tilde{\phi} : V \longrightarrow (V^*)^*$$
$$v \longmapsto \psi_v. \quad \text{with} \tag{14.23}$$
$$\psi_v : V^* \longrightarrow \mathbb{R}$$
$$\xi \longmapsto \psi_v(\xi) = \phi(\xi, v) = \xi(v). \tag{14.24}$$

This ψ_v is the same as the map $v^\#$ of Eq. (14.21) and we notice again the identification $\psi_v = v^\#$. With this preparation, it is easy to define a mixed tensor.

Definition 14.5 Mixed tensors.

Let V be a finite as usual dimensional vector space and k and l be nonnegative integrals. The space of mixed tensors on V of type (k, l) is given by

$$T_l^k(V) = \underbrace{V^* \otimes \cdots \otimes V^*}_{k\text{-times}} \otimes \underbrace{V \otimes \cdots \otimes V}_{l\text{-times}}. \tag{14.25}$$

Using again the reasoning of Proposition 14.1, we may denote a mixed tensor $S \in T_l^k(V)$ by:

$$S = \sigma^{j_1 \cdots j_l}_{i_1 \cdots i_k} \beta^{i_1} \otimes \cdots \otimes \beta^{i_k} \otimes b_{j_1} \otimes \cdots \otimes b_{j_l}. \tag{14.26}$$

With the corresponding coefficients of S given in the basis (B, B^*), we get

$$S = \sigma_{i_1 \cdots i_k}^{j_1 \cdots j_l} = S(b_{i_1}, \cdots, b_{i_l} \beta^{j_1}, \cdots, \beta^{j_k}) \in \mathbb{R}.$$

Mixed tensors are the most general case in which we need a change of basis formula. Using the notation of Example 8.1, we consider the bases $B = (b_s)$, $B^* = (\beta^s)$ and $C = (c_i)$, $C^* = (\gamma^i)$ of V and V^* correspondingly, with $\beta^r(b_s) = \delta^r_s$ and $\gamma^j(c_i) = \delta^j_i$, and $r, s, i, j \in I(n)$.

The change of basis transformation is given by the regular matrix:

$$
\begin{aligned}
T &= T_{CB} = (\tau^i_s) \in Gl(n) \\
\text{and} & \\
T^{-1} &= T_{BC} = (\bar{\tau}^s_i).
\end{aligned}
$$

We denote the coefficient of S in the basis (B, B^*) by $\sigma(B)_{r_1 \cdots r_k}^{s_1 \cdots s_l}$ and in the basis (C, C^*) by $\sigma(C)_{j_1 \cdots j_k}^{i_1 \cdots i_l}$.

So the transition map at the level of tensors is:

$$\sigma(C)_{j_1 \cdots j_k}^{i_1 \cdots i_l} = \tau^{i_1}_{s_1} \cdots \tau^{i_l}_{s_l} \bar{\tau}^{r_1}_{j_1} \cdots \bar{\tau}^{r_k}_{j_k} \sigma(B)_{r_1 \cdots r_k}^{s_1 \cdots s_l}.$$

14.4   Tensors on Semi-Euclidean Vector Spaces

In the last section, the vector space V was an abstract vector space without further structure. This section considers tensors on a vector space with additional structures. The most crucial additional structure on an abstract vector space in physics is a Euclidean or semi-Euclidean structure. We add to the vector space V a symmetric nondegenerate bilinear form $s \in T^2(V)$ which is, as we know, a special covariant tensor of rank 2. So we obtain (V, s), a Euclidean vector space if s is positive definite or a semi-Euclidean vector space (e.g. Minkowski space in special relativity) if s is symmetric and nondegenerate.

In both cases, the connection between V and V^* changes drastically. As we saw in Sect. 13.2, given the metric s, there exists a canonical isometry between V and V^* :

$$(V, s) \underset{can}{\overset{\cong}{=}} (V^*, s^*).$$

This means that we identify not only V with V^{**} but also V with V^*. The identification $V = V^*$ essentially indicates that we can forget the dual V^* and that the vector space V alone is relevant for the tensor formalism. This means that the distinction between covariant and contravariant tensors is obsolete: Given (V, s), we also have the identification $T^k(V^*) = T^k(V)$ or equivalently

$$\underbrace{V^* \otimes \cdots \otimes V^*}_{k-times} = \underbrace{V \otimes \cdots \otimes V}_{k-times}.$$

We therefore have only one kind of tensor. The distinction between covariant and contravariant tensors is only formal and refers to the representation (the coefficients) of a tensor relative to a given basis in V. This follows directly from our discussion in Sect. 13.2. The canonical isometry brings the basis $B^* = (\beta^1, \ldots, \beta^n)$ in V^* down to V (see Remark 13.2):

$$\check{s}(\beta^1) = b^1, \ldots, \check{s}(\beta^n) = b^n.$$

So we have two bases in V, the original $B = (b_1 \ldots, b_n)$ and $\check{B} = (b^1, \ldots, b^n)$ given by $b^j = \sigma^{ji} b_i$ (Proposition 13.2), with $s(b_i, b^j) = < b_i \mid b^j > = \delta_i^j$. We call B a covariant basis and $\check{B}$ a contravariant basis of V. For a given tensor $T \in T^k(V)$, we have two corresponding representations given by

$$T = \tau^{i_1 \cdots i_k} b_{i_1} \otimes \cdots \otimes b_{i_k} = \tau_{i_1 \cdots i_k} \otimes b^{i_1} \otimes \cdots \otimes b^{i_k}.$$

Accordingly, we may call $\tau^{i_1 \cdots i_k}$ a contravariant coefficient and $\tau_{i_1 \cdots i_k}$ a covariant coefficient for the only given tensor T. Similarly, we can continue with mixed coefficients of the type (k, l). If we considered V without a metric, as in a previous section, we would have for $T \in T_l^k(V)$

$$T = \tau_{i_1 \cdots i_k}^{j_1 \cdots i_l} \; b_{j_1} \otimes \cdots \otimes b_{j_l} \otimes \beta^{i_1} \otimes \cdots \otimes \beta^{i_k}.$$

Here, we consider V with metric (V, s), so we have

$$T = \tau_{i_1 \cdots i_k}^{j_1 \cdots j_l} \; b_{j_1} \otimes \cdots \otimes b_{j_l} \otimes b^{i_1} \otimes \cdots \otimes b^{i_k}.$$

In this case, T is a tensor of rank $m = k + l$. We further see that a fixed basis B, in fact, leads to different representations for the same tensor $T \in \underbrace{V \otimes \cdots \otimes V}$. So we

$$\text{(k+l)-times}$$

may set symbolically

$$\tau^{i_1 \cdots \tau_m} \longleftrightarrow \tau_{j_1 \cdots j_m} \longleftrightarrow \tau_{j_1 \cdots j_\varrho}^{i_1 \cdots i_k}.$$

with $i_1, \ldots, i_m \in I(m)$, $m = k + l$.

14.5   The Structure of a Tensor Space

In this section, we introduce the universal property of tensors.

14.5.1 Multilinearity and the Tensor Product

Suppose, as always, V is a vector space of dimension n. We start with a comparison between the vector space

$$T^k(V) = \underbrace{V^* \otimes \cdots \otimes V^*}_{k\text{-times}}$$

and an abstract vector space W of the same dimension (dim $W = n^k$). The two vector spaces $T^k(V)$ and W are isomorphic as abstract vector spaces because of their identical dimensions.

It should also be evident that a vector space like $V \otimes \cdots \otimes V$ has more structure than the vector space W. This is so for at least two reasons.

Firstly, the vector space $V \otimes \cdots \otimes V$ is different from W since its elements have the form of a linear combination of $v_1 \otimes \cdots \otimes v_k$, and secondly we have some kind of product on it (the tensor product). The identification $T^k(V) = V^* \otimes \cdots \otimes V^*$ and the explicit presence of the tensor product symbol $\otimes$ make this algebraic structure visible (see Comment 14.4).

On the other hand, such an algebraic structure is not available on the abstract vector space W. Furthermore, the comparison between the multilinear maps in $T^k(V^*)$ and the tensor product $V \otimes \cdots \otimes V$ or equivalently the multilinear maps $T^k(V)$ and $V^* \otimes \cdots \otimes V^*$, is an instructive one: Expressions like $V \otimes \cdots \otimes V$ or $V^* \otimes \cdots \otimes V^*$ are by definition purely algebraic objects. Since we start with an abstract vector space V, its elements are just vectors and not maps, the elements of $V \otimes \cdots \otimes V$ are vectors (tensors), not maps or even multilinear forms like the elements of $T^k(V^*)$. The same holds for V^* and $V^* \otimes \cdots \otimes V^*$ since by definition going from V^* to $V^* \otimes \cdots \otimes V^*$, we follow the same procedure as going from V to $V \otimes \cdots \otimes V$. This means that in such an algebraic construction, we discard in V^* any other structure different from the structure of an abstract vector space. With this in mind, we can say that $V^* \otimes \cdots \otimes V^*$ is a purely algebraic object as opposed to the multilinear forms $T^k(V)$.

14.5.2 The Universal Property of Tensors

We now return to Example 14.5 in Sect. 14.1 and once more discuss the relation of tensors as multilinear maps to linear maps we found there (see diagram in Example 14.5), from a more general point of view. A tensor space is also a special vector space that allows us to consider multilinear maps as linear maps. The domain of this special linear map is the tensor space which corresponds to the given multilinear map. This leads to the following proposition:

Proposition 14.2 *Universal property of tensor products.*

Let U and V be two real vector spaces. There exists a vector space $U \otimes V$ together with a bilinear map

$$\Theta : U \times V \longrightarrow U \otimes V, \qquad (14.27)$$

with the following "universal" property:
For every vector space Z, together with a bilinear map,

$$\varphi : U \times V \longrightarrow Z, \qquad (14.28)$$

there exists a unique linear map,

$$\tilde{\varphi} : U \otimes V \longrightarrow Z,$$

such that $\varphi = \tilde{\varphi} \circ \Theta$.

This corresponds to the commutative diagram

So we have $\dim U \otimes V = \dim U \cdot \dim V$ and the canonical isomorphism

$$T(U, V; Z) \cong \mathrm{Hom}(U \otimes V, Z). \qquad (14.29)$$

Proof The use of bases makes the proof quite transparent. Let

$$B = (b_1, \ldots, b_n) = (b_s)_n \quad \text{and} \quad C = (c_1, \ldots, c_m) = (c_i)_m$$

be the corresponding bases for U and V. We consider the set $B \times C = \{(b_s, c_i) : s \in I(n), i \in I(m)\}$. The two multilinear maps Θ and φ are, as we know, uniquely defined on the set $B \times C$. We define $U \otimes V$ to be the free vector space generated by $B \times C$,

$$U \otimes V := \mathbb{R}(B \times C)$$

with basis $b_s \otimes c_i := \Theta(b_s, c_i)$ (see Definition 8.1).

It is clear that $\dim U \otimes V = n \cdot m$ by construction. This corresponds to the $n \cdot m$ values $\varphi(b_s, c_i) := z_{si} \in Z$ of the bilinear map $\varphi \in T(U, V; Z)$.

There is a unique linear map $\tilde{\varphi} \in \mathrm{Hom}(U \otimes V, Z)$ defined on the basis elements by

$$
\begin{array}{ccccc}
U & \otimes & V & \xrightarrow{\tilde{\varphi}} & Z \\
b_s & \otimes & c_i & \mapsto & \tilde{\varphi}(b_s \otimes c_1) := z_{si}.
\end{array}
\tag{14.30}
$$

So we have

$$\tilde{\varphi}(\Theta(b_s, c_i)) = \varphi(b_s, c_i)$$

or equivalently

$$\tilde{\varphi} \circ \Theta = \varphi.$$

The bijectivity $\varphi \leftrightarrow \tilde{\varphi}$ is given by construction and we have a canonical isomorphism

$$T(U, V; Z) \cong \mathrm{Hom}(U \otimes V, Z).
\tag{14.31}$$

□

Remark 14.3 With the isomorphism $T(U, V; Z) \cong \mathrm{Hom}(U \otimes V, Z)$ we gave above and which depended on a choice of basis of U and V, it turns out that one can prove that all choices lead to the same isomorphism, and thus, that this isomorphism is a canonical isomorphism.

□

Corollary 14.1 *Universal property of tensors with $Z = \mathbb{R}$.*
The special case with $Z = \mathbb{R}$ leads to

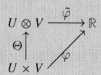

and to

$$T^2(U, V) \cong \mathrm{Hom}(U \otimes V, \mathbb{R}) \equiv (U \otimes V)^*.
\tag{14.32}$$

If we use the assertion of Example 14.5 in Sect. 14.1 and the isomorphism $V^* \otimes V^* \cong T^2(V)$ of Eq. (14.10), and after identification $T^2(U, V) = U^* \otimes V^*$, we obtain additionally the equation

$$(U \otimes V)^* = U^* \otimes V^*.
\tag{14.33}$$

The above proposition along with its proof give a good understanding of the properties of linear and multilinear maps. Tensor spaces as domains of linear maps of the form $V \otimes \cdots \otimes V$, corresponding to multilinear maps $T^k(V^*)$, can be considered as algebraic manifestations of multilinear maps. This makes some properties of multilinear maps more transparent and justifies once more the identification of, a priori different objects like

$$T^k(V) = (V \underbrace{\otimes \cdots \otimes}_{\text{k-times}} V)^* = V^* \underbrace{\otimes \cdots \otimes}_{\text{k-times}} V^k; \tag{14.34}$$

$$T^k(V^*) = (V^* \underbrace{\otimes \cdots \otimes}_{\text{k-times}} V^*)^* = (V \underbrace{\otimes \cdots \otimes}_{\text{k-times}} V). \tag{14.35}$$

14.6   Universal Property of Tensors and Duality

This section summarizes and discusses some of the essential identifications within tensor formalism which stem from duality and the universal properties of tensor products. We also comment shortly on the relevant proofs since this gives a better understanding of the key relations. When we consider duality, we use the identification $V^{**} = V$ and the expected identification $(U \times V)^* = U^* \otimes V^*$ for which, on this occasion, we prove the corresponding canonical isomorphism.

We first discuss the following isomorphisms.

> **Proposition 14.3** *Canonical isomorphisms in tensor formalism.*
>
> *(i)* $(U \otimes V)^* \cong T(U, V)$;
> *(ii)* $U^* \otimes V^* \cong T(U, V)$;
> *(iii)* $(U \otimes V)^* \cong U^* \otimes V^*$.

Proof Proof of (i)
The isomorphism (i) is best obtained from the commutative diagram

$$
\begin{array}{ccc}
U \otimes V & \xrightarrow{\tilde{\varphi}} & \mathbb{R} \\
{\scriptstyle \Theta} \uparrow & \nearrow {\scriptstyle \varphi} & \\
U \times V & &
\end{array}
$$

Proposition 14.2 in Sect. 14.5.2 leads to the bijection $\varphi \leftrightarrow \tilde{\varphi}$ with $\varphi \in T(U, V)$ and $\tilde{\varphi} \in (U \otimes V)^*$. It leads to the canonical isomorphism Θ:

$$T(U, V) \longrightarrow (U \otimes V)^*$$
$$\varphi \longmapsto \tilde{\varphi}$$

given by

$$T(U, V) \overset{\cong}{\underset{\tilde{\Theta}}{\to}} (U \otimes V)^*$$

and to the identification

$$T(U, V) = (U \otimes V)^*.$$

So (i) is proven.   □

Proof Proof of (ii)

The isomorphism (ii) is obtained again from Proposition 14.2 and the commutative diagram

Here, we need to show that the linear map $\tilde{\psi}$ is a bijection. Since $\dim((U^* \otimes V^*) = \dim T(U^*, V^*)$, we only have to show that $\tilde{\psi}$ is injective, see Example 14.5 in Sect. 14.1, with the corresponding notation $P = \psi$ and $\tilde{P} = \tilde{\psi}$. So we now have

$$\psi : U^* \times V^* \longrightarrow T(U, V),$$
$$\tilde{\psi} : U^* \otimes V^* \longrightarrow T(U, V),$$
$$\xi \otimes \eta \longmapsto \tilde{\psi}(\xi \otimes \eta) = \psi(\xi, \eta) = \xi\eta,$$

and with $\Theta, \tilde{\Theta}$ for the universal property:

$$T(U^*, V^*; T(U, V)) \overset{\tilde{\Theta}}{\longrightarrow} \mathrm{Hom}(U^* \otimes V^*, T(U, V)),$$
$$\psi \longrightarrow \tilde{\psi}.$$

Finally, we have to show the injectivity of $\tilde{\psi}$. We have for $(\beta^1, \ldots, \beta^n) = (\beta^s)_n$, basis of U^* and $(\gamma^1, \ldots, \gamma^m) = (\gamma^i)_m$, basis of V^*, with

$$\tilde{\psi}(\xi \otimes \eta)(\beta^s \otimes \gamma^i) = \xi\eta(\beta^s, \gamma^i) = \xi(\beta^s)\eta(\gamma^i) \overset{!}{=} 0.$$

This means $\tilde{\psi}(\xi \otimes \eta)(\beta^s, \gamma^i) = \xi(\beta^s)\eta(\gamma^i) = 0$ for all s and i. We obtain $\tilde{\psi}(\xi \otimes \eta) = 0^* \in U^* \otimes V^*$ and $\tilde{\psi}$ is indeed injective, and $\tilde{\psi}$ is the canonical isomorphism:

$$U^* \otimes V^* \cong T(U, V) \quad \text{(isomorphism (ii))}$$

corresponding to the identification

$$U^* \otimes V^* = T(U, V).$$

So (ii) is proven.   □

Proof Proof of (iii)
From the isomorphisms in (i) and (ii), we also obtain directly

$$(U \otimes V)^* \cong U^* \otimes V^* \quad \text{(isomorphism (iii))}.$$

                                                                                                       □

Summarizing, we have, altogether, identifications:

$$T(U, V) = (U \otimes V)^*, \quad T(U, V) = U^* \otimes V^*, \quad (U \otimes V)^* = U^* \otimes V^*$$

and, by taking duals, we obtain

$$T(U^*, V^*) = U \otimes V, \quad T(U^*, V^*) = U \otimes V, \quad (U^* \otimes V^*)^* = U \otimes V.$$

The following proposition is useful in general relativity, particularly in connection with curvature.

Proposition 14.4 *Canonical isomorphism between vector-valued multilinear maps and multilinear forms.*
Let V, Z be vector spaces and $k \in \mathbb{N}$. There exists the following canonical isomorphism.

$$T^k(V; Z) \cong T^{(k+1)}(V, Z^*).$$

with

$$T^{k+1}(V, Z^* *) = T \underbrace{(V, \ldots, V}_{k\text{-times}, Z^*}.$$

Proof This isomorphism is a generalization of the canonical isomorphism between vectors and bivectors: $Z \cong Z^{**}$. In addition, the proof here goes along the same line: We consider the dual pairing,

$$
\begin{array}{ccc}
Z^* \times Z & \longrightarrow & \mathbb{R} \\
(\xi, z) & \longmapsto & \xi(z) \in \mathbb{R}.
\end{array}
$$

The evaluation map $z^{\#} \in Z^{**}$ is given, as usual, by $z^{\#}(\xi) = \xi(z)$ whenever $\xi \in Z^*$. There is a similar relationship for tensor maps too:

$$
\begin{array}{ccc}
T^k(V; Z) & \overset{\mathcal{F}}{\longrightarrow} & T^{k+1}(V, Z^*) \\
\psi & \longmapsto & \psi^{\#},
\end{array}
$$

where

$$\psi^{\#}(v_1, \ldots, v_k, \xi) := \xi(\psi(v_1, \ldots, v_k)) \in \mathbb{R}. \qquad (14.36)$$

The analogy with $Z \tilde{=} Z^{**}$ becomes visible if we fix $(v_1, \ldots, v_k)$ and set $\tilde{\psi} := \psi(v_1, \ldots, v_k) \in Z$ and $\tilde{\varphi} := \varphi(v_1, \ldots, v_k, \xi) \in \mathbb{R}$. The canonical isomorphism $Z \tilde{=} Z^{**}$ corresponds to

$$T^k(V; Z) \cong T^{k+1}(V, Z^*).$$

So the proposition is proven.   □
    We get the identification

$$T^k(V; Z) = T^{k+1}(V, Z^*).$$

    □

Corollary 14.2 *From this, we can also obtain the following interesting special cases. When $k = 1$ we have*

$$\mathrm{Hom}(V, Z) \equiv T^1(V; Z)$$

and the identification

$$\mathrm{Hom}(V, Z) = T^{(2)}(V, Z^*)$$

or

$$\mathrm{Hom}(V, Z) = V^* \otimes Z.$$

When $k = 0$, we may write $T^{(0)}(V; Z) \equiv Z$ and we get

$$T^0(V; Z) = T^{(1)}(V, Z^*),$$

which is the same as

$$T^0(V; Z) = T^1(Z^*) \quad or$$
$$Z = Z^{**}.$$

14.7   Tensor Contraction

The tensor contraction, also known as the contraction or trace map, is a simple operation on tensor spaces that we already know. It is essentially the evaluation of a covector applied to a vector giving a scalar. It was used at the beginning of Sect. 14.3 about mixed tensors and the beginning of the proof in Proposition 14.4. The importance of this operation explains our repeating. We consider the map again:

$$\phi : \quad V^* \times V \quad \longrightarrow \quad \mathbb{R}$$
$$(\xi, v) \quad \longmapsto \quad \phi(\xi, v) := \xi(v).$$

or equivalently the map corresponding to the universal property of tensor products in Proposition 14.2:

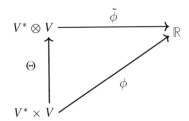

Here, we denote the relevant map by C.

$$C : \quad V^* \otimes V \quad \longrightarrow \quad \mathbb{R}$$
$$(\xi \otimes v) \quad \longmapsto \quad C(\xi \otimes v) = \xi(v).$$

The map C is called tensor contraction. As we see, both maps, ϕ and C above, are basis-independent. This C can also be considered as an operator on tensor spaces.

$$C : T_1^1(V) \longrightarrow T_0^0(V).$$

Comment 14.6 If we apply this reasoning to an operator $f \in \mathrm{Hom}(V, V)$, we may recognize that this corresponds to taking the trace of f.

From Corollary 14.2 in the previous section, there follows the canonical isomorphism:
$$\mathrm{Hom}(V, V) \underset{\iota}{\cong} V^* \otimes V$$

and the trace of f is given by the decomposition

$$\begin{aligned} \mathrm{Hom}(V, V) &\xrightarrow{tr} \mathbb{R} \\ f &\longmapsto tr(f) := C \circ \iota(f). \end{aligned}$$

With a given basis B of V, this takes the form $tr(f) := tr(f_{BB}) := tr(\varphi_s^i) = \varphi_i^i \in \mathbb{R}$ which is the well-known sum of the diagonal elements of the representation matrix f_{BB}. From the transformation properties of the matrix f_{BB}, we see again that $tr(f)$ does not depend on the chosen basis B.

We would now like to extend the tensor contraction C of the tensor of type (k, l):

$$C_j^i : T_l^k(V) \longrightarrow T_{l-1}^{k-1}(V).$$

For this purpose, we first consider the tensor space $V^* \otimes \cdots \otimes V^* \otimes V \otimes \cdots \otimes V$ of type (k, l). We have to specify the position of one covector space V_i^* with $i \in I(k)$ and one vector space V_j with $j \in I(l)$, and we can write

$$T_l^k(V) = V_1^* \otimes \cdots \otimes V_k^* \otimes V_1 \otimes \cdots \otimes V_l.$$

With appropriate notation and numeration, the operation C_j^i is given by:

$$V_1^* \otimes \cdots \otimes V_k^* \otimes V_1 \otimes \cdots \otimes V_l \xrightarrow{C_j^i} \mathbb{R} \otimes V_1^* \otimes \cdots \otimes V_{k-1}^* \otimes V_1 \otimes \cdots \otimes V_{l-1}$$
$$\cdots \otimes \xi^i \otimes \cdots \quad \cdots \otimes v_j \otimes \longmapsto \xi^i(v_j)\xi^1 \otimes \cdots \otimes \xi^{k-1} \otimes v_1 \otimes \cdots \otimes v_{l-1}.$$

Evidently, $C_j^i(\cdots \otimes \xi^i \otimes \cdots \otimes v_j \otimes \cdots)$ is a tensor of type $(k - 1, l - 1)$.

Example 14.7 $T = \xi^1 \otimes \xi^2 \otimes v_1 \otimes v_2 \otimes v_3$
If we take $T = \xi^1 \otimes \xi^2 \otimes v_1 \otimes v_2 \otimes v_3$, we obtain

$$C_3^1(\xi^1 \otimes \xi^2 \otimes v_1 \otimes v_2 \otimes v_3) = \xi^1(v_3)\xi^2 \otimes v_1 \otimes v_2 \in T_2^1(V).$$

Similarly, we get

$$C_2^1(\xi^1 \otimes \xi^2 \otimes v_1 \otimes v_2 \otimes v_3) = \xi^1(v_2)\xi^2 \otimes v_1 \otimes v_3 \in T_2^1(V).$$

We can also consider $T_l^k(V)$ as a multilinear form. In this case, the analogous procedure for the application of contraction C_j^i leads to the following development: Consider $T \in T_l^k(V)$, with

$$T : V \underbrace{\times \cdots \times V}_{k\text{-times}} \times V^* \underbrace{\times \cdots \times V^*}_{l\text{-times}} \longrightarrow \mathbb{R}.$$

We first fix the vectors $v_1, \ldots, v_{k-1}$ and the covectors $\xi^1, \ldots, \xi^{l-1}$ and we then consider a basis $B = (b_1, \ldots, b_n)$ in V and its dual $\mathcal{B} = (\beta^1, \ldots, \beta')$ so that $(\beta^i(b_s) = \delta^i_s i, s \in I(n))$. We define:

$$C^i_j T(v_1, \ldots, v_{k-1}, \xi^1, \ldots, \xi^{l-1})$$

$$:= \sum_{s=1}^n T(v_1, \ldots, v_{j-1}, b_s, v_{j+1}, \ldots v_k, \xi^1, \ldots, \xi^{i-l}, \beta^s, \xi^{i+1}, \ldots, \xi^l).$$

Taking the vector $b_s \in V$ in position j, and the covector $\beta^s \subset V^*$ in position i.

With these definitions, we have $(C^i_j T)(v_1, \ldots, v_{k-1}, \xi^1, \ldots, \xi^{l-1}) \in \mathbb{R}$ so that $C^i_j T$ is evidently a multilinear form of type $(k-1, l-1)$ and C^i_j a contraction of T corresponding to (i, j):

$$T^k_l(V) \xrightarrow{C^i_j} T^{k-1}_{l-1}(V).$$

Example 14.8 $T(v_1, v_2, v_3, \xi^1, \xi^2)$
For $T(v_1, v_2, v_3, \xi^1, \xi^2)$ and $i = 1$ and $j = 3$:

$$C^1_3 T(v_1, v_2, \xi^2) = \sum_{s=1}^n T(v_1, v_2, b_s, \beta^s, \xi^2).$$

If instead $i = 1$ and $j = 2$, then :

$$C^1_2 T(v_1, v_3, \xi^2) = \sum_{s=1}^n T(v_1, b_s, v_3, \beta^s, \xi^2).$$

Example 14.9 Example 14.8 in coefficients.
For the coefficients relative to the bases $B = (b_s)$ and $B^* = (\beta^r)$, with $s, r, i_1, i_2, j_3 \in I(n)$ in the above examples, we have:

$$C_3^1 T(b_{i_1}, b_{i_2}, \beta^{j_1}) = \sum_{s=1}^{n} T(b_{i_1}, b_{i_2}, b_s, \beta^s, \beta^{j_1})$$

$$= T_{i_1 i_2 s}^{s j_1}$$

and similarly:

$$C_2^1 T(b_{i_1}, b_{i_2}, \beta^{j_1}) = \sum_{s=1}^{n} T(b_{i_1}, b_s, b_{i_2}, \beta^s, \beta^{j_1})$$

$$= T_{i_1 s i_2}^{s j_1}.$$

Summary

This was the third time we delved into tensors. The first two approaches in Sect. 3.5 and Chap. 8 were basis-dependent to facilitate understanding. Simultaneously, our index notation, which we have used throughout linear algebra, significantly facilitated this understanding.

In this chapter, it was time for the basis-free and coordinate-free treatment of tensor formalism. In this sense, we could affirm that a tensor is a multilinear map. Since we initially considered abstract vector spaces, the distinction between covariant, contravariant, and mixed tensors was necessary. Following that, we introduced and discussed an inner product vector space, establishing and discussing the corresponding tensor quantities.

After this, the universal property of the tensor product was introduced, allowing for a deeper understanding of the concept of tensors. Finally, using the universal property, several commonly used relationships, essentially involving the dual space of a tensor product of two vector spaces, were proven.

References and Further Reading

1. M. DeWitt-Morette, C. Dillard-Bleick, Y. Choquet-Bruhat, *Analysis, Manifolds and Physics* (North-Holland, 1978)
2. G. Fischer, B. Springborn, *Lineare Algebra. Eine Einführung für Studienanfänger*. Grundkurs Mathematik (Springer, 2020)
3. K.-H. Goldhorn, H.-P. Heinz, M. Kraus, *Moderne mathematische Methoden der Physik. Band 1* (Springer, 2009)
4. P. Grinfeld, *Introduction to Tensor Analysis and the Calculus of Moving Surfaces* (Springer, 2013)

5. S. Hassani, *Mathematical Physics: A Modern Introduction to its Foundations* (Springer, 2013)
6. N. Jeevanjee, *An Introduction to Tensors and Group Theory for Physicists* (Springer, 2011)
7. N. Johnston, *Advanced Linear and Matrix Algebra* (Springer, 2021)
8. J.M. Lee, *Introduction to Smooth Manifolds*. Graduate Texts in Mathematics (Springer, 2013)
9. T. Levi-Civita, *The Absolute Differential Calculus. (Calculus of Tensors)* (Dover Publications, 1977)
10. P. Renteln, *Manifolds, Tensors, and Forms: An Introduction for Mathematicians and Physicists* (Cambridge University Press, 2013)
11. A.J. Schramm, *Mathematical Methods and Physical Insights: An Integrated Approach* (Cambridge University Press, 2022)
12. B.F. Schutz, *Geometrical Methods of Mathematical Physics* (Cambridge University Press, 1980)
13. C. Von Westenholz, *Differential Forms in Mathematical Physics* (Elsevier, 2009)
14. R. Walter, *Lineare Algebra und analytische Geometrie* (Springer, 2013)

Index

A
Abelian group, 19, 20, 54, 81, 99
Affine space, 17, 55, 56, 138
Algebra, 20
Algebra End(V), 267
Algebraic multiplicity, 256
Alternating tensors, 239
Angle, 47
Annihilators, 184

B
Basis, 94, 95, 106, 380
Basis isomorphism, 109
Bijective, 108
Bilinear form, 40, 49, 355

C
Canonical basis, 72
Canonical dual basis, 176
Canonical isometry, 366
Canonical isomorphism, 365
Canonical map, 4, 60
Cauchy-Schwarz, 43
Cayley-Hamilton, 282
Change of basis, 157, 235
Characteristic polynomial, 254, 287
Colist, 86, 87, 147, 362
Column rank, 148
Column space, 148
Complementary subspace, 116
Complement in set theory, 116
Complexification, 329, 330
Complex spectral theorem, 314, 316, 327
Contravariant tensor, 382
Covariant k-tensor, 372
Covector, 53, 71, 87, 97, 175, 362

Covector fields, 71

D
Decomposition, 117, 260, 261, 281
Determinant, 203
Determinant function, 204
Diagonalizability, 259, 260, 262, 265, 266, 287

Dimension, 96, 108, 115, 125
Direct product, 114, 118
Direct sum, 62, 118, 252
Disjoint composition, 221
Dual, 37, 50, 71, 86
Dual basis, 176
Dual map, 178
Dual space, 37

E
Effective action, 11
Eigen element, eigenelement, 243
Eigenspace, 243
Eigenvalue, 243
Eigenvector, 243
Elementary matrices, 199
Elementary row operations, 201
Equivalence classes, 2, 3, 6, 57, 58, 106, 111, 152, 220
Equivalence relation, 1–4, 6, 11, 12, 14, 58, 90, 112, 152, 203, 218, 219, 356
Equivariant map, 1, 13, 14, 100, 102
Equivariant vector space, 101
Euclidean space, 138
Euclidean vector space, 17, 39, 44, 47, 48, 54, 74, 135, 186, 190, 215, 299, 333, 336, 346, 365, 385

© The Editor(s) (if applicable) and The Author(s), under exclusive license
to Springer Nature Switzerland AG 2024
N. A. Papadopoulos and F. Scheck, *Linear Algebra for Physics*,
https://doi.org/10.1007/978-3-031-64908-0

Printed in the United States
by Baker & Taylor Publisher Services